Alternating Current (AC) Heating for Lithium-Ion Batteries in Electric Vehicles

Jiangong Zhu · Ranjun Huang · Haifeng Dai

Alternating Current (AC) Heating for Lithium-Ion Batteries in Electric Vehicles

Heating Principles, Modeling, and Implementation

Springer

Jiangong Zhu
School of Automotive Studies
Tongji University
Shanghai, China

Ranjun Huang
School of Automotive Studies
Tongji University
Shanghai, China

Haifeng Dai
School of Automotive Studies
Tongji University
Shanghai, China

ISBN 978-981-96-9070-1 ISBN 978-981-96-9071-8 (eBook)
https://doi.org/10.1007/978-981-96-9071-8

This Springer imprint is published by the registered company Springer Nature Singapore Pte Ltd.
The registered company address is: 152 Beach Road, #21-01/04 Gateway East, Singapore 189721, Singapore

Preface

With the continuous advancement of global industrialization, the demand for energy is steadily increasing, accompanied by a surge in fossil fuel consumption and the worsening of environmental issues such as global warming and air pollution. In response to these challenges, countries around the world are increasingly embracing sustainable energy strategies, with a particular focus on the development and deployment of clean energy technologies. Against this backdrop, electric vehicles (EVs) have emerged as a pivotal platform for driving technological transformation in the transportation sector. In recent years, the combined influence of supportive policies and growing market demand has accelerated the rapid growth of the EV industry. Consequently, lithium-ion batteries (LIBs)—an essential component of EVs—have experienced significant advancements and widespread adoption.

However, the performance degradation of LIBs in low-temperature environments remains a major obstacle to the widespread adoption of EVs in cold regions and a key factor limiting user experience. At low temperatures, EVs suffer a sharp decline in driving range and a noticeable drop in power performance. Moreover, charging under such conditions becomes significantly slower and poses greater safety risks. The root cause of these issues lies in the deterioration of LIB performance at low temperatures. This degradation is primarily characterized by reduced usable capacity and diminished power output, which are mainly attributed to the increase in internal resistance under low-temperature conditions. Prolonged operation in cold environments further accelerates battery aging, shortening the service life of the battery. More critically, low-temperature charging increases the risk of lithium plating. On one hand, lithium plating consumes active lithium, leading to capacity loss; on the other hand, it can form lithium dendrites that may penetrate the separator and cause internal short circuits, posing serious safety hazards. Therefore, overcoming the performance limitations of LIBs at low temperatures is not only a pressing research focus in academia but also a key technical challenge that the industry must address.

This book focuses on the alternating current (AC) heating of automotive LIBs, systematically establishing a comprehensive knowledge framework that spans from fundamental performance analysis to engineering applications. Chapters 1 and 2 examine the low-temperature performance and aging mechanisms of LIBs, elucidating the underlying causes of performance degradation in cold environments. Chapter 3 provides a detailed analysis of the theory and principles of AC heating, laying a solid scientific foundation for subsequent research on heating technologies. Chapters 4 and 5 explore the methodologies for designing AC heating strategies, shedding light on the electro-thermal coupling mechanisms at low temperatures and the constraints involved in strategy development. Chapter 6 presents the practical implementation of AC heating technology within vehicle systems. Through interdisciplinary research, this book offers low-temperature battery management solutions that combine theoretical rigor with engineering practicality.

The completion of this book is the result of the collective efforts of our team and the invaluable feedback from colleagues both at home and abroad. We extend our heartfelt gratitude to Prof. Dr. Xuezhe Wei for his meticulous guidance on key theoretical issues. We also wish to express our sincere thanks to those who contributed to the preparation of this book, including Ph.D. students Yihan Zhu, Jin Yang, Renjie Zhang, Zhehang Xu, Zhen Yin, and Jiajun Liu.

This book presents a comprehensive summary of our team's research on low-temperature AC heating conducted over the past decade. It is our hope that this work will offer new insights into the development of battery thermal management technologies for researchers, engineers, educators, and students engaged in the field of new energy. By doing so, we aim to contribute to the advancement of EVs and the improvement of energy efficiency in cold regions. Despite multiple rounds of revision, some oversights may remain. We sincerely ask for readers' understanding and warmly welcome any feedback or suggestions for improvement. We also hope that the methodologies proposed in this book will inspire further innovative exploration and help drive the industry toward greater safety and energy efficiency.

Shanghai, China

Jiangong Zhu
Ranjun Huang
Haifeng Dai

Contents

Chapter 1
Low-Temperature Performance of Lithium-Ion Batteries for Electric Vehicles

Abstract The performance of electric vehicles (EVs) is largely determined by the properties of lithium-ion batteries (LIBs), particularly in terms of range, charging efficiency, and usage safety. Ambient temperature plays a critical role in influencing these properties, with low temperatures causing a notable decline in energy availability and power output. Moreover, prolonged exposure to such conditions accelerates battery degradation, ultimately reducing its lifespan. The commonly used LIBs for EVs have a noticeable decrease in capacity and operating voltage at −10 °C. At −20 °C, the performance significantly declines, marked by a drastic reduction in discharge capacity, retaining only approximately 30% of the specific capacity observed at normal temperatures. Meanwhile, in low-temperature environments, it is difficult to charge LIBs. Metal lithium is prone to accumulate on the anode's surface during charging. The lithium dendrites formation can penetrate the battery separator, leading to an internal short circuit. This not only results in irreversible damage to the battery but also triggers thermal runaway. Research data shows that the battery power characteristics deterioration and the charge–discharge coulombic efficiency's decrease caused by low temperature are one of the objective factors that restrict the play of electric vehicle technical advantages and practical application. Therefore, the following will be through battery single cell testing research, and carry out low-temperature charge–discharge tests on the currently mainstream nickel–cobalt-manganese system ($LiNi_{1/3}Co_{1/3}Mn_{1/3}O_2/C$) and lithium iron phosphate system ($LiFePO_4/C$) batteries, to explore the lithium-ion batteries characteristics under low temperature and the key factors that restrict their low-temperature performance.

Keywords Electric vehicles · Lithium-ion batteries · Electrode materials · Low-temperature performance · Charge–discharge tests

J. Zhu et al., *Alternating Current (AC) Heating for Lithium-Ion Batteries in Electric Vehicles*, https://doi.org/10.1007/978-981-96-9071-8_1

1.1 Lithium-Ion Batteries for Electric Vehicles

Lithium-ion batteries, with their high energy density, high power density, reliable safety features, and long cycle life, have already occupied an important position in consumer electronics, energy storage power supply, electric transportation, and other fields. The main shapes of vehicle-used lithium-ion batteries include cylindrical, square, and pouch types. Regardless of the shape, their basic composition is the same. Taking a typical $LiCoO_2$/graphite battery as an example, its basic structure is as Fig. 1.1 shown: the main components include the positive electrode (cathode), negative electrode (anode), electrolyte, separator, current collector, and battery shell. The positive and negative electrodes are usually composed of active material, binder, and conductive additive. In the entire battery, only the active materials in the positive and negative electrodes have electrochemical activity and can provide capacity. The rest of the parts do not have electrochemical activity and cannot provide capacity. Therefore, under the premise of ensuring that battery performance is not affected, reducing the mass of inactive components can improve the battery's specific energy density.

1.1.1 The Types of Lithium-Ion Batteries

The selection of cathode materials determines the energy density, lifespan, and safety of batteries. Lithium Cobalt Oxide ($LiCoO_2$, LCO) has emerged as the dominant choice in consumer electronics owing to its superior energy density. However, its

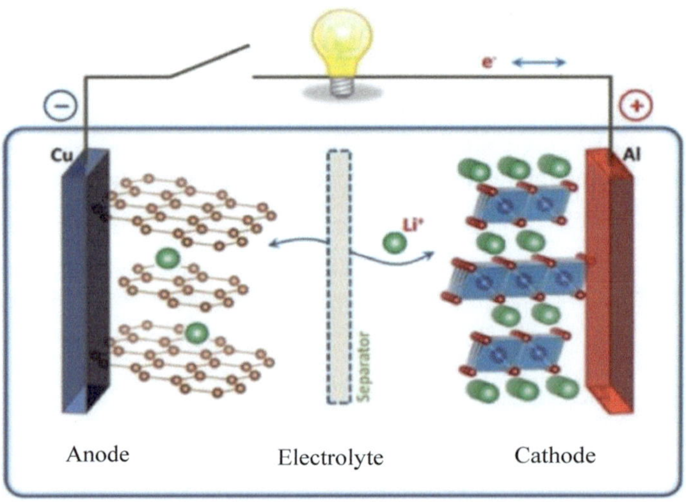

Fig. 1.1 Illustration of the structure of $LiCoO_2$ battery [1]

high cobalt cost and thermal instability have driven the industry to seek alternatives. Ternary batteries (e.g., $LiNi_xCo_yMn_{1-x-y}O_2$(Nickel Cobalt Manganese Oxide, NCM), $LiNi_xCo_yAl_{1-x-y}O_2$(Nickel Cobalt Aluminum Oxide, NCA)), leveraging the synergistic effects of Ni, Co, Mn, or Al, dominate EV sector. High-nickel formulations like NCM811 further reduce cobalt dependency, though safety challenges at high temperatures persist. In contrast, Lithium Iron Phosphate ($LiFePO_4$, LFP) achieves breakthroughs in cycle life and safety through its cost-effective iron-phosphorus composition. Despite lower energy density, LFP has gained traction in energy storage systems and affordable EVs. Lithium Manganese Oxide ($LiMn_2O_4$, LMO) is used in short-cycle applications due to its cost advantage, while cutting-edge materials like lithium-rich manganese-based cathodes remain under laboratory exploration to push energy density limits.

Traditional liquid LIBs rely on organic electrolytes for ion conduction, offering mature technology and low cost. However, flammability and leakage risks persist. Solid-state batteries, utilizing solid electrolytes, eliminate leakage risks and promise energy densities exceeding 500 Wh/kg, positioning them as next-generation solutions. Yet, challenges such as interfacial impedance between solid electrolytes and electrodes, along with high manufacturing costs, hinder commercialization. Semi-solid-state batteries, blending liquid and solid electrolytes as a transitional solution, partially enhance safety and have been piloted in premium EVs like the NIO ET7's "semi-solid-state" battery pack.

Battery form factors also cater to diverse applications: Cylindrical cells (e.g., Tesla's early 18650) suit standardized production; Prismatic cells optimize space utilization; Pouch cells adopt flexible designs for wearable devices. Innovations like Tesla's 4680 tables cylindrical cell reduce internal resistance for higher power output, while CATL's CTP (Cell-to-Pack) technology improves volumetric efficiency by 15–20% through modularization elimination.

LCO batteries use $LiCoO_2$ as the cathode and graphite as the anode. When it's charging, lithium-ions deintercalate from the cathode, migrate through electrolyte, and embed into the graphite anode. Discharging reverses this "rocking-chair" mechanism. LCO batteries exhibit high energy density (up to 280 Wh/kg or 700 Wh/L) and a stable discharge plateau, enabling extended EV range and consistent power output. However, safety concerns arise from thermal runaway risks under overcharge, overdischarge, or short-circuit conditions. Additionally, cobalt's scarcity and high cost elevate production expenses, while limited cycle life leads to gradual capacity degradation.

The cathodes of Ternary batteries typically combine nickel (Ni), cobalt (Co), manganese (Mn), or aluminum (Al). Common formulations include $LiNi_xCo_yMn_{1-x-y}O_2$ (NCM) and $LiNi_xCo_yAl_{1-x-y}O_2$ (NCA). Adjusting Ni/Co/Mn ratios (e.g., NCM523, NCM622) balances energy density, cycle life, and safety. Higher nickel content increases reversible capacity but exacerbates lattice oxygen release risks. NCA substitutes manganese with aluminum, achieving high energy density (~280 Wh/kg) and thermal stability, albeit with complex manufacturing and elevated costs. The layered α-$NaFeO_2$ structure facilitates lithium-ion transport, outperforming LCO and lithium iron phosphate (LFP) in electrochemical behavior.

However, nickel's high reactivity during cycling induces phase transitions and microcracks, accelerating capacity fade. Compared to LFP, ternary batteries offer superior energy density but inferior cycle life and thermal stability.

LFP cathodes ($LiFePO_4$) undergo reversible lithium extraction/insertion:

Charging: $LiFePO_4 \rightarrow FePO_4 + Li^+ + e^-$

Discharging: $FePO_4 + Li^+ + e^- \rightarrow LiFePO_4$

The stable olivine structure ensures flat charge/discharge plateaus and intrinsic safety. LFP batteries withstand extreme conditions (e.g., overcharge, short circuit, needle penetration) without combustion. With cycle lives exceeding 2000 cycles at 1 C rate—double that of ternary batteries—LFP excels in longevity. However, its energy density (~140–180Wh/kg) lags behind ternary systems.

The mainstream lithium-ion batteries currently are ternary lithium-ion batteries and LFP batteries. Therefore, the two sample batteries used in the low-temperature test experiment in Sect. 1.3 were selected as ternary lithium-ion batteries (NCM system) and LFP lithium-ion batteries (LFP system).

1.1.2 The Working Principles

LIBs, as one of the classic electrochemical energy storage devices, can convert electrical energy and chemical energy into each other. This is achieved through the intercalation and de-intercalation of lithium ions and electrons between the active materials of the positive and negative electrodes. The essence is the electrochemical oxidation–reduction reaction between the active materials of the positive and negative electrodes. With the intercalation and de-intercalation of lithium-ions and electrons, the microscopic intrinsic parameters of the active materials of the cathode and anode, such as Gibbs free energy and Fermi surface, will also change. The macroscopic manifestation is the change of battery capacity and voltage.

Taking the $LiCoO_2$/graphite battery as example, the charge–discharge principle is shown in the Fig. 1.2, and the electrode reactions are as follows:

Anode: $C + xLi^+ + xe^- \rightleftharpoons Li_xC$

Cathode: $LiCoO_2 \rightleftharpoons Li_{1-x}CoO_2 + xLi^+ + xe^-$

Total reaction: $C + LiCoO_2 \rightleftharpoons Li_{1-x}CoO_2 + Li_xC$

While charging, lithium-ions de-intercalate from the cathode material LiCoO2, and simultaneously release one electron. The cobalt oxidation state increases from +3 to +4. The de-intercalated lithium-ions intercalate into the graphite negative electrode through the electrolyte. The released electrons transfer to the cathode via the external circuit, thereby maintaining charge balance.

During the discharging process, lithium-ions de-intercalate from the anode, and intercalate into the cathode side through electrolyte. The oxidation state of cobalt decreases from +4 to +3, and simultaneously releases one electron. The electron transfers to the cathode side via the external circuit to achieve charge balance.

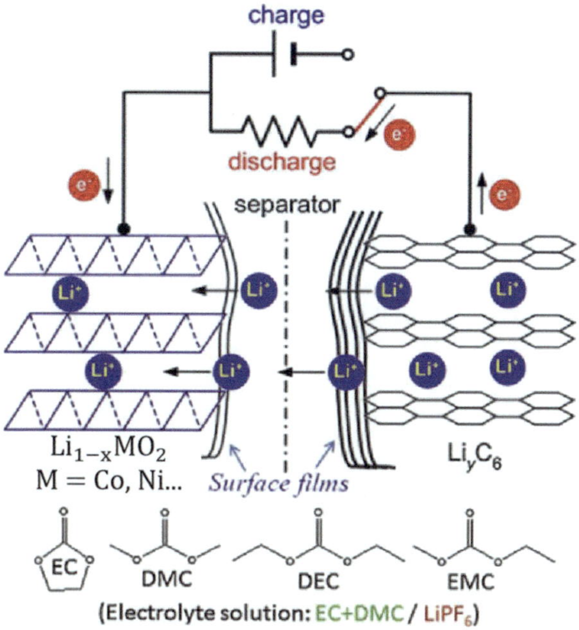

Fig. 1.2 Illustration of the working principle of $LiCoO_2$ battery [2]

1.2 The Impact of Low Temperature on Electrode Materials

In a low-temperature environment, the operation of EVs faces multiple challenges, the most notable of which is the decline in battery performance. First, low temperatures can significantly reduce the driving range of EVs. For example, in an environment of −10 °C, the driving range of some EVs may decrease by 30–50%. Second, the power performance will also significantly decline, manifested as slower acceleration and weakened climbing ability. In addition, the battery's charging speed will obviously slow down in a low-temperature environment, and the charging time may be extended to twice or even more than that under normal temperature. More seriously, low temperatures may also trigger the instability of chemical reactions inside the battery, increasing the risk of thermal runaway and thus bringing safety hazards.

The battery performance deterioration at low temperatures mainly stems from the slowdown of electrochemical reactions. As the temperature decreases, the batteries' impedance increases. Simultaneously, the discharge voltage platform drops. The batteries' terminal voltage also falls rapidly. This causes a significant decay in their available capacity and power characteristics. At low temperatures, achieving high-current discharge becomes difficult. Additionally, due to the increased impedance, the charging voltage rises quickly. As a result, the time needed to reach the charging protection termination voltage shortens. This leads to charging difficulties and low charging efficiency. All of these phenomena are caused by the impact of low temperatures on electrode materials. First, low temperatures will make the crystalline structure of the cathode material more rigid, slowing down the lithium ions' diffusion speed in it, which results in a reduction in the battery's discharge performance. Second, anode materials are more likely to experience lithium metal precipitation at low temperatures, forming lithium dendrites. These dendrites decrease the battery's capacity and may pierce the separator, leading to short circuits or even fires and explosions. In addition, low temperatures will also increase the interfacial impedance between electrode materials and electrolytes, further affecting the battery's performance. Consequently, conducting thorough investigations into the effects of low temperatures on anodes, cathodes, and electrolytes holds substantial importance for enhancing the operational efficiency of electric vehicles in cold climates.

1.2.1 The Impact on Anode

Among the numerous anode materials, graphite, graphene, and LTO have received considerable focus because of their unique characteristics, particularly their characteristics in low-temperature environments, which are particularly remarkable. The following section will provide a detailed introduction to the low-temperature characteristics of these three typical anode materials.

GRAPHITE: Graphite, the most extensively utilized anode material in LIBs, is renowned for its affordability, extended service life, and adaptable structure. During the charging process of LIBs, lithium ions move from the cathode to the anode and become inserted between the graphite layers. The characteristics of the solid electrolyte interphase (SEI) film covering the anode surface play a crucial role in lithium-ion movement, which in turn affects the low-temperature conductivity of graphite electrodes. The formation of a stable SEI film is essential for achieving desirable electrochemical performance in LIBs. Research indicates that surface modification of graphite during initial cycles (for instance, through oxidation) can boost interfacial stability by facilitating SEI development. In harsh conditions, such as extremely low temperatures (below −40 °C) or high charging rates, the lithium ions' insertion into the graphite structure slows down, resulting in lithium plating. This process is shown in Fig. 1.3 as follows. This phenomenon, which is driven by the growth of lithium dendrites, is a primary aging mechanism in LIBs and significantly shortens their lifespan. Moreover, unmodified graphite electrodes are highly prone to lithium plating because their redox potential is close to that of Li+/Li. Experimental data shows rapid capacity loss when temperatures fall below 0 °C. In addition, an increase in anode overpotential can trigger electrode polarization reactions, potentially leading to internal short circuits and catastrophic thermal events, such as fires or explosions [3]. Enhancing the graphite electrodes performance at low temperatures continues to be a key area of research. One promising approach involves incorporating carbon nanotubes (CNTs) into porous graphite (PGN/CNT composites). The CNTs act as

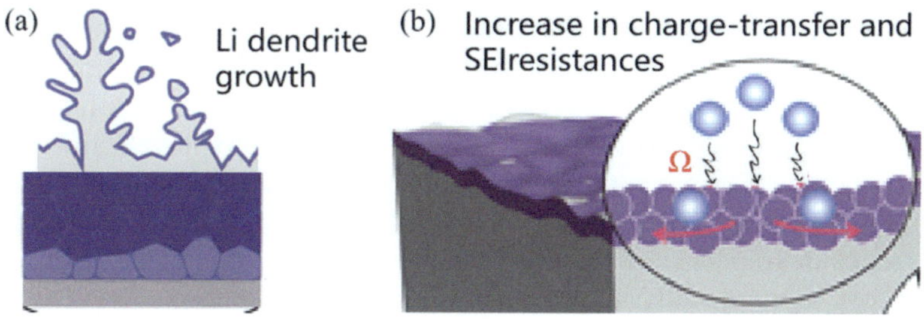

Fig. 1.3 Challenges imposed by temperature on the performance of graphite anodes. At the low temperature extreme, lithium plating becomes very problematic (**a**), and reaction kinetics are negatively affected by the increased Rct and the SEI's blocking character (**b**) [5]

conductive pathways, facilitating Li+ transport between the electrode and electrolyte, thereby mitigating interfacial Li+ blockage caused by lithium plating [4]. Commonly used methods include the surface coating modification method, which involves coating the surface of graphite with a layer of material, such as amorphous carbon, metals, and metal oxides, to improve the performance of graphite. For example, amorphous carbon coating can enhance the high-current charge–discharge performance of graphite materials, while metal and metal oxide coatings can reduce charge transfer resistance and increase the lithium's diffusion coefficient. Another method is the doping modification method, which involves adding appropriate elements, such as B, N, P, K, and S, to alter the electronic environment of carbon atoms, thereby changing the structure and properties of graphite materials. For instance, boron doping can increase the lithium-insertion capacity of graphite, and the doping of elements like Si and Sn can have a synergistic effect, enhancing lithium-storage performance.

GRAPHENE: Graphene, an allotrope of graphite, exhibits a more stable structure and higher charge–discharge capacity compared to commercial graphite electrodes. It also demonstrates superior low-temperature performance, making it a promising substitute for graphite in LIBs. As a novel conductive material, graphene is a two-dimensional crystal with perfectly sp^2-hybridized carbon atoms and boasts a theoretical capacity of 4100 mAh/g—about 10 times higher than conventional graphite. It also possesses exceptional electron mobility (up to 15,000 $cm^2/(V \cdot s)$) and outstanding mechanical and thermal properties. The integration of graphene with other materials to form composites can reduce charge transfer resistance and establish a continuous conductive network, thereby mitigating electrochemical performance degradation (e.g., due to volume expansion) under low-temperature conditions [6]. These composites offer a high specific surface area and robust electrical conductivity, creating ample pathways for Li+ diffusion within the electrode to enhance battery performance [7, 8]. Despite graphene's exceptional electronic conductivity, it tends to agglomerate due to van der Waals forces, potentially reverting to graphite-like structures. This aggregation impedes lithium intercalation and restricts Li+ movement across graphene layers [9, 10]. Furthermore, at temperatures below 0 °C, The ability of lithium-ions to reach the interface between electrolyte and active material is essential for electrochemical performance, particularly since the electrolyte ionic conductivity decreases significantly [11, 12]. Thus, achieving better low-temperature performance necessitates methods to improve Li+ penetration into the electrolyte-active material interface. Current research on graphene electrodes under low-temperature conditions has yet to yield ideal results. The existence of oxygen-containing functional groups in graphene can trigger excessive SEI film formation and irreversible capacity loss. Consequently, future efforts to optimize graphene's low-temperature electrochemical performance will likely focus on: Developing low-temperature-compatible composite materials; Designing electrolytes tailored for low-temperature operation; Engineering structural features (e.g., pore size, defect density) to enhance Li+ kinetics.

Spinel Lithium Titanium Oxide: At lower temperatures, LTO's high lithiation voltage (~1.55 V vs. Li+/Li) enhances its safety advantage over graphite, as lithium plating is less probable. Established methods for synthesizing nanoparticle LTO have effectively preserved power density below 0 °C by shortening lithium ions' diffusion path. LTO's lower activation energy for charge transfer may also boost its sub-zero performance. While graphite's SEI grows more obstructive at low temperatures, any surface layers on LTO show no significant electrochemical signature, thereby not raising the energy barrier for Li+ transport. Huang and his team used a simple hydrothermal method combined with ion-exchange and calcination treatment. They prepared dual-phase hierarchical porous $Li_4Ti_5O_{12}$–TiO_2 (HP LTO-TO) microspheres. These microspheres served as anode materials for low-temperature lithium-ion batteries (LIBs). [13] The HP LTO-TO microspheres are made of extremely thin nanosheets with a large specific surface area. This structure can be used for interface intercalation reactions and offers a shorter diffusion path for electrons and lithium ions (Li+). As a result, HP LTO-TO electrode demonstrated remarkable low-temperature cycling ability (143 $mA \cdot h \cdot g^{-1}$ at 0 °C, 130 $mA \cdot h \cdot g^{-1}$ at − 10 °C, and 94 $mA \cdot h \cdot g^{-1}$ at − 40 °C under 0.2 C) and rate capability (77 $mA \cdot h \cdot g^{-1}$ at − 20 °C under 5 C).

1.2.2 The Impact on Cathode

Because of their high energy density, long cycle life, and environmental-friendliness, LIBs have garnered significant attention. However, their performance at low temperatures remains a significant challenge, particularly for the cathode materials. Although the exact mechanisms behind LIBs' decreased performance at low temperatures are not fully understood, intrinsic kinetic limitations may be partly responsible for the technical difficulties. The following sections will delve into the low temperature's impact on cathode materials.

It has been reported that the LFP charge transfer resistance at −20 °C is 300% higher than that at room temperature, which affects the rate performance. In the study by Zhang et al., it was found that for cathode half-cells at −30 °C, the charge transfer resistance (Rct) exhibits a strong dependence on the state-of-charge(SOC), with much elevated resistance in the fully-lithiated state. [14] This means that even when the rate is moderate, charging batteries in cold conditions is more difficult than discharging them.

The challenges in sustaining battery performance under 0 °C also stem from elevated grain boundary resistance and sluggish solid-state lithium-ion diffusion within the active material. Certain materials have low activation energy for diffusion, which is advantageous for use in sub-zero conditions. For example, LVP has an activation energy of 6.57, over 7 times smaller than LFP's, and its relative capacity retention rate at −20 °C is almost 300% higher [15]. More broadly, decreasing particle size to nanoscale dimensions and applying conformal coatings of highly conductive carbon/polymer materials on particle surfaces can effectively reduce the diffusional path length and inter-grain resistance, respectively. For instance, $Li_{0.99}La_{0.01}Fe_{0.9}Mg_{0.1}PO_4$/carbon aerogel composites demonstrated exceptional performance at −20 °C, achieving successful operation from 1 to 50 C rate with specific capacities of $120.3\ mA \cdot h \cdot g^{-1}$ and $56.6\ mA \cdot h \cdot g^{-1}$, respectively. [16] While creating these porous structures is scientifically valuable for investigating the fundamental performance boundaries of cathode materials, their production substantially decreases the cell's volumetric energy density, thereby restricting their real-world applicability.

In addition, the performance of cathode materials with different structures also varies at low temperatures. Layered-structured materials, which boast unparalleled rate performance due to their one-dimensional lithium-ion diffusion channels and structural stability from three-dimensional channels, were the first to be commercially used as cathodes in lithium-ion batteries. Their representative compounds include $LiCoO_2$, $Li(Co_{1-x}Ni_x)O_2$, and $Li(Ni, Co, Mn)O_2$, etc. Xie et al. [17] used $LiCoO_2$/MCMB as the research subject and tested its low-temperature charge–discharge characteristics. The results showed that as the temperature decreased, the discharge platform dropped from 3.762 V (at 0 °C) to 3.207 V (at −30 °C), and the battery capacity also sharply decreased from 78.98 mA•h (at 0 °C) to 68.55 mA · h (at −30 °C). [18] For ternary materials, the battery capacity also shows decay with the decrease of temperature.

Compared with the layered structure that needs to use Co to stabilize the structure and improve electrical conductivity, the spinel-structured $LiMn_2O_4$ cathode material, which does not contain Co, offers the benefits of being inexpensive and harmless. However, due to the multiple valence states of Mn and the Jahn–Teller effect of Mn^{3+}, this component has problems such as structural instability and poor reversibility. Regarding low-temperature characteristics, $LiMn_2O_4$ also faces issues like reduced specific capacity at low temperatures, inferior cycling stability, and significant polarization.

1.2.3 The Impact on Electrolytes

Given the relatively limited variety of electrolytes currently used in lithium-ion batteries, the following section will primarily focus on introducing the characteristics of electrolytes as they perform in low-temperature conditions.

From an electrolyte standpoint, lithium-ion battery failure at low temperatures stems from multiple factors: (1) bulk electrolyte limitations, (2) high Li+ desolvation energy barriers, (3) fragile, high-resistance electrolyte/electrode interfaces, and (4) excessive lithium plating and lithium dendrite formation. However, the impact mechanisms of most of these factors on battery performance are similar to those of low temperatures on the cathode and anode. In order to more thoroughly explore this issue, we have selected three unique impacts of low-temperature conditions on the electrolyte for detailed explanation.

Low Bulk Ionic Conductivity Currently, in the realm of LIBs, non-aqueous electrolytes continue to hold a dominant position. Ethylene carbonate (EC) serves as a crucial solvent in advanced lithium-ion battery electrolytes, primarily due to its high dielectric constant and its effectiveness in forming a stable SEI layer on graphite anodes. Because of its viscosity (1.9 cp at 40 °C) and high melting point (38 °C), EC needs to be mixed with linear carbonates like diethyl carbonate (DEC), ethyl methyl carbonate (EMC), or dimethyl carbonate (DMC). While a high proportion of linear carbonates can reduce the electrolyte's viscosity and melting point, thus expanding its temperature range, a minimum bulk EC content of 30 wt.% is essential. This ensures stable SEI layer formation via the EC-dominated solvation structure [19]. In order to maintain compatibility with graphite, the EC content in conventional electrolytes is typically high, ranging from 30 to 50%. This is acceptable at room temperature because these electrolytes possess a high ionic conductivity of 10. However, the situation reverses under low-temperature conditions. When the temperature falls below 0 °C, electrolytes with a high EC content experience a sharp rise in viscosity. This significant "viscosity effect" impedes the movement of the Li+ solvation structure within the bulk electrolyte, resulting in a decrease in ionic conductivity. At room temperature, aqueous electrolytes typically demonstrate greater ionic conductivity than non-aqueous ones. However, water's freezing point (0 °C) is much higher than that of organic solvent mixtures (like cyclic and acyclic carbonates). As a result, aqueous electrolytes experience a more significant drop in ionic conductivity when the temperature decreases [20, 21]. However, ionic conductivity is not the only factor related to the low-temperature performance of LIBs, particularly when there's enough electrolyte to facilitate the movement of lithium ions between the electrodes. Rather than focusing on other properties, the liquid range at low temperatures is crucial for low-temperature electrolytes. This is because electrolyte crystallization can severely harm battery performance.

Freezing Temperature Change: This issue is one that both aqueous and non-aqueous systems face. Therefore, EC-based electrolyte will be taken as an example in this section. On the one hand, the eutectic effect broadens the freezing temperature range of the EC and linear carbonate mixture [22]. On the other hand, the "red line" of 30% EC makes it difficult for the eutectic effect to be fully utilized. For instance, although EMC has the freezing point as low as −53 °C, the EC:EMC (3:7 v/v) solvent mixture's freezing temperature is only 0 °C (Fig. 1.4a). Even though the EC:DMC ratio of 3:7 v/v is at the eutectic point, the freezing temperature is still only 7.6 °C (Fig. 1.4b). The electrolyte's freezing temperature will be lowered by adding lithium salt. Within a certain range, the higher the lithium salt concentration, the lower the freezing point of the electrolyte [23, 24]. However, high salt concentration will bring additional problems such as increased viscosity of lithium salt and precipitation at low temperatures. Currently, most traditional EC-based electrolytes have a freezing temperature exceeding 20 °C. For instance, when the temperature is 20 °C, the classic electrolyte with 1 M in EC/DMC (3:7 v/v) becomes fully solid. Once the temperature drops below 20 °C, EC-based electrolytes experience a complete halt in ionic transport. The Li+ bulk conductivity is nearly zero, making it impossible to speed up the desolvation process and adjust the interface through solvation structure regulation. In conclusion, achieving a broad liquid range at low temperatures is essential for the following development and refinement of low-temperature electrolytes.

Interphase Degradation: Most of the solvents that can work below 20 °C in low-temperature electrolytes are essentially unstable to the anode. Taking graphite as an example, some linear carbonates and propylene carbonate (PC) will jointly insert into graphite, resulting in its exfoliation. Esters engage in condensation reactions with lithiated graphite, generating a substantial volume of gas [25, 26]. Even at low

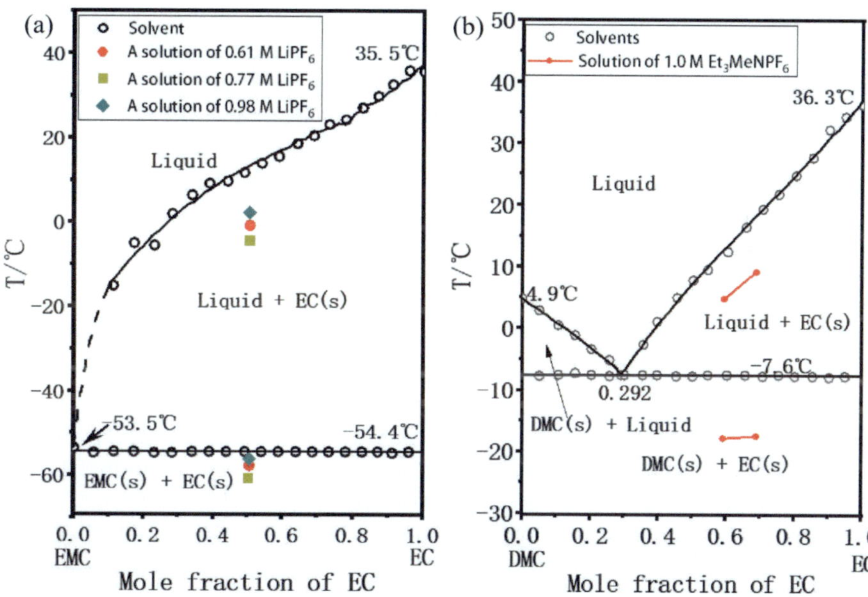

Fig. 1.4 **a** Liquid/solid phase diagram of EMC/EC with different concentration of $LiPF_6$. **b** Liquid/solid phase diagram of DMC/EC. Reproduced with permission [22]

temperatures, co-intercalation and gas generation from low-temperature solvents can cause extreme instability at the electrode/electrolyte interface, leading to a rapid drop in LIBs' cycling life [25]. Besides the SEI, the CEI's chemical composition and spatial structure also greatly affect battery low-temperature performance. This impact is particularly significant when enhancing the low-temperature LIBs' baseline capacity by introducing high-voltage cathodes. [27, 28]

1.3 Battery Low Temperature Performance

At present, the cathode materials of lithium-ion power batteries are mainly lithium-embedded metal oxides, such as lithium manganate $LiMn_2O_4$, lithium iron phosphate $LiFePO_4$, ternary materials like nickel cobalt manganese $LiNi_{1/3}Co_{1/3}Mn_{1/3}O_2$ and nickel cobalt aluminum $LiNi_{1/3}Co_{1/3}Al_{1/3}O_2$, etc., and the anodes are mainly carbon-based materials, such as artificial graphite, natural graphite, etc., with the electrolyte mostly being organic lithium salt solutions (such as $LiPF_6$).

Macroscopically, the low-temperature performance of lithium-ion power batteries is manifested as an increase in the battery's impedance with decreasing temperature, a decrease in the discharge voltage platform, and a rapid drop in the terminal voltage of the battery, leading to a significant reduction in its available capacity and power characteristics. It's hard to discharge at high current in the cold environment, and this is worsened by the battery's increased impedance [29]. The charging voltage rises rapidly at low temperatures, and the time to reach the charging protection termination voltage is shortened, resulting in difficulties in charging and low charging efficiency [29, 30]. On a microscopic scale, the activity of lithium ions inside the battery decreases at low temperatures, and the conductivity of the electrolyte and the solid–liquid phase interface decreases [31–34]. The impedance of the SEI film inside the battery increases, and the charge transfer rate at the reaction interface decreases [14, 35]. All of these cause the battery's impedance to rise and a significant reduction in its power characteristics [36]. Zhang et al. [14, 37] concluded through low-temperature experiments on lithium-ion power batteries that the electrochemical charge transfer resistance increases significantly with decreasing temperature. Wei et al. [38] believe that the low conductivity of the SEI film at the solid–liquid phase interface of the anode graphite at low temperatures leads to poor low-temperature performance of the battery. In addition, at lower temperatures, the solid-phase diffusion coefficient of lithium ions decreases [39, 40] and the impedance of the SEI film increases, resulting in an increase in the polarization of the electrode reaction interface. Especially under high-rate charging conditions, it is easy for lithium ions to deposit on the surface of the anode and form lithium metal dendritic crystals [41–46]. Moreover, as the charging time and number of charges increase, lithium dendrites also grow. Eventually, they may pierce the separator, causing an internal short circuit in the battery, leading to the end of the battery's service life and a significant decrease in its usage safety.

Table 1.1 Specifications and parameters of test battery samples

Parameters	Sample A	Sample B
Type of battery	18650	8 A · h prismatic
Material system	$LiNi_{1/3}Co_{1/3}Mn_{1/3}O_2$	$LiFePO_4$
Quality of battery/g	45.0	340.0
Capacity/A · h	2.3	8
Rated voltage/V	3.6	3.2
End of charge voltage/V	4.2	3.7
End of discharge voltage/V	2.7	2.5
Battery size (diameter × height or length × width × thickness)	18 × 65	135 × 70 × 18
Manufacture	CBAK	LTD

1.3.1 Low Temperature Test for Typical Cells

Two different specifications and materials of batteries were selected for testing, including discharge, charging, and AC impedance testing at low-temperature. The battery sample parameters are shown in Table 1.1.

Sample A is a 18650-type NCM system ($LiNi_{1/3}Co_{1/3}Mn_{1/3}O_2/C$) battery, and Sample B is a LFP system ($LiFePO_4/C$)battery. The charge–discharge testing was conducted using an Arbin (BT200) charge–discharge loading tester. The electrochemical impedance spectroscopy (EIS) testing was performed using a Solartron (SI 1287, 1255B) electrochemical workstation. Before the tests, both batteries were placed in a 25 °C environment and charged and discharged to achieve the same state of charge (SOC). For example, in the discharge testing, both batteries were charged to 100%, and then they were placed in a Patna constant-temperature chamber at different temperatures for 4 h. Once the battery temperature matched the set temperature, the corresponding tests were carried out.

1.3.2 Discharge Analysis

From the discharge curves of the two power batteries shown in Fig. 1.5, it can be observed that as the ambient temperature decreases, the batteries' discharge platform voltage plummets, and the batteries' power characteristics decrease. Moreover, as the temperature drops, the time it takes for the power batteries to reach the cut-off voltage is shortened, and the available capacity decays severely. Comparing the two, it is evident that at the same temperature, Sample B $LiFePO_4/C$ battery has a much higher decay rate than Sample A 18650 $LiNi_{1/3}Co_{1/3}Mn_{1/3}O_2/C$ battery. The main reason stems from the material properties. The lithium iron phosphate material's naturally poor low-temperature conductivity leads to a severe decay in the low-temperature characteristics of the battery [47]. The lower the temperature, the greater the initial terminal voltage drop of the battery. This happens because when the temperature drops, the battery's impedance rises. This increase causes a bigger voltage drop over the battery's internal resistance, making the terminal voltage decreases It is noticed that as indicated by the black circle in Fig. 1.5a, at the beginning of the low-temperature discharge of the battery, the terminal voltage of the battery rises.

This phenomenon is mainly due to the heat generated during the discharge process of the battery. Figure 1.6 shows the discharge curves of the two lithium-ion power batteries at different discharge rates under an ambient temperature of −10 °C. It can be seen that under low-temperature conditions, as the discharge rate increases, there are serious problems of power and capacity characteristic decay in the power batteries.

To show how temperature and discharge rate affect power and capacity traits of power batteries in a more intuitive way, Fig. 1.7 presents the available capacity ratio of the two power batteries at various discharge rates and temperatures. The capacity

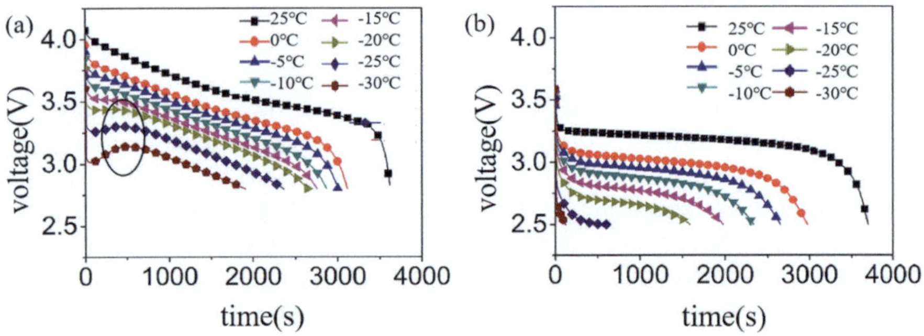

Fig. 1.5 The discharge voltage curves of **a** Sample A $LiNi_{1/3}Co_{1/3}Mn_{1/3}O_2/C$ battery and **b** Sample B $LiFePO_4/C$ battery at different temperatures when the discharge rate is 1 C

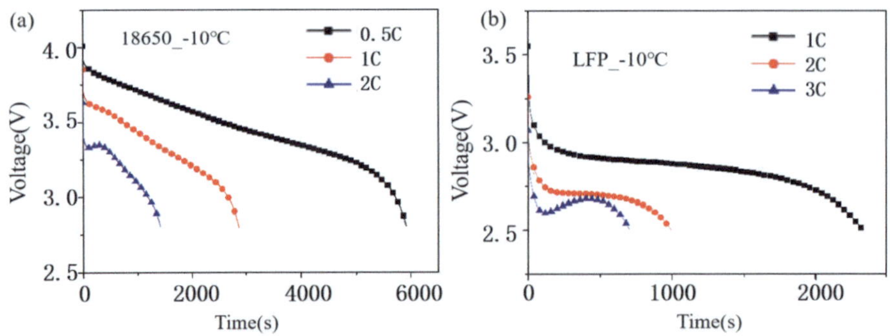

Fig. 1.6 The voltage curves of **a** Sample A $LiNi_{1/3}Co_{1/3}Mn_{1/3}O_2/C$ battery and **b** Sample B $LiFePO_4/C$ battery at different discharge rates at an ambient temperature of −10 °C

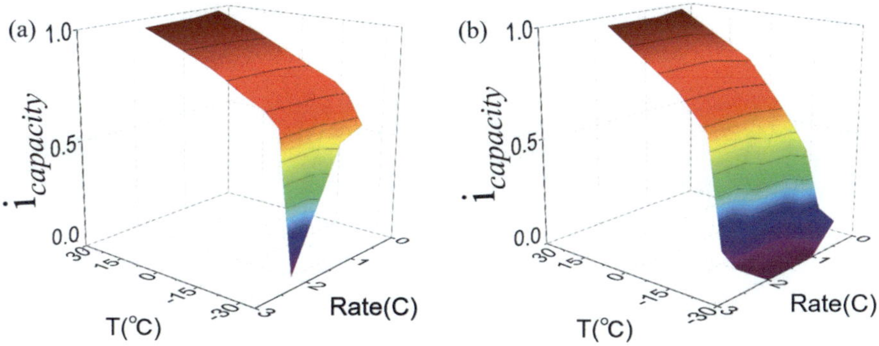

Fig. 1.7 The discharge capacity ratio of **a** Sample A $LiNi_{1/3}Co_{1/3}Mn_{1/3}O_2/C$ battery and **b** Sample B $LiFePO_4/C$ battery under different discharge rates and different discharge temperatures

percentage in the figure is based on the discharge capacity of the LIB at a discharge rate of 0.5 °C at 25 °C, which is defined as:

$$i_{cap} = \frac{\text{Current measured capacity}}{\text{Discharge capacity at } 25\,^\circ\text{C and } 0.5\,\text{C}} \tag{1.1}$$

From Fig. 1.7, it can be seen that the capacity of the batteries significantly decreases with the reduction of ambient temperature. For Sample A 18650 $LiNi_{1/3}Co_{1/3}Mn_{1/3}O_2/C$ batteries, the discharge capacity at a 0.5 C rate and a 1 C rate at −30 °C has already decayed to about 50% of the discharge capacity at 25 °C, and the 2 C constant current discharge capacity is zero. It is evident that low-temperature environments severely restrict the use of batteries and are the main reason for the significant reduction in the range of EVs. Moreover, the impact of the discharge rate on the capacity of batteries becomes significantly more pronounced as the temperature decreases.

In view of the fact that the discharge curve of power battery A does not have an obvious voltage plateau, and there is a voltage recovery phenomenon during low-temperature discharge due to its own heat generation, as shown in Fig. 1.6. Therefore, the relationship between the power characteristics of the power battery, temperature, and discharge rate is evaluated by comparing the initial terminal voltages. Figure 1.8 presents the initial terminal voltage of the two LIBs under various ambient temperatures and discharge rates. The voltage percentage in the figure is referenced to the initial terminal voltage of the LIB during discharge at a rate of 0.5 C at 25 °C, which is defined as:

$$i_{voltage} = \frac{\text{Current initial terminal voltage}}{\text{Discharge initial terminal voltage at } 25\,^\circ\text{C and } 0.5\,\text{C}} \tag{1.2}$$

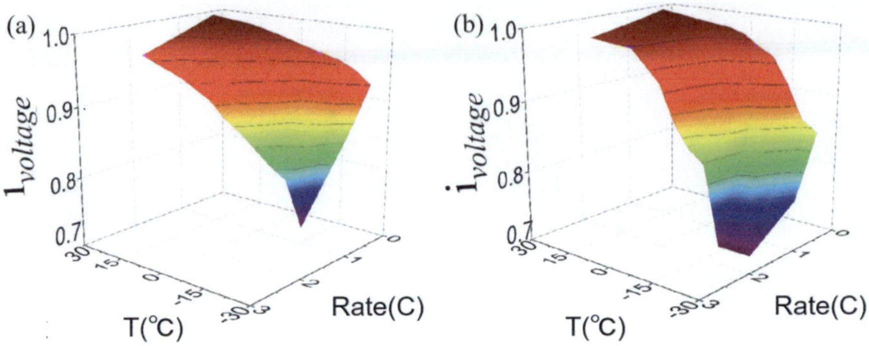

Fig. 1.8 The ratio of **a** Sample A $LiNi_{1/3}Co_{1/3}Mn_{1/3}O_2/C$ battery's and **b** Sample B $LiFePO_4/C$ battery's initial terminal voltage during discharge under different discharge rates and temperatures

From Fig. 1.8, it can be seen that the initial terminal voltage of the battery during discharge is significantly affected by both temperature and discharge rate. As the temperature decreases, the initial discharge terminal voltage of the battery continues to drop. This is mainly because lower temperatures cause greater internal resistance, which in turn leads to a larger voltage drop within the battery. Additionally, when the temperature falls, the disparity in the battery's initial discharge terminal voltage at varying rates becomes more noticeable. At −30 °C, the initial voltage during 0.5 C rate discharge only decreased by 6.8% compared to that at 25 °C, the initial voltage during 1 C rate discharge decreased by nearly 12.7% compared to that at 25 °C, and the initial voltage during 2 C rate discharge decreased by nearly 22.8% compared to that at 25 °C. The significant decay of the voltage platform of the battery during low-temperature high-rate discharge severely affects the power output of the battery. In automotive applications, this mainly impacts the vehicle's acceleration and hill-climbing performance.

1.3.3 Charge Analysis

The charging time, depth of charge, and charging safety of batteries have always been key concerns in the field of EV applications. Figure 1.9 shows the CC–CV (Constant Current–Constant Voltage) charging curves of the two batteries at different temperatures with a charging current rate of 0.5 C. Considering the differences in low-temperature characteristics of the two batteries, for Sample A 18650 $LiNi_{1/3}Co_{1/3}Mn_{1/3}O_2/C$ battery, tests were conducted at −15 and −25 °C, while for Sample B $LiFePO_4/C$ battery, tests were conducted at −10 and −20 °C. It can be seen that when charging just starts, the terminal voltage of the battery experiences an immediate rise, with lower temperatures resulting in a higher initial charging voltage for the battery. At low temperatures, the terminal voltage soars quickly, soon reaching the cut-off voltage and entering the constant voltage charging phase. As the

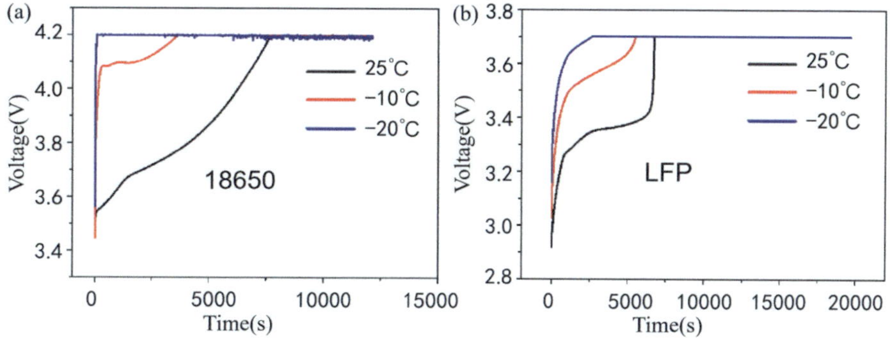

Fig. 1.9 Low-temperature CC–CV charging curves of **a** Sample A $LiNi_{1/3}Co_{1/3}Mn_{1/3}O_2/C$ battery and **b** Sample B $LiFePO_4/C$ battery test samples

temperature drops, the battery's constant current charging time shortens, while the constant voltage charging stage lasts longer. To charge the same amount of charge, the required charging time for the battery will significantly increase.

Figure 1.10 shows the charging capacity test results of the two batteries at different temperatures, divided into the charging capacity during the constant current stage and the constant voltage stage. From the analysis of the figure, it can be seen that when the same charging termination conditions are set, as the temperature decreases, the overall charging capacity of the battery decreases. The 18650 $LiNi_{1/3}Co_{1/3}Mn_{1/3}O_2/C$ battery's charging capacity at −25 °C has already dropped to 50% of the normal temperature charging capacity. In the set CC–CV charging mode, the charging capacity during the battery's constant voltage stage rises as the temperature drops. This indicates that as the temperature drops, the battery's constant current charging capacity decreases, with most charging then occurring in the constant voltage stage. The long duration of the constant voltage stage charging leads to an extension of the overall charging time of the battery, reducing charging time efficiency. In addition, prolonged low-temperature constant voltage charging is also a factor in battery performance degradation due to side reactions.

Figure 1.11 compares the charging curves of a certain manganese acid lithium system ($LiMn_2O_4/C$) power battery at different temperatures and charging rates [48]. Figure 1.9a and b show the charging current values of 10 A and 35 A, respectively. Unlike the low-temperature charging characteristics of $LiNi_{1/3}Co_{1/3}Mn_{1/3}O_2/C$ batteries and $LiFePO_4/C$ batteries, which are not significantly affected, the initial charging voltage in low-temperature environments is significantly higher than that in normal temperature environments. Low-temperature charging can cause "lithium deposition" on the anode graphite. As the "lithium deposition" in the battery gradually crystallizes, it can pierce the separator, posing a safety hazard in use. Therefore, in the technical manuals provided by most power battery manufacturers, it is required that the temperature of the power battery must be above 0 °C before charging operations can be performed.

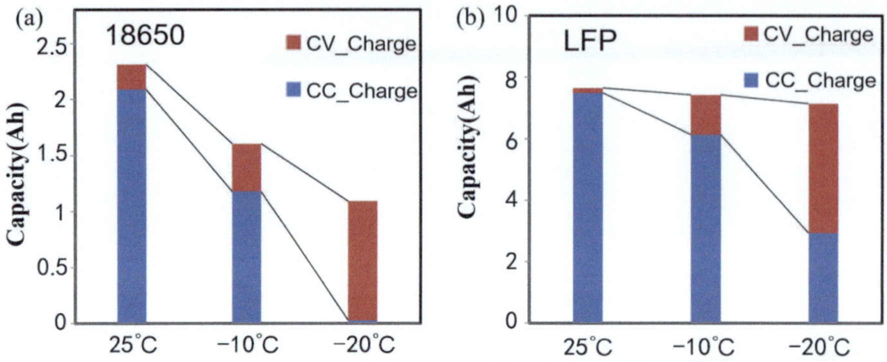

Fig. 1.10 Comparison chart of charged capacity for **a** Sample A $LiNi_{1/3}Co_{1/3}Mn_{1/3}O_2/C$ battery and **b** Sample B $LiFePO_4/C$ battery at different temperatures

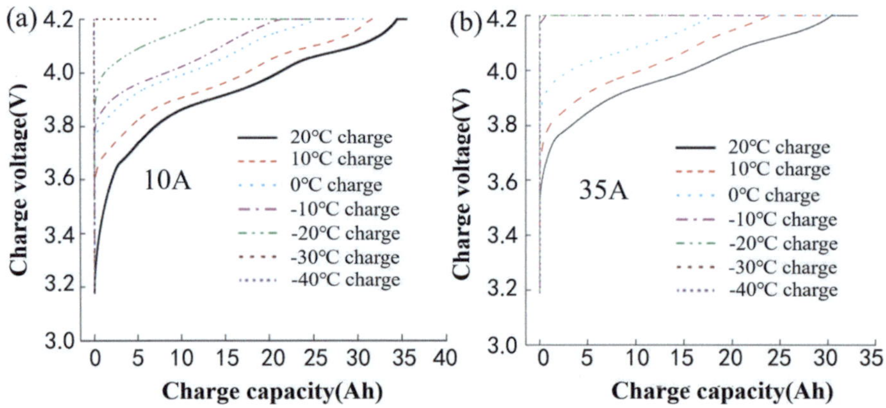

Fig. 1.11 Comparison of charging curves for LIBs with a lithium manganese oxide system at different temperatures and rates. The charge curve is 10 A (**a**) and 35 A (**b**)

1.3.4 Impedance Change

Electrochemical Impedance Spectroscopy (EIS), or AC impedance spectroscopy, involves applying a small AC signal (voltage or current) with a sinusoidal waveform within a specific frequency range to an electrochemical system and measuring how the AC voltage and current signals change with frequency. This method offers greater insight into the electrode interface structure and kinetics compared to other conventional electrochemical methods, making it a popular tool for studying the internal mechanisms of LIBs. [48–52] Next, EIS will be used to study the effect of temperature on the electrochemical impedance characteristics of power batteries.

Figure 1.12 shows the AC impedance spectra of the two test samples at different temperatures when SOC is 50%. In the plot, the x-axis represents the real part of the

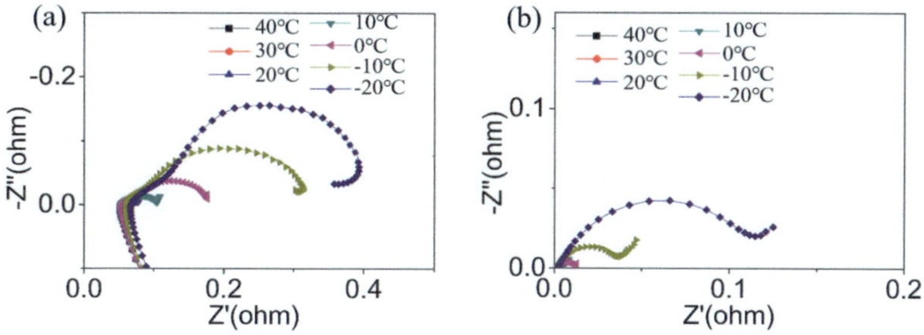

Fig. 1.12 The coupling relationship between **a** Sample A $LiNi_{1/3}Co_{1/3}Mn_{1/3}O_2/C$ battery and **b** Sample B $LiFePO_4/C$ battery AC impedance spectrum and temperature [53]

impedance, while the y-axis represents the imaginary part, with the plot drawn from high to low frequency (the frequency range is 100–0.01 kHz; the excitation current amplitude is 1.5 A). As the temperature decreases, the AC impedance arc (diameter) significantly increases. The entire spectrum shifts rightward, signaling higher ohmic impedance in the ultra-high frequency region. As the temperature drops, the high-frequency and medium-frequency impedance arcs gradually expand, indicating a rise in the SEI film impedance and charge transfer resistance. In the low-frequency region, the slope of the line that represents the diffusion process also undergoes a significant alteration. At low temperatures, for Sample A battery, the diffusion line has already disappeared from the spectrum, and at −20 °C, the impedance has increased to several times that of the normal temperature.

To further analyze the variation of impedance with temperature, a quantitative analysis is conducted using an equivalent impedance model based on EIS analysis. Figure 1.13a illustrates the correspondence between the equivalent impedance model and the impedance spectra [54–57], which is explained as follows:

Ultra-high frequency region (above 10kHz): This covers electron movement through active material particles and lithium-ion migration in the electrolyte within the pores of these particles [58, 59]. On the EIS spectrum, the electrochemical ohmic resistance is represented by the intersection with the real axis.

High frequency region: The diffusion and migration of lithium ions through the SEI film on the surface of the active material particles are represented as a semicircular arc on the impedance spectrum [60–62]. This process is equivalently substituted by a RSEI/CSEI parallel structure in the impedance model.

Medium frequency region: This is the impedance arc related to the electrochemical reaction, which includes the processes of charge transfer and the charging and discharging of the electric double layer. The charge transfer happens at the contact surface between the solid-phase electrode and the electrolyte. [63, 64] This process obeys Faraday's law, so it is also known as the Faraday process. The charge transfer's speed in this process is reflected by the Faraday current, and it can generally be equivalently represented as a pure resistor, known as the charge transfer resistance or

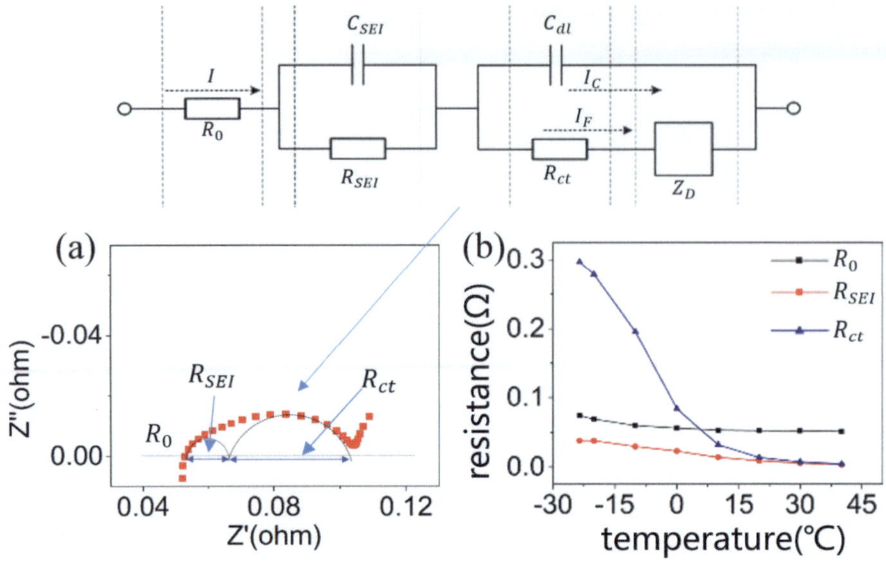

Fig. 1.13 **a** The EIS spectrum of Sample A battery with R_0, R_{SEI}, R_{ct} when the SOC is 50%. **b** The relationship between R_0, R_{SEI}, R_{ct} and temperature of Sample A battery with 50% SOC [53]

Faraday resistance, represented by Rct. The charging and discharging of the electric double layer, occurring at the interface between the solid-phase electrode and liquid-phase electrolyte, forms a capacitor—like physical structure, creating the electrode's interface electric double layer. This process, represented by a capacitor Cdl, is also known as the non-Faraday process.

Low frequency region: The diffusion process of lithium ions mainly occurs in this region, which is associated with the active material particles. [57, 65, 66] During the electrochemical reaction, the buildup of products and depletion of reactants at the interface create a concentration gradient at the interface of the solid and liquid phases. This is due to the Faradaic current passing through the solid-phase electrode and electrolyte interface. According to the porous electrode theory, the solid-phase electrode is considered as spherical particles with specific porosity. As the reaction proceeds, substances build up inside these particles. This accumulation reduces the concentration gradient between the interior and exterior of the interface and slows down the diffusion rate. [67–69] A stable concentration polarization, which results from the lithium ion concentration difference within the battery, occurs once the diffusion of material on the electrode reaches a steady state. This diffusion process can generally be represented by the impedance of a semi-infinite diffusion, known as the Warburg impedance ZW. Taking into account the geometric factors of the electrode surface, such as porosity and roughness, as well as the presence of adsorption, this is typically modeled using a phase component, represented by the symbol ZD.

Very low frequency region (<0.001 Hz): This involves the buildup and depletion of lithium ions within the active material, leading to alterations in the active material particles' crystal structure or the new phases' emergence.

Since the testing range of EIS is 100–0.01 kHz, the semi-circles related to the changes in the active material particles' crystal structure or the new phases' emergence in the very low frequency region are not observable in the EIS spectrum. As shown in Fig. 1.13a, the EIS analysis software ZView fits and identifies the battery impedance parameters R_0, R_{SEI} and R_{ct}. These three impedance values can be determined separately based on the horizontal axis of the impedance spectrum. Figure 1.13b shows the variation curves of the electrochemical ohmic resistance (R0), SEI film resistance (RSEI), and charge transfer resistance (R_{ct}) of Sample A battery with temperature at SOC of 50%.

As shown in Fig. 1.13, the temperature decrease leads to an increase in impedance. Among them, R_0 and R_{SEI} are relatively stable with temperature changes, increasing only slightly as the temperature drops. However, R_{ct} increased significantly with the temperature decrease. The reason is that R_0 and R_{SEI} are mainly influenced by the electrolyte's ionic conductivity, and their variation with temperature resembles the pattern of the electrolyte's ionic conductivity. Of the three internal resistance components, R_{ct} was the most significantly affected by temperature. With the temperature's decrease, the R_{ct}'s proportion in the whole battery's impedance gradually increases. When the temperature drops below −20 °C, R_{ct} in the proportion of the whole battery impedance is basically nearly 80%. Therefore, it can be considered that the LIB's low temperature performance is primarily restricted by the large charge migration internal resistance R_{ct}.

Based on sample B, the battery focuses on the internal resistance of charge migration in the impedance data of 50% SOC, and the results are shown in Fig. 1.14. In the impedance map, R_{ct} is mostly used to represent the power battery's electrochemical reaction capacity, and the smaller the value indicates that the faster the electrochemical reaction rate is. The impedance of the battery showed a significant increase with the battery temperature, indicating that the chemical reaction rate of the battery decreased significantly. The dependence of charge migration internal resistance on temperature can be described by the Arrhenius equation for [70]:

$$R_{ct} = A_1 \cdot \exp\left(-\frac{E_a}{RT}\right) \tag{1.3}$$

where A_1 is the pre-exponential factor of R_{ct}, E_a is the electrochemical reaction's activation energy, R is the gas constant, T is the temperature.

Figure 1.14b shows the relationship between $\ln(R_{ct})$ and the absolute temperature's reciprocal. From the figure, it can be seen that $\ln(R_{ct})$ and 1/T have a significant linear relationship. Using this relationship to plot according to the Arrhenius law, the the reaction's activation energy can be determined.

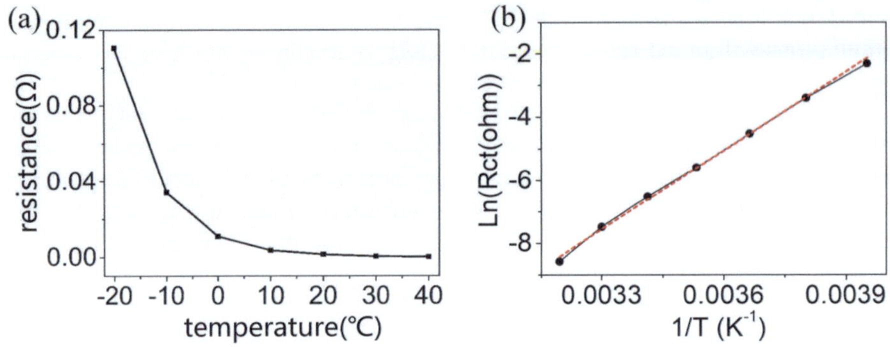

Fig. 1.14 **a** The relationship between the charge transfer internal resistance and temperature of sample B battery with SOC of 50%. **b** The relationship between $\ln(R_{ct})$ and the absolute temperature's reciprocal [53]

1.4 Summary of Key Factors Affecting Battery Low-temperature Characteristics

LIBs confront severe degradation in energy and power characteristics when used at low temperatures. Referring to the literature and combining the experimental research mentioned earlier, macroscopically, the LIBs' low-temperature performance is characterized by an increase in impedance and a decrease in the discharge voltage platform as the temperature drops. The battery's terminal voltage decreases rapidly, leading to a significant reduction in available capacity and power. Additionally, at low temperatures, the batteries not only struggle with high-current discharge but also experience increased impedance [29]. This leads to a rapid rise in charging voltage, shortening the time to reach the charge protection termination voltage. As a result, charging becomes difficult with low efficiency [29, 30]. On a microscopic scale, the low-temperature characteristics of LIBs can be summarized as being mainly constrained by the following key factors:

(1) The electrolyte's ionic conductivity inside the battery is too low at low temperatures [36, 71, 72];
(2) The SEI film's conductivity on the surface of the anode graphite particles decreases at low temperatures [73–75];
(3) The electrochemical reaction's rate at the battery electrodes decreases at low temperatures [70, 76–78];
(4) The lithium ions' solid-phase diffusion coefficient in the anode graphite material particles is too low at low temperatures [14, 36, 37, 74, 79].

Specifically, the battery electrolyte closely relates to the low-temperature performance of LIBs. The battery electrolyte solvent not only directly affects the electrolyte's liquid-phase line temperature range but also participates directly in the reactions that form the SEI film. At low temperatures, the electrolyte conductivity decreases. Lithium plating, shown in Fig. 1.15, is the main battery deterioration

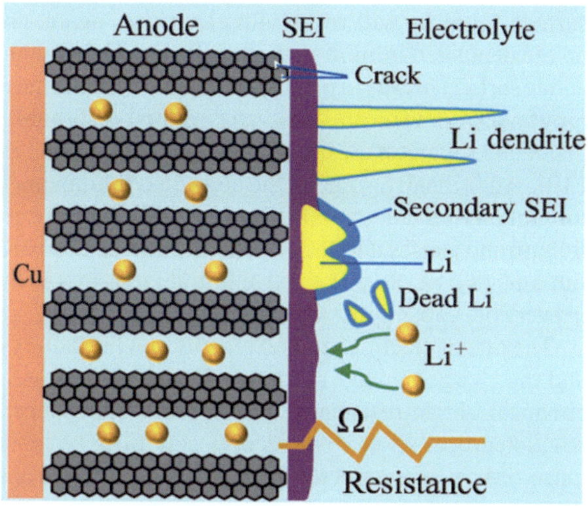

Fig. 1.15 The degradation mechanism caused by lithium plating on anode under low temperature [80]

mechanism at low temperatures. Currently, the majority of LIBs' anode materials are carbon-based, particularly graphite. When charging, lithium ions move from the cathode to the anode and insert into the graphite layers. Nevertheless, at low temperatures, the electrolyte's viscosity increases and its conductivity decreases, slowing the charge transfer kinetics of lithium ions. [81] When the anode potential drops below 0 V versus Li/Li+, lithium deposition reaction becomes dominant [46, 82–84]. At this point, some lithium ions are reduced to metallic lithium, depositing on the anode surface instead of being inserted into graphite layers. Usually, deposited lithium is re-oxidized at about 0.1 V, which is significantly below the anode delithiation potential, so the plating process is reversible [85]. During the discharging process, the deposited lithium dissolves, producing lithium ions. However, the plated lithium is highly reactive, directly reacting with electrolyte, and consuming part of the lithium inventory [43, 84].

Additionally, when the active material is covered by plated lithium, local volume expansion occurs, leading to an increase in mechanical stress [86]. During cycling, mechanical stress rises, causing active material cracking and subsequent loss. This loss decreases active sites, thereby lowering the anode's capacity.

The increase in the impedance of the SEI film on the internal electrodes of the battery at low temperatures is another factor contributing to the deterioration of battery performance. At low temperatures, the SEI film's impedance on the battery's internal electrodes rises, leading to reduce the battery's available power. Meanwhile, the SEI ionic conductivity plummets, heavily hampering the lithium ions intercalation into the graphite layers [87]. Especially when charging at low temperatures, metallic lithium precipitates on the anode particles' surface, and the reaction between lithium metal and electrolyte causes the SEI film to thicken. On one hand, lithium plating can harm the SEI film, causing it to regenerate and become thicker [86]. Thus, the SEI film impedance increases. Besides, the plated lithium can be deposited on the SEI film surface. Lithium will react with electrolyte again, creating the secondary SEI film on the deposited lithium surface [88]. The plated lithium's uneven deposition on the anode surface results in protrusions on the SEI film surface [89]. As the lithium plating progresses, some protrusions can develop into lithium dendrites. These dendrites can pierce the separators, causing internal short circuits and compromising the safety of LIBs. Additionally, some of the deposited lithium may break away and detach during lithium plating. On the other hand, when the SEI film completely covers the plated lithium and causes it to lose electrical contact, this part of lithium turns into the "dead lithium" [84, 88, 89]. The active lithium ions loss in the anode results in the battery's irreversible capacity loss.

The rate of electrochemical reactions in the battery decreases at low temperatures, and the charge transfer resistance increases substantially. Compared to the electrochemical ohmic resistance and SEI film impedance, the battery's electrochemical reaction process is more noticeably controlled by temperature. Charge transfer resistance rises at an exponential rate as temperature drops. It can be considered that the dramatic increase in low-temperature charge transfer resistance is the primary cause of the battery's power performance deterioration.

The reduction in the lithium ions' solid-phase diffusion coefficient in the anode is also significantly contributes to the battery performance deterioration at low temperatures. The drop in lithium ions' solid-phase diffusion coefficient in graphite at low temperatures is the main rate—controlling step causing the battery's capacity characteristic deterioration. When charging a battery at low temperature, a smaller diffusion coefficient can slow down the lithium ions diffusion in the anode graphite. This increases the likelihood of "lithium deposition" on the anode particles' surface, causing permanent battery damage.

1.5 Conclusions

This chapter delves into the impact of low temperatures on lithium-ion battery materials, including both the cathode and anode materials as well as electrolyte. It also examines the lithium-ion batteries performance in low-temperature environments and the factors that constrain their performance. It initially points out that the range, charging time, and safety of EVs are mainly affected by the characteristics of the batteries. The characteristics of LIBs significantly decline in cold environments. For example, the battery capacity and operating voltage markedly decrease. Charging becomes difficult, and the formation of lithium dendrites can lead to internal short circuits, reducing the battery's lifespan and safety.

At low temperature, the performance degradation of lithium-ion batteries mainly stems from the deterioration of the physical and chemical properties of electrode materials and electrolytes. For anode materials, graphite faces capacity decay and safety hazards due to lithium dendrite formation and insufficient SEI film stability at low temperatures, while graphene and LTO show potential but still need to address issues of aggregation and interface compatibility. The low-temperature performance of cathode materials (such as LFP and LVP) is limited by increased charge transfer resistance and slow solid-state diffusion of lithium-ions, which requires optimization through nanosization, conductive coatings, and other means. In terms of electrolytes, the decline of ionic conductivity, the narrowing of the liquid range, and the intensification of interfacial side reactions (such as co-intercalation and gas generation) at low temperatures further restrict battery performance. It is necessary to balance the solvent components (such as EC content), lithium salt concentration, and additive design to expand low-temperature adaptability.

Through comparative discharge, charge, and EIS tests on two different material systems of batteries (nickel–cobalt–manganese and lithium iron phosphate), the chapter provides a detailed analysis of the impact of low temperatures on battery performance. The test results show that as the temperature drops, the batteries' discharge platform voltage rapidly declines, and the available capacity significantly decays. Moreover, the decay rate of lithium iron phosphate system batteries is higher than that of nickel–cobalt–manganese system batteries. Under low-temperature conditions, the initial terminal voltage drop of the batteries increases, and the increase in discharge rate further exacerbates the decay of power and capacity characteristics.

In terms of charging performance, low temperatures cause the initial charging voltage of the batteries to rise. The constant current charging time shortens, while the constant voltage charging time extends. As a result, the overall charging time increases, and charging efficiency decreases. In addition, low-temperature charging can cause "lithium plating" on the negative graphite electrode, posing safety hazards.

Subsequently, the chapter analyzes the temperature impact on the electrochemical impedance characteristics of batteries using EIS. As the temperature decreases, the AC impedance of the batteries significantly increases. This is manifested as an increase in ohmic resistance, an increase in SEI film resistance and charge transfer resistance, and a change in the slope of the diffusion process. Equivalent impedance model analysis indicates that the charge transfer resistance rises dramatically as temperature drops, becoming the primary reason restricting batteries' low-temperature performance.

Finally, the chapter summarizes the key factors affecting the low-temperature characteristics of LIBs. These include the low ionic conductivity of the electrolyte, the reduced conductivity of the SEI film on the negative electrode particles, the decreased rate of electrochemical reactions, and the lithium ions' lowered solid-phase diffusion coefficient in the graphite particles of the anode. These factors collectively lead to a significant decay in the energy and power characteristics of batteries at low temperatures.

References

1. Goodenough J B, Park K-S. The Li-ion rechargeable battery: a perspective [J]. Journal of the American Chemical Society, 2013, 135(4): 1167–1176.

2. Yoo H D, Markevich E, Salitra G, et al. On the challenge of developing advanced technologies for electrochemical energy storage and conversion [J]. Materials Today, 2014, 17(3): 110–121.
3. Billaud D, Nabais C, Mercier C, et al. Electrochemical lithium insertion in graphite containing dispersed tin–antimony alloys [J]. Energy Conversion and Management, 2008, 49(9): 2447–2454.
4. Park M-S, kang Y-M, Kim J-H, et al. Effects of low-temperature carbon encapsulation on the electrochemical performance of SnO2 nanopowders [J]. Carbon, 2008, 46(1): 35–40.
5. Wang Y, Liu F, fan G, et al. Electroless formation of a fluorinated Li/Na hybrid interphase for robust lithium anodes [J]. Journal of the American Chemical Society, 2021, 143(7): 2829–2837.
6. Zhu X, zhu Y, Murali S, et al. Reduced graphene oxide/tin oxide composite as an enhanced anode material for lithium ion batteries prepared by homogenous coprecipitation [J]. Journal of Power Sources, 2011, 196(15): 6473–6477.
7. Grishanov D A, Mikhaylov A A, Medvedev A G, et al. Graphene Oxide-Supported β-Tin Telluride Composite for Sodium- and Lithium-Ion Battery Anodes [J]. Energy Technology, 2018, 6(1): 127–133.
8. Dubin S, Gilje S, Wang K, et al. A One-Step, Solvothermal Reduction Method for Producing Reduced Graphene Oxide Dispersions in Organic Solvents [J]. ACS Nano, 2010, 4(7): 3845–3852.
9. Reddy M J K, Ryu S H, Shanmugharaj A M. Synthesis of SnO2 pillared carbon using long chain alkylamine grafted graphene oxide: an efficient anode material for lithium ion batteries [J]. Nanoscale, 2015, 8(1): 471–482.
10. Shivhare S, Vyas S, Bagal V S, et al. An electrode comprising of graphene nanopowder inserted in an enclosed structure in anodic aluminium oxide coated with PANI by using low temperature hydrothermal process [J]. AIP Conference Proceedings, 2018, 1942(1): 140009.
11. He J, Li Q, Chen Y, et al. Self-assembled cauliflower-like FeS2 anchored into graphene foam as free-standing anode for high-performance lithium-ion batteries [J]. Carbon, 2017, 114: 111–116.
12. Gupta A, Manthiram A. Designing Advanced Lithium-Based Batteries for Low-Temperature Conditions [J]. Advanced Energy Materials, 2020, 10(38): 2001972.
13. Li Z, Huang J, Yann Liaw B, et al. A review of lithium deposition in lithium-ion and lithium metal secondary batteries [J]. Journal of Power Sources, 2014, 254: 168–182.
14. Zhang S S, Xu K, Jow T R. The low temperature performance of Li-ion batteries [J]. Journal of Power Sources, 2003, 115(1): 137–140.
15. Rui X H, jin Y, Feng X Y, et al. A comparative study on the low-temperature performance of LiFePO4/C and Li3V2(PO4)3/C cathodes for lithium-ion batteries [J]. Journal of Power Sources, 2011, 196(4): 2109–2114.
16. Zhang H, Xu Y, Zhao C, et al. Effects of carbon coating and metal ions doping on low temperature electrochemical properties of LiFePO4 cathode material [J]. Electrochimica Acta, 2012, 83: 341–347.
17. Xie X H, Xie J Y, Xia B J. Study on the Charge and Discharge Performance of Lithium Ion Batteries at Low Temperature [J]. Chemical World, 2008(10): 581–583+614.
18. Zhao S X, Guo S T, Zhao J W, et al. Development on Low-temperature Performance of Lithium Ion Batteries [J]. Journal of The Chinese Ceramic Society, 2016, 44(1): 19–28.
19. Xu K. "Charge-Transfer" Process at Graphite/Electrolyte Interface and the Solvation Sheath Structure of Li + in Nonaqueous Electrolytes [J]. Journal of The Electrochemical Society, 2007, 154(3): A162.
20. Huang J, Dong X, Wang N, et al. Building low-temperature batteries: Non-aqueous or aqueous electrolyte? [J]. Current Opinion in Electrochemistry, 2022, 33: 100949.
21. Yan L, Zhang S, Kang Q, et al. Iodine conversion chemistry in aqueous batteries: Challenges, strategies, and perspectives [J]. Energy Storage Materials, 2023, 54: 339–365.
22. Ding M S, Xu K, Jow T R. Liquid-Solid Phase Diagrams of Binary Carbonates for Lithium Batteries [J]. Journal of The Electrochemical Society, 2000, 147(5): 1688.
23. Plichta E J, Behl W K. A low-temperature electrolyte for lithium and lithium-ion batteries [J]. Journal of Power Sources, 2000, 88(2): 192–196.

24. Hubble D, Emory Brown D, Zhao Y, et al. Liquid electrolyte development for low-temperature lithium-ion batteries [J]. Energy & Environmental Science, 2022, 15(2): 550–578.
25. Xu G, Huang S, Cui Z, et al. Functional additives assisted ester-carbonate electrolyte enables wide temperature operation of a high-voltage (5 V-Class) Li-ion battery [J]. Journal of Power Sources, 2019, 416: 29–36.
26. Ein-Eli Y, Thomas S R, Chadha R, et al. Li-Ion Battery Electrolyte Formulated for Low-Temperature Applications [J]. Journal of The Electrochemical Society, 1997, 144(3): 823.
27. Wang J, nie P, Xu G, et al. High-Voltage LiNi0.45Cr0.1Mn1.45O4 Cathode with Superlong Cycle Performance for Wide Temperature Lithium-Ion Batteries [J]. Advanced Functional Materials, 2018, 28(4): 1704808.
28. Chen J, Xing L, Yang X, et al. Outstanding electrochemical performance of high-voltage LiNi1/3Co1/3Mn1/3O2 cathode achieved by application of LiPO2F2 electrolyte additive [J]. Electrochimica Acta, 2018, 290: 568–576.
29. Zhang S S, Xu K, Jow T R. Charge and discharge characteristics of a commercial LiCoO2-based 18650 Li-ion battery [J]. Journal of Power Sources, 2006, 160(2): 1403–1409.
30. Pesaran A, Vlahinos A, Stuart T. Cooling and preheating of batteries in hybrid electric vehicles [J]. The 6th ASME-JSME Thermal Engineering Joint Conference, 2003.
31. Li S, li X, Liu J, et al. A low-temperature electrolyte for lithium-ion batteries [J]. Ionics, 2015, 21(4): 901–907.
32. Fan J. On the discharge capability and its limiting factors of commercial 18650 Li-ion cell at low temperatures [J]. Journal of Power Sources, 2003, 117(1): 170–178.
33. Smart M C, Ratnakumar B V, Surampudi S, et al. Irreversible Capacities of Graphite in Low-Temperature Electrolytes for Lithium-Ion Batteries [J]. Journal of The Electrochemical Society, 1999, 146(11): 3963–3969.
34. Fu Q M. Research on low temperature performance of Lithium-Ion batteries[D/OL]. Harbin Engineering University, 2005[2025–01–21].
35. Wang C, Appleby A J, Little F E. Low-Temperature Characterization of Lithium-Ion Carbon Anodes via Microperturbation Measurement [J]. Journal of The Electrochemical Society, 2002, 149(6): A754.
36. Ji Y, Zhang Y, Wang C-Y. Li-Ion Cell Operation at Low Temperatures [J]. Journal of The Electrochemical Society, 2013, 160(4): A636–A649.
37. Zhang S S, Xu K, Jow T R. Electrochemical impedance study on the low temperature of Li-ion batteries [J]. Electrochimica Acta, 2004, 49(7): 1057–1061.
38. Wei Z H. Research on The Modification of Graphite as Anode Material for Low Temperature Li-Ion Battery[D/OL]. Harbin Institute of Technology, 2012[2025–01–21].
39. Smart M C, ratnakumar B V, Whitcanack L D, et al. Improved low-temperature performance of lithium-ion cells with quaternary carbonate-based electrolytes [J]. Journal of Power Sources, 2003, 119–121: 349–358.
40. Huang C-K, Sakamoto J S, Wolfenstine J, et al. The Limits of Low-Temperature Performance of Li-Ion Cells [J]. Journal of The Electrochemical Society, 2000, 147(8): 2893.
41. Uhlmann C, Illig J, Ender M, et al. In situ detection of lithium metal plating on graphite in experimental cells [J]. Journal of Power Sources, 2015, 279: 428–438.
42. Ely D R, García R E. Heterogeneous Nucleation and Growth of Lithium Electrodeposits on Negative Electrodes [J]. Journal of The Electrochemical Society, 2013, 160(4): A662–A668.
43. Akolkar R. Modeling dendrite growth during lithium electrodeposition at sub-ambient temperature [J]. Journal of Power Sources, 2014, 246: 84–89.
44. Bhattacharyya R, Key B, Chen H, et al. In situ NMR observation of the formation of metallic lithium microstructures in lithium batteries [J]. Nature Materials, 2010, 9(6): 504–510.
45. Khurana R, Schaefer J L, Archer L A, et al. Suppression of Lithium Dendrite Growth Using Cross-Linked Polyethylene/Poly(ethylene oxide) Electrolytes: A New Approach for Practical Lithium-Metal Polymer Batteries [J]. Journal of the American Chemical Society, 2014, 136(20): 7395–7402.
46. Petzl M, Danzer M A. Nondestructive detection, characterization, and quantification of lithium plating in commercial lithium-ion batteries [J]. Journal of Power Sources, 2014, 254: 80–87.

47. Xie X H, Zhang J, Li J, et al. Study on the low temperature performance of the LiFePO4/Li batteries [J]. Journal of South China Normal University (Natural Science Edition), 2009(S1): 243–244.
48. Lei Z, Zhang C, Li J, et al. Preheating method of lithium-ion batteries in an electric vehicle [J]. Journal of Modern Power Systems and Clean Energy, 2015, 3(2): 289–296.
49. Zhang S S, Xu K, Jow T R. EIS study on the formation of solid electrolyte interface in Li-ion battery [J]. Electrochimica Acta, 2006, 51(8): 1636–1640.
50. Gomez J, Nelson R, Kalu E E, et al. Equivalent circuit model parameters of a high-power Li-ion battery: Thermal and state of charge effects [J]. Journal of Power Sources, 2011, 196(10): 4826–4831.
51. Orazem M E, Tribollet B. Electrochemical Impedance Spectroscopy[M]. John Wiley & Sons, 2011.
52. Cao C N, Zhang J Q. Introduction to Electrochemical Impedance Spectroscopy[M]. Science Press, 2002.
53. Zhu J G, Sun Z C, Wei X Z, et al. Research Progress on Low-temperature Characteristics and Heating Techniques of Vehicle Lithium-ion Battery [J]. Automotive Engineering, 2019, 41(5): 571–581+589.
54. Ciucci F, Lai W. Electrochemical impedance spectroscopy of phase transition materials [J]. Electrochimica Acta, 2012, 81: 205–216.
55. Gaberscek M, Jamnik J. Impact of electrochemical wiring topology on the kinetics of insertion electrodes [J]. Solid State Ionics, 2006, 177(26): 2647–2651.
56. Gaberscek M, Moskon J, Erjavec B, et al. The Importance of Interphase Contacts in Li Ion Electrodes: The Meaning of the High-Frequency Impedance Arc [J]. Electrochemical and Solid-State Letters, 2008, 11(10): A170.
57. schmidt J P, Chrobak T, Ender M, et al. Studies on LiFePO4 as cathode material using impedance spectroscopy [J]. Journal of Power Sources, 2011, 196(12): 5342–5348.
58. Bai S J, Song Y S. Correlation between Internal Structure and Electrochemical Impedance Spectroscopy of Multiphase Slurry Systems [J]. Analytical Chemistry, 2013, 85(8): 3918–3925.
59. Lohmann N, Wesskamp P, Haussmann P, et al. Electrochemical impedance spectroscopy for lithium-ion cells: Test equipment and procedures for aging and fast characterization in time and frequency domain [J]. Journal of Power Sources, 2015, 273: 613–623.
60. Boukamp B A. Electrochemical impedance spectroscopy in solid state ionics: recent advances [J]. Solid State Ionics, 2004, 169(1): 65–73.
61. Li S E, Wang B, Peng H, et al. An electrochemistry-based impedance model for lithium-ion batteries [J]. Journal of Power Sources, 2014, 258: 9–18.
62. Harada T, Tanaka J, Nakazawa A, et al. AC Impedance Spectroscopy of Lithium-Ion Secondary Cell Exposed to Elevated Temperature [J]. Electrochemistry, 2010, 78(5): 435–437.
63. Xiao M, Choe S-Y. Impedance model of lithium ion polymer battery considering temperature effects based on electrochemical principle: Part I for high frequency [J]. Journal of Power Sources, 2015, 277: 403–415.
64. Xie Y, Li J, Yuan C. Mathematical modeling of the electrochemical impedance spectroscopy in lithium ion battery cycling [J]. Electrochimica Acta, 2014, 127: 266–275.
65. Gaberscek M, Dominko R, Jamnik J. The meaning of impedance measurements of LiFePO4 cathodes: A linearity study [J]. Journal of Power Sources, 2007, 174(2): 944–948.
66. Malik R, Abdellahi A, Ceder G. A Critical Review of the Li Insertion Mechanisms in LiFePO4 Electrodes [J]. Journal of The Electrochemical Society, 2013, 160(5): A3179–A3197.
67. Doyle M, Newman J, Gozdz A S, et al. Comparison of Modeling Predictions with Experimental Data from Plastic Lithium Ion Cells [J]. Journal of The Electrochemical Society, 1996, 143(6): 1890–1903.
68. Newman J, Balsara N P. Electrochemical Systems[M]. John Wiley & Sons, 2021.
69. Newman J, Tiedemann W. Porous-electrode theory with battery applications [J]. AIChE Journal, 1975, 21(1): 25–41.
70. Abraham D P, Heaton J R, Kang S-H, et al. Investigating the Low-Temperature Impedance Increase of Lithium-Ion Cells [J]. Journal of The Electrochemical Society, 2008, 155(1): A41.

71. Hu L X, Wang C, Chen X Q, et al. Low temperature properties research of lithium ion battery electrolyte [J]. New Chemical Materials, 2011, 39(7): 62–64+81.
72. Li J, Yuan C F, Guo Z H, et al. Limiting factors for low-temperature performance of electrolytes in LiFePO4/Li and graphite/Li half cells [J]. Electrochimica Acta, 2012, 59: 69–74.
73. Qiao Y Q, Tu J P, Wang X L, et al. The low and high temperature electrochemical performances of Li3V2(PO4)3/C cathode material for Li-ion batteries [J]. Journal of Power Sources, 2012, 199: 287–292.
74. Wang H W, Du C L, Wang C B. Study of low temperature performance of Li-ion battery [J]. Battery Bimonthly, 2009, 39(4): 208–210.
75. Nagasubramanian G. Electrical characteristics of 18650 Li-ion cells at low temperatures [J]. Journal of Applied Electrochemistry, 2001, 31(1): 99–104.
76. Zhang S S, Xu K, Jow T R. Low temperature performance of graphite electrode in Li-ion cells [J]. Electrochimica Acta, 2002, 48(3): 241–246.
77. Zhang S S, Xu K, Jow T R. A new approach toward improved low temperature performance of Li-ion battery [J]. Electrochemistry Communications, 2002, 4(11): 928–932.
78. Zhang S S, Xu K, Jow T R. An inorganic composite membrane as the separator of Li-ion batteries [J]. Journal of Power Sources, 2005, 140(2): 361–364.
79. Fan J, Tan S. Studies on Charging Lithium-Ion Cells at Low Temperatures [J]. Journal of The Electrochemical Society, 2006, 153(6): A1081.
80. Zhang G, Wei X, Han G, et al. Lithium plating on the anode for lithium-ion batteries during long-term low temperature cycling [J]. Journal of Power Sources, 2021, 484: 229312.
81. Hu X, Zheng Y, Howey D A, et al. Battery warm-up methodologies at subzero temperatures for automotive applications: Recent advances and perspectives [J]. Progress in Energy and Combustion Science, 2020, 77: 100806.
82. Legrand N, Knosp B, Desprez P, et al. Physical characterization of the charging process of a Li-ion battery and prediction of Li plating by electrochemical modelling [J]. Journal of Power Sources, 2014, 245: 208–216.
83. Mendoza-Hernandez O S, Hosono E, Asakura D, et al. Effect of the Charge Process on the Performance of Li-ion Cells during Charge-Discharge Cycling at 0°C [J]. Electrochemistry, 2020, 88(3): 230–235.
84. Ecker M, Shafiei Sabet P, Sauer D U. Influence of operational condition on lithium plating for commercial lithium-ion batteries – Electrochemical experiments and post-mortem-analysis [J]. Applied Energy, 2017, 206: 934–946.
85. Burow D, Sergeeva K, Calles S, et al. Inhomogeneous degradation of graphite anodes in automotive lithium ion batteries under low-temperature pulse cycling conditions [J]. Journal of Power Sources, 2016, 307: 806–814.
86. Koleti U R, Rajan A, Tan C, et al. A Study on the Influence of Lithium Plating on Battery Degradation [J]. Energies, 2020, 13(13): 3458.
87. Ouyang M, Chu Z, Lu L, et al. Low temperature aging mechanism identification and lithium deposition in a large format lithium iron phosphate battery for different charge profiles [J]. Journal of Power Sources, 2015, 286: 309–320.
88. Zhao X, Yin Y, Hu Y, et al. Electrochemical-thermal modeling of lithium plating/stripping of Li(Ni0.6Mn0.2Co0.2)O2/Carbon lithium-ion batteries at subzero ambient temperatures [J]. Journal of Power Sources, 2019, 418: 61–73.
89. Wu X, Wang W, Du J. Effect of charge rate on capacity degradation of LiFePO4 power battery at low temperature [J]. International Journal of Energy Research, 2020, 44(3): 1775–1788.

Chapter 2
Battery Low-Temperature Degradation Mechanisms

Abstract This chapter delves into the intricate degradation mechanisms of lithium-ion batteries (LIBs) under low-temperature conditions, emphasizing the effects of low-temperature charging and alternating current (AC) heating. At low temperatures, LIBs exhibit significant performance degradation, characterized by decreased capacity, slower charging rates, and increased likelihood of lithium plating. The primary factors contributing to this degradation include elevated solid–liquid interface impedance, charge transfer impedance, solid-phase diffusion impedance, and the formation of lithium plating on the graphite anode. Lithium plating, a detrimental side reaction during charging, not only leads to irreversible capacity loss but also poses safety risks by potentially causing internal short circuits and thermal runaway. This chapter first investigates the aging mechanism of LIBs in charge/discharge cycles at low temperatures. Experimental results reveal that the degradation is more pronounced at high charging rates and low temperatures, with lithium plating being the dominant mechanism driving capacity fade. The study also investigates the effectiveness of AC heating methods in mitigating low-temperature effects. AC heating demonstrates uniform temperature distribution and minimal battery degradation after extensive cycling, making it a promising strategy for enhancing LIB performance in cold environments. The findings underscore the importance of optimizing battery management strategies and material designs to improve the low-temperature performance of LIBs. These insights are particularly valuable for the development of energy storage technologies in electric vehicles (EVs) and other applications where reliable operation under varying environmental conditions is crucial. Overall, this chapter summarizes the low-temperature degradation mechanism of LIBs, which is an important guide for the design of future low-temperature batteries and the development of battery management system (BMS).

Keywords Lithium-ion batteries · Low-temperature cycling · Postmortem characterization · Alternating current heating · Degradation mechanisms

J. Zhu et al., *Alternating Current (AC) Heating for Lithium-Ion Batteries in Electric Vehicles*, https://doi.org/10.1007/978-981-96-9071-8_2

2.1 Battery Degradation at Low Temperature

The low-temperature performance impairment of LIBs is predominantly characterized by multiple interrelated phenomena: diminished discharge capacity, lowered output power density, compromised cycle stability, prolonged charging duration, and notably enhanced propensity for metallic lithium deposition during charge cycles. The underlying reason is that the internal electrode processes of LIBs change with decreasing temperature and increasing cycle number. During operation, the internal processes of LIBs mainly involve electronic conduction, solid-phase diffusion of lithium ions, charge transfer, and mass transfer. Figure 2.1 illustrates the process of lithium-ion intercalation into the electrode, which can be generally divided into three stages: Stage 1 represents the migration of solvated lithium-ion in the electrolyte to the surface of the active material; Stage 2 indicates the process of lithium-ion overcoming the energy barrier to detach from the solvation layer and pass through the liquid–solid interface; Stage 3 involves the migration of desolvated lithium-ion through the cathode electrolyte interphase (CEI) or solid electrolyte interphase (SEI), culminating in their intercalation into the electrode's host structure. Conversely, lithium-ion extraction from the active material constitutes the reverse reaction, representing the deintercalation counterpart to this interfacial transport mechanism [1]. These processes are strongly coupled. At low temperatures, the electrolyte's ionic conductivity, the solid–liquid interfacial film's conductivity, the electrode's electrochemical reaction rate, and the lithium-ion solid-phase diffusion coefficient all diminish. These factors lead to a reduced rate of lithium-ion deintercalation and intercalation in the electrode, thereby causing the performance of LIBs to deteriorate. To explore the mechanism underlying LIBs' low-temperature performance decline, the subsequent analysis will be carried out from the perspectives of the positive and negative electrodes, as well as electrolyte properties.

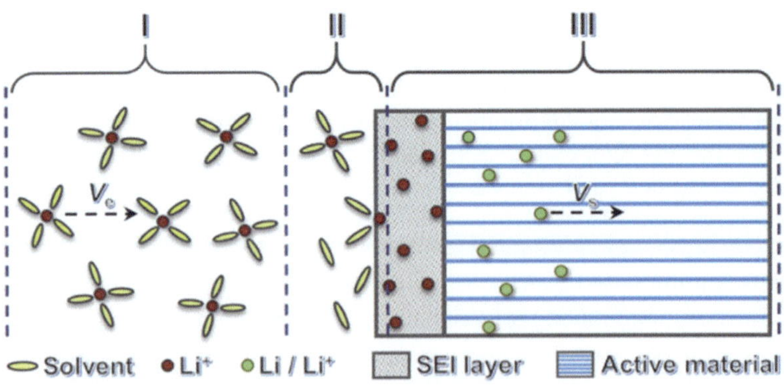

Fig. 2.1 The process of lithium-ion intercalation into the electrode [1]

The current commercial cathode materials for LIBs mainly include lithium-rich materials and ternary materials. Lithium-rich cathode chemistries are principally represented by olivine-structured lithium-ion phosphate ($LiFePO_4$) and spinel-phase lithium manganese oxide ($LiMn_2O_4$). Ternary cathode systems, conversely, are dominated by layered oxide configurations typified by lithium nickel cobalt manganese oxide ($LiNiCoMnO_2$, NCM) and lithium nickel cobalt aluminum oxide ($LiNiCoAlO_2$, NCA). These materials exhibit distinct crystallographic frameworks: $LiMn_2O_4$ adopts a three-dimensional (3D) spinel lattice, ternary oxides feature two-dimensional (2D) layered arrangements, and $LiFePO_4$ stabilizes in a one-dimensional (1D) olivine structure. Such structural divergences directly govern their ionic transport properties, with lithium-ion diffusion pathways dimensionally constrained to 3D, 2D, and 1D spatial configurations, respectively [2, 3]. The three-dimensional structure has fewer barriers to lithium intercalation. Thus, lithium manganese oxide batteries with a spinel structure exhibit relatively better rates and low-temperature characteristics compared to the other two types of batteries. Under low-temperature conditions, the charge transfer impedance, CEI impedance, and solid-phase diffusion impedance of the cathode materials all increase simultaneously. Experimental investigations on $LiFePO_4$ cathodes demonstrate that low-temperature operation induces more pronounced variations in solid-phase lithium-ion diffusion coefficients relative to charge transfer resistance deviations [4], with the temperature sensitivity of charge transfer impedance significantly surpassing that of the CEI impedance [5]. To a certain extent, this suggests to a certain extent that the decline in the low-temperature performance of the cathode materials is primarily influenced by the solid-phase diffusion coefficient. Symmetric cathode electrode experiments have revealed a direct positive correlation between solid-phase diffusion impedance and capacity degradation at reduced temperatures [6]. Furthermore, elevated intergranular impedance within cathode active materials under cryogenic conditions emerges as a critical limiting factor for LIBs performance under such thermal constraints, attributable to restricted lithium-ion percolation pathways across crystallite boundaries [7]. Low-temperature cycling induces cathode active material delamination from metallic current collectors, manifesting as progressive elevation of interfacial contact resistance. This mechanical degradation is further compounded by the disruption of electrical continuity between the cathode composite and current collector, impeding electron transport kinetics and accelerating the capacity fade of LIBs. Moreover, secondary particles composed of cathode materials may crack due to fatigue stress [8], increasing the instability of the electrode. The cracking of secondary particles can also cause the dissolution of transition metals, which can adhere to the graphite anode through the separator and induce side reactions [9], thereby accelerating the degradation of the battery. Li et al. [10] also discovered that extremely low temperatures might result in the cracking of cathode active material particles because of excessive low-temperature stress. Meanwhile, the locations of the binder and conductive agent could be altered, thus degrading the performance of the cathode active material.

Under low-temperature conditions, graphite anodes exhibit significantly elevated charge transfer resistance, SEI impedance, and solid-state diffusion resistance. [7]

Compared to the SEI impedance, the graphite anode is typically more affected by the charge transfer impedance [3], which is also more sensitive to temperature. Under cryogenic charging conditions, this impedance triad (charge transfer/SEI/diffusion resistances) promotes substantial Li^+ accumulation at graphite-electrolyte interfaces, ultimately driving metallic lithium deposition through kinetically-limited plating reactions. Such irreversible lithium plating phenomena not only compromise Coulombic efficiency but also pose critical safety risks through dendritic growth [11]. The precipitated lithium further undergoes irreversible reactions with the electrolyte, generating a new SEI film, which in turn increases the SEI impedance. During low-temperature cycling, lithium plating and the development of a new SEI film can damage the existing SEI film. This exposes the graphite anode surface to various side reaction products and SEI films with different densities and morphologies [9, 12]. This process not only continues to increase the SEI impedance but also consumes the electrolyte and reduces the number of active lithium ions, leading to battery capacity fade [13]. Moreover, as the precipitated lithium continues to accumulate, lithium dendrites can form. The growth of lithium dendrites can "pierce" the separator and come into contact with the cathode, forming an internal short circuit, which induces thermal runaway [14]. Additionally, once the precipitated lithium detaches from the graphite anode, it loses electrical contact and forms "dead lithium," which also results in a decrease in battery capacity.

Understanding the state of lithium-ion in the electrolyte at a microscopic level is crucial for conducting a comprehensive investigation into the impact of electrolytes on the low-temperature performance of LIBs. Figure 2.2a illustrates the trajectory of lithium-ion in the electrolyte. As depicted in Fig. 2.2b, when lithium-ion intercalate into the electrode, they need to first detach from the two solvation layers and then pass through the SEI/CEI film to intercalate into the active electrode material. Throughout this process, a certain energy barrier must be overcome, as illustrated in Fig. 2.2c. As the temperature drops, the electrolyte's ionic conductivity decreases while its viscosity goes up, which results in a slower ion diffusion rate within the electrolyte, thus impacting the process illustrated in Fig. 2.2a. Moreover, low temperatures also result in an increased energy barrier required for lithium-ion desolvation, thus affecting the intercalation rate of lithium ions. Moreover, the electrolyte plays a role in the formation of interface films on the active materials of both electrodes. Electrolytes with different formulations can influence the stability and impedance of these films [15], thereby affecting their low-temperature performance. Furthermore, the electrolyte can also affect the impedance value of the charge transfer process, which may be related to the slow desolvation of lithium-ion at the solid–liquid interface [1]. The electrolyte in LIBs influences not only the transport of lithium-ion in the liquid phase but also their intercalation into electrodes. Driven by the need to boost the low-temperature performance of LIBs, many researchers have directed their efforts towards investigating electrolytes, which are characterized by low freezing points and high ionic conductivity.

In low-temperature conditions, the binder's morphology in LIBs alters, particularly for those with styrene-butadiene rubber, which experience glass transition

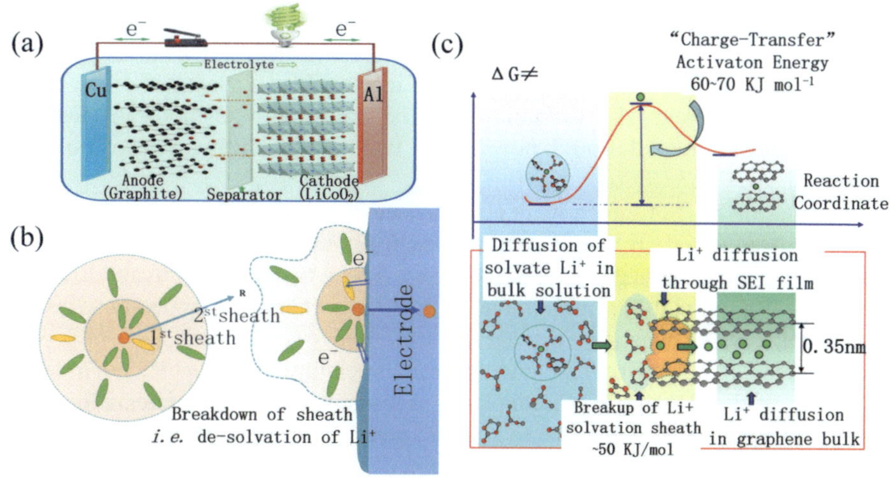

Fig. 2.2 The state of lithium-ion within the electrolyte: **a** the movement state of lithium-ion in the electrolyte [16]; **b** the microscopic process of lithium-ion intercalation into the electrode [15]; **c** the energy change during the process of lithium-ion intercalation into the electrode [17]

at lower temperatures, affecting adhesion [18]. Moreover, after repeated charge–discharge cycles, deposits from battery side reactions adhere to the separator, blocking its pores, which impedes electrolyte distribution and ion movement [9, 19].

2.2 Lithium Plating Mechanism

When charging LIBs at low temperatures, lithium plating tends to develop. This phenomenon not only results in irreversible capacity loss of the battery [20–23], but also poses a risk of inducing internal short circuits, which may subsequently trigger thermal runaway. Consequently, this issue warrants particular attention.

2.2.1 Lithium Plating Reactions

Lithium plating, an undesirable side effect during the lithium intercalation process, presents a significant challenge in LIB technology. Equation (2.1) describes the full insertion of Li^+ ions into the graphite anode, a process limited by diffusion. At a constant temperature, only a finite number of Li^+ ions can be inserted into the graphite structure per unit time [24]. The operating potential range for Li^+ ion insertion within the graphite matrix is 65–200 mV versus Li^+/Li [25]. In contrast, Eq. (2.2) illustrates the partial or complete deposition of lithium on the anode surface. The charging current is divided into two parts: the intercalation current and the lithium plating curren [26]. Ideally, The charging current determines the rate at which Li^+ ions arrive at the anode surface. However, the intercalation and plating currents compete with each other. As charging continues, the number of available sites in the graphite layers decreases. This results in a drop in the intercalation current and a rise in the lithium plating current [27]. Lithium plating becomes thermodynamically feasible when the anode potential drops below 0 V. At this point, the lithium deposition rate exceeds that of intercalation.

The overpotential of the graphite electrode mainly comes from three aspects: (i) charge transfer, (ii) electrolyte concentration (mass transfer), and (iii) lithium solid-phase diffusion in the negative electrode [28]. These factors collectively drive the kinetics of lithium plating. Lithium plating can occur thermodynamically when the local potential at the negative electrode drops below 0 V (vs. Li^+/Li) due to high state of charge (SoC), high charging C-rate, or low temperatures, all of which polarize the electrode [29, 30]. However, lithium plating is less favorable than intercalation from a thermodynamic perspective due to its more positive reaction enthalpy. The occurrence of lithium plating cannot be fully explained by kinetic factors alone, as charging is a dynamic, non-equilibrium process, especially at high C-rates. The cell voltage can be calculated using the Nernst equation when the system is in thermodynamic equilibrium [31, 32]. The equilibrium potential differential provides a measure of the thermodynamic driving force governing the competitive mechanisms of metallic plating and intercalation reactions, since these two electrochemical pathways contend for shared electron transfer and lithium ions [33]. The equilibrium electrochemical potential demonstrates temperature-dependent behavior in both metallurgical lithium deposition and intercalative insertion processes, thereby inducing spatial heterogeneity in anode surface potential distribution [32]. This heterogeneity promotes non-uniform lithium plating [33]. Furthermore, the deposited lithium undergoes reactions with the electrolyte, resulting in its consumption and potentially causing localized lithium electroplating owing to the heterogeneous characteristics of the graphite electrode [34].

$$xLi^+ + Li_\delta C_6 + xe^- \rightarrow Li_{x+\delta}C_6 \tag{2.1}$$

$$(1 - X)Li^+ + (1 - X)e^- \rightarrow (1 - X)Li \tag{2.2}$$

Lithium electrodeposition exhibits three principal morphological variants: (i) electrically isolated metallic lithium (denoted as dead Li), (ii) electrochemically recoverable lithium phases, and (iii) SEI-derived interfacial phases [35]. The dead lithium configuration constitutes metallic lithium deposits that become electrically isolated from the host graphite matrix during charge/discharge cycling [14]. This reaction creates a tortuous path for lithium-ion transport, reducing the active surface area for intercalation [35]. The secondary SEI film forms via the reduction of solvent electrolyte molecules (R) by deposited lithium, as per Eq. (2.4). Both dead lithium and the secondary SEI film are irreversible. They cause lithium loss and a gradual drop in battery capacity over time [36, 37]. Electrochemically recoverable lithium phases undergo system reintegration through the stripping mechanism (Eq. 2.5), occurring during relaxation or resting periods (Eq. 2.3) [38]. This lithium reinsertion persists throughout the relaxation period until achieving complete intercalation within the anode matrix. The stripping pathway exhibits enhanced kinetic preference relative to lithium extraction from host structures (Eq. 2.3). During discharge, electrochemically recoverable lithium species liberated through the stripping mechanism follow two competing pathways: (1) intercalation into graphite crystalline structures, or (2) migration toward the cathode to participate in current generation [38].

$$\varepsilon Li + Li_{\delta+x}C_6 \rightarrow Li_{\delta+x+\varepsilon}C_6 \tag{2.3}$$

$$R + Li \rightarrow R - Li \tag{2.4}$$

$$Li \rightarrow Li^+ + e^- \tag{2.5}$$

Figure 2.3 illustrates that lithium deposit morphologies differ based on varying operating and charging conditions, falling into three main categories: mossy, granular (particle-like), and dendritic (needle-like) [29, 39]. Current rate is the main factor influencing these morphologies' formation. Specifically, mossy and granular deposits usually develop under low current rates, while dendritic structures appear under high current rates [40]. The morphological evolution of lithium dendrites critically compromises cell structural integrity through separator breach and cathode contact, inducing internal short-circuit phenomena accompanied by localized thermal runaway. While transient thermal effects from dendrite melting may create self-terminating short-circuit pathways, this energy dissipation paradoxically accelerates degradation mechanisms including SEI formation and electrolyte decomposition [36, 41]. From a safety perspective, dendritic lithium poses a greater risk than mossy or granular lithium. Internal short circuits are categorized into soft and hard shorts [14]. Soft shorts usually disappear upon discharge and don't cause catastrophic failure. Instead, they may lower the cell's current and voltage while raising the local temperature [42, 43]. The thermal energy released by soft shorts can trigger exothermic reactions between the electrolyte and other cell components, with the potential to cause separator melting. By contrast, hard shorts exhibit lower resistance, resulting in substantially larger short-circuit currents between the anode and cathode [43]. This leads to more pronounced local temperature elevation, increasing the likelihood of thermal runaway events [43]. It should be emphasized that lithium plating formation is governed by a multifaceted interplay of factors, encompassing actual operating parameters, cell balancing status, material characteristics, electrode architecture, and comprehensive cell design considerations [30].

2.2.2 *Main Factors Affecting Lithium Plating*

Numerous in-depth studies have been carried out to clarify the mechanisms of lithium plating during standard and fast charging. However, the complexity of lithium plating

Fig. 2.3 The various morphology of lithium deposits. **a** Granular; **b** Moss; **c** Dendritic [39]

has made it difficult to fully understand its mechanisms [44]. As reported in numerous prior investigations, lithium plating primarily results from three key factors, which encompass: (i) hazardous operating conditions, (ii) cell defects, and (iii) cell aging (Table 2.1).

Table 2.1 Factors causing lithium plating [45]

Factors	Causes and conditions
Hazardous operating conditions	Low temperatures
	High charging C-rates
	High SoCs
Cell defects	Cell properties and poor design, such as an imbalanced negative to positive ratio
Aging of the cell	Leading to cell unbalance
	Kinetic degradation (Capacity fade, energy fade, CE decrease, energy efficiency fade, and resistance increase)

2.2.2.1 Hazardous Operating Conditions

Lithium plating emerges under non-optimal charging protocols, such as high C-rate charging, high SoC charging, and low-temperature charging [46–48]. Such electrochemical stressors compromise ionic transport dynamics across both liquid electrolyte and solid-phase interfaces, establishing conditions where the electrochemical potential of the anode descends below lithium's redox equilibrium. This thermodynamic instability drives preferential metallic deposition rather than intercalation [49, 50]. As critical operational thresholds governing lithium deposition kinetics constitute principal determinants of battery degradation, the following systematic analysis identifies and evaluates governing parameters that exacerbate this phenomenon under specified electrochemical constraints.

(1) Low-Temperature Effects

Low temperatures significantly hinder fast charging and can cause substantial capacity loss when charging at high C-rates. For example, charging a 7.5 Ah cell at 0 °C with a 1 C rate can cause a 3.6% capacity loss [48]. Generally, low temperatures reduce LIBs' power and energy densities, especially during charging, mainly due to three factors: 1. reduced ionic conductivity in the electrolyte; 2. impaired solid-state diffusivity of lithium ions in the electrode; 3. sluggish charge-transfer kinetics [51, 52]. According to the Arrhenius equation, low temperatures increase the cell's internal resistance due to decreased ionic conductivity in the electrolyte. However, reduced ionic conductivity is not the primary issue in low-temperature charging. Studies reveal that during low-temperature operation, limited Li^+ mobility in electrode materials is a major cause of lithium plating. In this scenario, lithium ions build up at the interface between carbon particles and the electrolyte [35, 48]. Once the surface concentration of lithium ions in graphite particles peaks, lithium plating is initiated. Additionally, the sluggish charge-transfer kinetics is another key limiting factor in low-temperature charging of LIBs. When current is applied, significant overpotential is generated. As anode polarization intensifies, the anode potential drops below 0 V (vs. Li^+/Li). Consequently, lithium plating is triggered since Li^+ ions accumulate at the anode interface instead of undergoing intercalation [40, 44]. This process is illustrated in Fig. 2.4a. Furthermore, the composite anode's potential decline near the separator is more pronounced than elsewhere, suggesting lithium plating initiates on the anode adjacent to the separator [53]. Thus, low-temperature charging currents must be carefully regulated. To alleviate low-temperature lithium deposition, preheating the cell prior to charging or employing low charging rates are viable strategies. [54] Recently, Yang et al. [54] proposed a cell design with thin nickel (Ni) foils inside. The Ni foil can generate heat in under 10 s as an internal heating element. This innovative cell structure enables 15 min charging to 80% SoC without lithium plating at −50 °C, even at 3.5 C rate. The same team later presented an asymmetric temperature modulation (ATM) method that charges a cell by heating it to 60 °C during charging. They showed that exposing the cell to 60 °C for 10 min

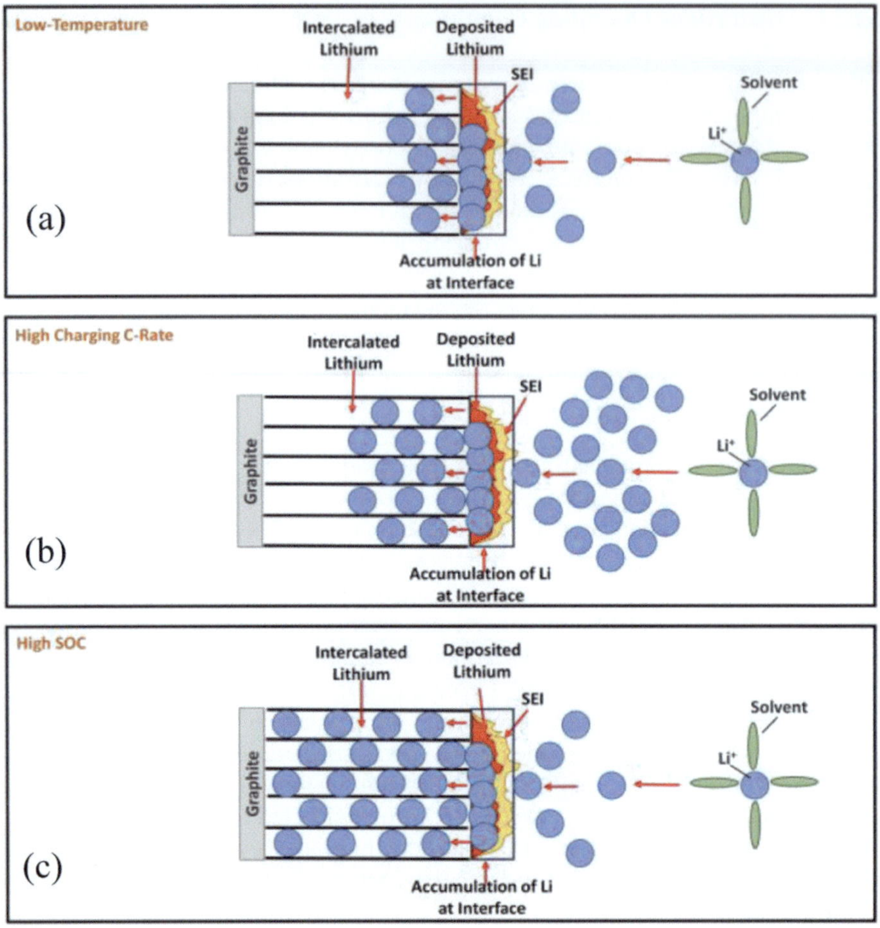

Fig. 2.4 Lithium plating behavior under different operational conditions. **a** Low temperatures; **b** High charging C-rates; **c** High SoCs [45]

per cycle can prevent lithium plating [55]. For EVs, most modern models have effective thermal management systems to avoid extreme operating temperatures. Also, at low temperatures, grid electricity is often utilized to preheat the cells.

(2) High Charging C-rates

Fast charging is becoming increasingly important for EVs and many other applications. It depends on high charging currents, which can significantly affect battery performance and cyclic life by accelerating aging. During fast charging, the charging rate is more likely to exceed the intercalation rate. At a high C-rate, more Li^+ ions are transported from the cathode to the anode per unit time during charge-transfer [14]. Higher charging rates often lead to greater polarization from transport and kinetic overpotentials, favoring lithium plating [56, 57]. For instance, recharging a cell within 10 min requires a charge rate of 6 C. At this rate, Li^+ ions begin to accumulate at the anode surface, as shown in Fig. 2.4b. As high-rate charging continues, the accumulated Li^+ ions lead to a high concentration of Li^+ ions on the graphite surface. If the concentration at the anode surface reaches saturation, lithium plating occurs [27, 58]. Additionally, the application of high charge currents in fast charging leads to increased heat generation. Notably, lithium plating and temperature elevation are two widely recognized concerns that emerge during fast charging procedures. [59]

(3) High SoC

Cells feature an upper cutoff voltage set by manufacturers. A BMS's failure to cease charging when cell voltage surpasses this preset threshold during charging is the main reason for cell overcharging (high SoC). High SoC in LIBs indicates a fully charged state where charging current still flows into the battery [60]. As charging continues at high SoCs, the Li^+ ion concentration on the anode surface is more likely to reach saturation beyond the allowable maximum, as depicted in Fig. 2.4c. Once saturated, lithium deposition on the anode surface begins [27]. Juarez-robles et al. [61] explored the high SoC impact on graphite/LCO 5 Ah pouch cells across cutoff voltages from 4.2 to 4.8 V. Cells charged above 4.5 V showed marked capacity loss, lithium plating, electrolyte breakdown, and swelling. Those charged to 4.6, 4.7, and 4.8 V exhibited dendritic structures [61]. Dendrites traverse the porous separator, leading to micro-internal short-circuits. Furthermore, at elevated SoCs, side reactions extend beyond the anode; cathode electrolyte decomposition also takes place.

2.2.2.2 Cell Defects

Numerous studies have demonstrated that cell manufacturing processes and characteristics significantly impact lithium plating. As an example, Liu et al. [62] showed that cells with a negative to positive (N/P) ratio of 1.19 have a slower aging rate and smaller impedance rise than those with an N/P ratio of 1.06. Local cell defects can also impact lithium plating. For instance, separator deformation (pore closure) may happen during cell manufacturing or operation due to internal mechanical stress buildup during charging or aging, which can trigger lithium plating [63]. Furthermore, the material-level kinetics can be explained by the activation energy barrier. The kinetics of interfacial Li^+ ion transfer is a key factor in charge transfer [64]. A correlation exists between intercalation kinetics and lithium plating behavior [65]. Xu et al. [66] found that in the charge-transfer process, desolvation is the most energy-intensive step (50 kJ/mol). They also determined that the overall activation energy barrier at the graphite/electrolyte interface is roughly 60–70 kJ/mol. [67] In another study, Yao et al. [68] found that due to the difference in the energy barrier for lithium de-solvation on the edge plane and the basal plane of graphite, the intercalation process tends to occur at the edge plane of the graphite rather than the basal plane. Lithium deposition shows a propensity for accumulating on pre-existing lithium deposits during lithium plating.

2.2.2.3 Degradation of the Cell

Lithium plating may develop even under normal operating conditions as a result of cell aging. As previously mentioned, the literature widely reports that the predominant degradation mechanisms are loss of lithium inventory (LLI) and loss of active material (LAM). LAM is classified into four categories depending on the impacted electrode and lithiation level: loss of active material on the delithiated negative electrode (LAMdeNE), on the delithiated positive electrode (LAMdePE), on the lithiated negative electrode (LAMliNE), and on the lithiated positive electrode (LAMliPE) [69]. After lithium plating happens, the plated lithium and electrolyte's side reaction creates a new SEI, which results in capacity loss and higher impedance [36]. Researchers can use changes in degradation mechanisms to detect lithium plating. Studying capacity fade curve shapes offers insights into aging mechanisms and lithium plating signs. Capacity fade curves fall into three categories: linear, decelerated, and accelerated capacity fade, all closely related to the number of cycles [14]. In commercial batteries, a two-stage capacity fade is common. In the initial stage, capacity fade remains stable, with LLI being the dominant degradation mechanism [69]. Accelerated capacity fade usually happens in the latter degradation stage, often after 500 cycles, and is generally attributed to LAMdeNE. Ansean et al. [70] discovered that LAMdeNE progresses four times faster than LLI, creating cell imbalance and over-lithiation of the negative electrode, which results in lithium plating. They noticed that at the sudden capacity loss point (curve shape), lithium deposition turns irreversible. The latter stage is not typically caused by changes in cell usage but may stem from silent degradation mechanisms present from the start [70, 71]. Cell chemistry and form factors (like pouch and cylindrical cells) influence the type of silent degradation. These silent degradations have an incubation period without causing capacity loss [71]. Hence, studying degradation modes, particularly those leading to lithium plating, is crucial. In addition to the LAMNE to LLI ratio, the plating threshold (LAMNE, PT) can also forecast an accelerated degradation phase [72]. Cell design parameters (such as mass ratio), capacity loss, and the two degradation modes (LLI and LAMPE) affect this value. Values exceeding this threshold cause lithium plating [72]. In another study, Schuster et al. [73] found a significant decrease in capacity at moderate temperatures and charging rates due to lithium plating. They demonstrated that lithium plating arises due to the significant decline in graphite's ionic kinetics, caused by SEI growth and loss of graphite active material. It is also important to note that in the literature, the capacity curve is frequently plotted against temperature in an Arrhenius plot, derived from cycling experiments at different temperatures. While lithium plating results in accelerated aging at lower temperatures, batteries without lithium plating typically age more rapidly at higher temperatures [74].

2.3 Degradation Mechanism During Low-Temperature Cycling

To fully understand the mechanism of low-temperature cycling attenuation, this chapter explores the issue from two perspectives: one based on external characteristics and the other combined with disassembly experiments. The subjects of these two investigations are sample C and sample D, which are detailed as follows.

2.3.1 Experimental Design

2.3.1.1 Experimental Subjects and Equipment

(1) Experimental Subjects

1) Sample C

In this study, 18650 ternary LIBs with NCA (Nickel–Cobalt–Aluminum) as the cathode material were selected as the experimental subjects and referred to as sample C. The experimental samples are detailed in Fig. 2.5, and their fundamental characteristics are summarized in Table 2.2.

2) Sample D

In this study, sample D denotes commercial prismatic LIBs with a rated capacity of 37 Ah. The cathode is composed of $LiNi_{0.5}Mn_{0.2}Co_{0.3}O_2$ (NMC523), while the anode is made of graphite.

(2) Experimental Equipment

- A lithium-ion battery charge–discharge testing device, as shown in Fig. 2.6a, for conducting battery cycling experiments.
- A high- and low-temperature environmental simulation chamber, as shown in Fig. 2.6b, for adjusting the environmental temperature during battery charge discharge processes.

Fig. 2.5 A 18650 ternary battery

Table 2.2 Fundamental parameters of the sample C

Rated capacity	2900 mAh (@0.2 C)
Typical capacity	2850 mAh (@0.2 C)
Nominal voltage	3.62 V (@0.2 C)
Electrode material	NCA (Nickel Cobalt Aluminum)
Maximum charging voltage	4.2 V ± 0.05 V
Charging current	Standard charging: 1375 mA
Maximum charging current	2750 mA (not applicable for cycle life)
Maximum discharging current	2850 mA
Minimum discharging voltage	2.5 V
Operating temperature	Charging: 0~45 °C
	Discharging: −20~60 °C
Storage temperature	1 year: −20~25 °C
	3 months: −20~45 °C
	1 month: −20~60 °C

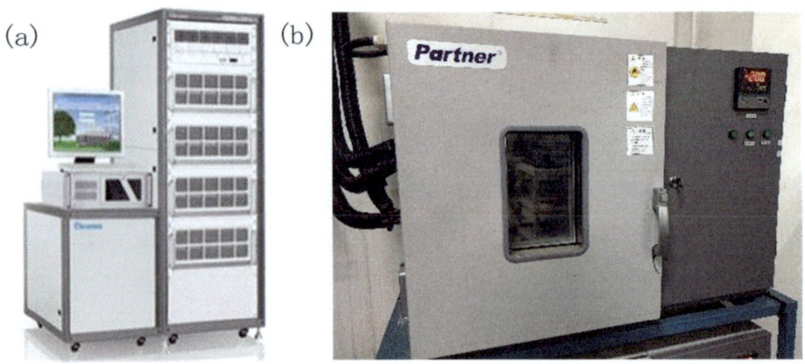

Fig. 2.6 Experimental setup. **a** Charging and discharging equipment; **b** High and low temperature environment simulation experiment box

- HIOKI wireless receiver module and a HIOKI wireless transmitter module for temperature data acquisition and recording, are used to monitor the surface temperature of the batteries.
- Several cylindrical battery fixtures for linking the positive and negative terminals of the batteries to the corresponding terminals of the charge–discharge testing device.

2.3.1.2 Experimental Protocol

(1) Experimental Protocol for Sample C

For sample C, two low-temperature charging experiment types were designed. The first, a single-factor control experiment, assesses the impacts of charging temperature, rate, and cutoff voltage on battery aging [75]. The second is an orthogonal experiment considering interaction effects. It uses range and variance analysis to determine influencing factor importance and explore factor interactions. The factor levels selected for each type of experiment are described below.

1) Single-Factor Control Experiment Design

The factor levels for the single-factor control experiment are listed in Table 2.3. The selection of factor levels is based on two criteria. The first is that the levels chosen should span the practical range of the factor. The second is that the selected levels should be distributed as evenly as possible. In Table 2.3, the temperature levels range from 0 to −20 °C, with a uniform interval of 5 °C. The charging rate levels encompass both low-rate and high-rate charging scenarios. Specifically, 0.5 C is the battery manufacturer's recommended charging rate, while 1 C represents the maximum charging rate. The charging cutoff voltage levels cover a broad range, with 4.25 V being the highest allowable voltage for charging.

2) Orthogonal Experimental Design

Orthogonal experimental design is a commonly used method for analyzing multifactorial experiments. In multifactorial experiments, factors not only act independently but also influence the experimental outcomes through their combined interactions, known as interaction effects [76]. When arranging experiments with interaction effects using an orthogonal array, the interaction between two factors is treated as a

Table 2.3 Design of factor levels for single-factor contrast experiments

Group	Temperature (°C)	Charging rate (C-rate)	Charging cutoff voltage (V)
S01	−20	0.5	4.2
S02	−15	0.5	4.2
S03	−10	0.5	4.2
S04	−5	0.5	4.2
S05	0	0.5	4.2
S06	−15	0.2	4.2
S07	−15	1	4.2
S08	−15	0.5	4.0
S09	−15	0.5	4.05
S10	−15	0.5	4.1
S11	−15	0.5	4.25

new factor and occupies a column referred to as the interaction column [76]. In this study, factor A, factor B, and factor C respectively represent the charging temperature, charging rate, and charging cutoff voltage [77]. The interactions between these factors are denoted as A × B for charging temperature and rate, A × C for charging temperature and cutoff voltage, and B × C for charging rate and cutoff voltage. An $L_8(2^7)$ orthogonal array was selected for this study, as listed in Table 2.4, with the numbers in the table representing the level distribution of the factors. The factor levels selected for the $L_8(2^7)$ orthogonal array are detailed in Table 2.5.

3) Capacity Characterization and Analysis

Before the low-temperature charging aging experiments, all battery samples underwent activation stability testing and reference performance testing (RPT). The activation stability testing process involved the batteries undergoing five cycles of constant current–constant voltage (CC–CV) charging followed by constant current (CC) discharging at a 0.5 C rate. This procedure took place in a temperature-controlled chamber maintained at 25 °C. This procedure ensured that the batteries were in a stable performance state. In this study, the RPT exclusively focused on capacity calibration. During battery aging, RPT was periodically performed to evaluate the aging state by calibrating the actual capacity, with experimental steps in Table 2.6.

(2) Experimental Protocol for Sample D

1) Battery Performance Tests

The evaluation of Sample D is divided into three stages: standard capacity test I, hybrid pulse power characteristic (HPPC) test II, and low-temperature high-rate

Table 2.4 $L_8(2^7)$ orthogonal experiment considering interaction effects

Group	Factor A	Factor B	Interaction A × B	Factor C	Interaction A × C	Interaction B × C	Dummy column
L01	1	1	1	1	1	1	1
L02	1	1	1	2	2	2	2
L03	1	2	2	1	1	2	2
L04	1	2	2	2	2	1	1
L05	2	1	2	1	2	1	2
L06	2	1	2	2	1	2	1
L07	2	2	1	1	2	2	1
L08	2	2	1	2	1	1	2

Table 2.5 Factor levels for the $L_8(2^7)$ orthogonal experiment

Factor	A (°C)	B (C-rate)	C (V)
Level 1	−10	0.5	4.1
Level 2	−15	1	4.2

Table 2.6 Experimental procedure for battery capacity calibration

Step number	Procedure	Current (C_rate)	End condition
1	Rest	0	1 h
2	CC charge	0.5	Charging cutoff voltage of 4.2 V
3	CV charge	–	Charging current drops to 0.02 C
4	Rest	0	1 h
5	CC discharge	0.5 C	Discharging cutoff voltage of 2.5 V
6	Repeat steps 1–5		

charging test III. Through these stages, the cell degradation characteristics are obtained.

As shown in Tables 2.7, 2.8, and 2.9, cell performance tests include three stages. Stage one is the standard capacity test I, in which the cell undergoes three CC−CV charging and CC discharging cycles, with the last discharging capacity regarded as the cell's standard capacity. Then, the cell is discharged to 50% SoC via 1 C CC discharging for the HPPC test. Both tests I and II are performed at 23 °C. Finally, the cell undergoes ten CC−CV charging and discharging cycles at −25 °C.

2) Postmortem Tests

The tested cells, after undergoing different cycles, are disassembled within an argon-filled glovebox (BRAUN, MB20G) at a fully discharged state (0% SoC). The glovebox maintains oxygen and water content levels below 5 ppm and 3 ppm respectively [11]. A cutting machine is used to slice the cell casing. Subsequently, the cells are extracted from the casings and unfolded.

To examine lithium deposition's structure and morphology, prepare surface and cross-section samples of different anode parts. For surface observation, cut 1 cm × 0.5 cm samples directly from the anode at two locations: near tab edge and center. Obtain cross-section samples by cutting jellyrolls with the Cross-section Polisher

Table 2.7 Profiles of standard capacity test

	Step	State	Duration	Condition	Temperature (°C)
Standard capacity test I	1	Rest	30 min	1 C until 4.2 V 4.2 V until I < C/20	23
	2	CC−CV charging			
	3	Rest	30 min	1 C until 3 V	
	4	CC discharging			
	5	Rest	30 min		
	6	Cycle	2	Step 2−Step 5	
	7	CC−CV charging		1 C until 4.2 V 4.2 V until I < C/20	

Table 2.8 Profiles of HPPC test

	Step	State	Duration	Condition	Temperature (°C)
HPPC test II	8	Rest	30 min		23
	9	CC discharging		1 C until 4.2 V 4.2 V until I < C/20	
	10	CC discharging	18 s	150 A	
	11	Rest	40 s		
	12	CC charging	10 s	100 A	
	13	Rest	40 s		
	14	CC discharging		1 C until 3 V	

Table 2.9 Profiles of low-temperature high-rate charging test

	Step	State	Duration	Condition	Temperature (°C)
Low-temperature high-rate charging test III	15	Rest	30 min		−25
	16	CC−CV charging		2 C until 4.2 V 4.2 V until I < C/20	
	17	Rest	30 min		
	18	CC−CV discharging		2 C until 2 V 2 V until I < C/20	
	19	Rest	30 min		
	20	Cycle	9	Step 16−Step 19	

(JEOL, IB-19520CCP) at −120 °C to prevent heat damage; the size is 1 cm × 1 cm, also from near-tab edge and center. Capture SEM images using a JEOL JSM-7610FPLUS, transferring samples via vacuum transfer box to avoid air contact.

XPS analysis of anode electrodes was conducted using Thermal K-ALPHA+. Anode samples were taken from two locations: near-tab edge and center. They were transferred to the XPS chamber via vacuum transfer box. To detect elemental lithium, samples were etched using XPS's Monatomic mode with a 3000 eV argon ion beam for 120 s of single-layer etching.

2.3.2 Results and Analysis

This chapter conducts a thorough analysis of LIBs' low-temperature charging-induced aging, focusing on charging temperature, rate, and cutoff voltage's impact on capacity degradation. For sample C, analysis shows the capacity degradation rate increases with lower charging temperature, higher charging rate, or higher cutoff voltage. For sample D, the study explores degradation mechanisms under low-temperature high-rate charging. Results indicate a significant initial capacity drop followed by slower degradation with cycling. Post-mortem examination shows severe anode lithium plating, significant lithium dendrite growth, and SEI film thickening. These findings imply that lithium plating and SEI film formation are the main degradation mechanisms under these conditions.

2.3.2.1 Analysis of Sample C

(1) Analysis of Capacity Degradation Characteristics

1) Analysis of the Effect of Charging Temperature on Capacity Degradation Rate

Figure 2.7a–f show the linear fitting curves for capacity degradation rate k at different temperatures.

As mentioned earlier, low temperatures reduce the solid-phase diffusion coefficient of lithium-ions, worsening lithium plating and loss of recyclable lithium-ions, thus hastening capacity degradation. Therefore, the capacity degradation rate k increases with decreasing charging temperature. The capacity degradation rate k is fitted using a piecewise linear function for each condition. The battery's capacity degrades linearly throughout its life cycle when charged at 0 °C, rendering piecewise treatment unnecessary for this condition. Additionally, a repeat experiment was performed for the −10 °C, 0.5 C, 4.2 V condition, denoted as Re_−10 °C 0.5 C 4.2 V.

The fitting parameters for each condition are listed in Tables 2.10 and 2.11.

The analysis of the fitting parameters for the two repeat experiments at −10 °C reveals that the capacity degradation rate in the second experiment is significantly higher than that in the first experiment. There are two possible reasons for this discrepancy. First, since the two experiments were not conducted simultaneously, the second experiment was approximately three months later than the first one. During this period, the battery experienced calendar aging, resulting in a poorer health state, which in turn accelerated capacity degradation [30]. The second factor might be experimental error. Thus, the mean value from two experiments is used to reflect capacity degradation at −10 °C. After averaging, the k value for the first stage at −10 °C is 0.5995, and for the second stage, it is 2.0557.

This study extends its investigation to the relationship between the capacity degradation rate k and charging temperature, conducting further analysis to understand their connection. Figure 2.8 shows that k in the second stage is higher than in the first stage, rising with decreasing temperature, indicating accelerated capacity degradation. The Arrhenius equation, a classic empirical formula illustrating the temperature-capacity degradation relationship, is employed to describe this, as shown in Eq. (2.6).

$$k = A \cdot \exp\left(-\frac{Ea}{RT}\right) \tag{2.6}$$

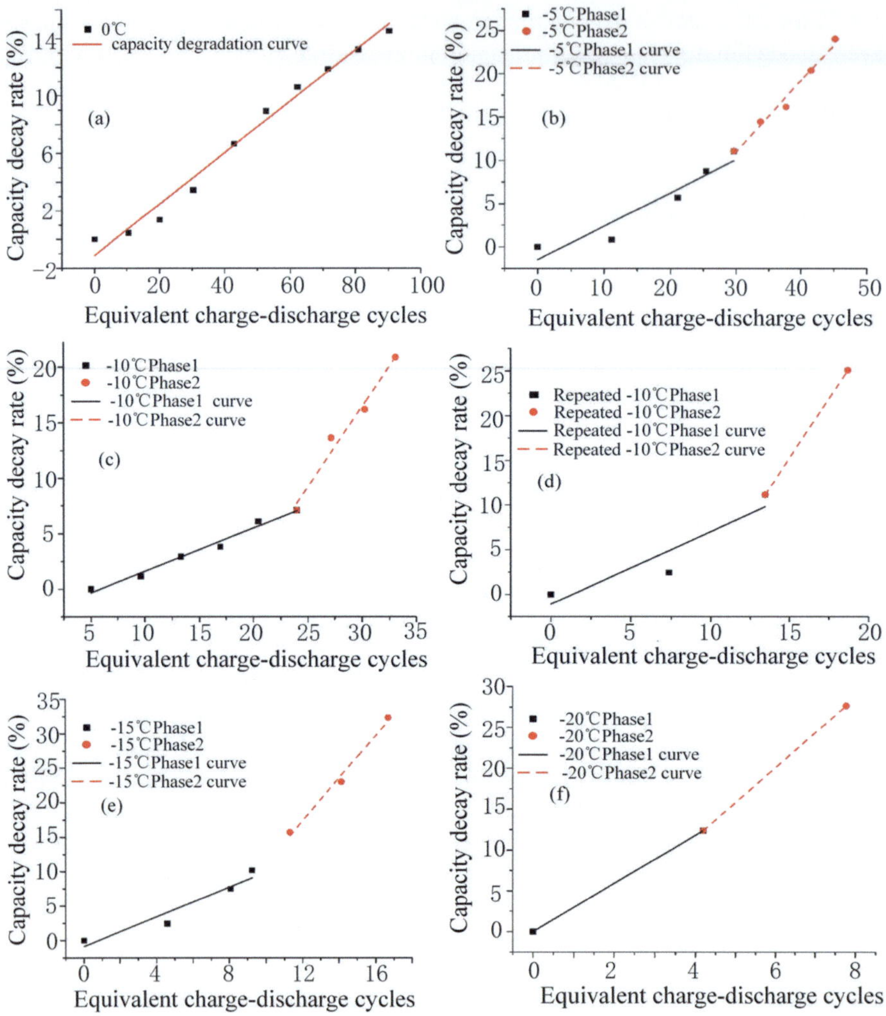

Fig. 2.7 Linear fitting curves for capacity degradation rate k at different charging temperatures. **a** 0 °C; **b** −5 °C; **c** −10 °C; **d** Repeated −10 °C; **e** −15 °C; **f** −20 °C

Table 2.10 Linear fitting parameters for capacity decline at different temperatures in the first stage

Condition	Intercept (a)	Slope (k)	Adj. R-square
0 °C 0.5 C 4.2 V	−1.1106	0.1791	0.9819
−5 °C 0.5 C 4.2 V	−1.4813	0.3862	0.8773
−10 °C 0.5 C 4.2 V	−0.3309	0.3883	0.9795
Re_−10 °C 0.5 C 4.2 V	−1.0961	0.8107	0.7407
−15 °C 0.5 C 4.2 V	−0.8698	1.0738	0.8889
−20 °C 0.5 C 4.2 V	1.2561E−15	2.9503	–

Table 2.11 Linear fitting parameters for capacity decline at different temperatures in the second stage

Condition	Intercept (a)	Slope (k)	Adj. R-square
−5 °C 0.5 C 4.2 V	−13.5455	0.8194	0.9742
−10 °C 0.5 C 4.2 V	−19.8536	1.4536	0.9641
Re_−10 °C 0.5 C 4.2 V	−24.5866	2.6578	0.9903
−15 °C 0.5 C 4.2 V	−19.6954	3.0912	0.9787
−20 °C 0.5 C 4.2 V	−5.5105	4.2653	–

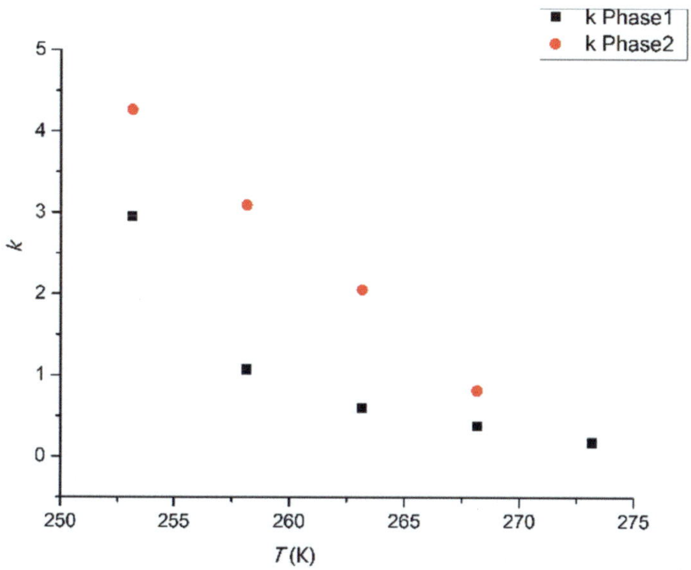

Fig. 2.8 Variation of capacity decline rate k with temperature

where k is the capacity degradation rate, A is the pre-exponential factor, Ea is the activation energy, R is the universal gas constant, and T is the thermodynamic temperature.

Linearizing Eq. (2.6) yields Eq. (2.7).

$$\ln k = \ln A + \left(-\frac{Ea}{R}\right) \cdot \frac{1}{T} \tag{2.7}$$

The relationship between ln k and $1/T$ is plotted and fitted linearly, as shown in Fig. 2.9.

As shown in Eq. (2.7) and Fig. 2.9, the capacity degradation rate k during the initial aging stage is characterized by:

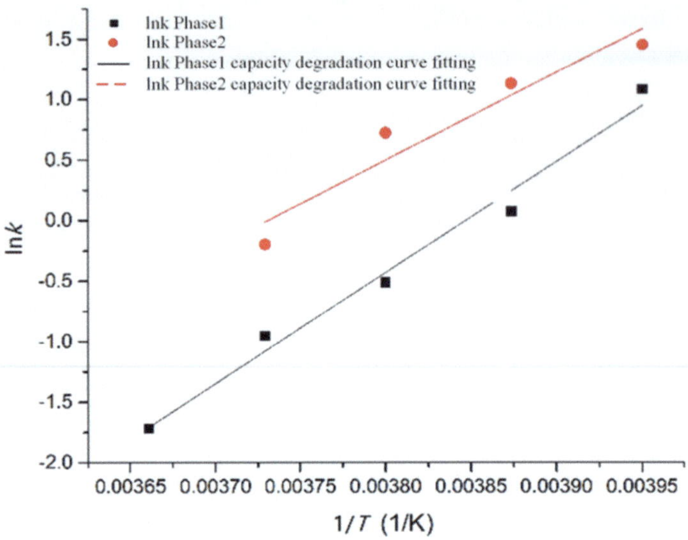

Fig. 2.9 Linear fit of ln(k) versus $1/T$

$$\ln k_1 = -35.3076 + 9177.7251 \cdot \frac{1}{T} \quad R^2 = 0.9786 \tag{2.8}$$

which can be rewritten as:

$$k_{\text{Phase 1}} = \exp(-35.3076) \cdot \exp\left(9177.7251 \cdot \frac{1}{T}\right) \tag{2.9}$$

For the second stage of aging, the capacity degradation rate k is described by:

$$\ln k_2 = -27.0158 + 7240.3602 \cdot \frac{1}{T} R^2 = 0.8914 \tag{2.10}$$

which can be rewritten as:

$$k_2 = \exp(-27.0158) \cdot \exp\left(7240.3602 \cdot \frac{1}{T}\right) \tag{2.11}$$

The relatively high fitting correlation coefficients for both stages suggest that the Arrhenius equation effectively describes capacity degradation characteristics during low-temperature charging. Nevertheless, the activation energy *Ea* for low-temperature aging differs from room-temperature aging mechanisms by being negative. This difference reveals the distinct aging mechanisms between low and high temperatures. In the literature [74, 78], attempts have been made to explain the negative activation energy. Thomas et al. [78] attributed the negative activation energy to the negative electrode potential, which causes lithium plating on the negative electrode surface. Prior to the chemical reaction between the electrolyte and plated lithium, there is a rapid exothermic adsorption/desorption process on the lithium surface [74].

2) Analysis of the Effect of Charging Rate on Capacity Degradation Rate

Figure 2.10 presents the capacity degradation curves during charging at −15 °C at various rates up to 4.2 V. At −15 °C, capacity degradation rate rises with increasing charging rates, likely due to worsening lithium plating. Notably, 0.5 C and 1 C show similar degradation rates, yet 0.2 C's rate is much lower. This suggests the existence of a critical turning point in the charging rate range between 0.2 and 0.5 C. When the charging rate exceeds this turning point, the capacity degradation rate rises sharply. OuYang et al. [19] observed similar trends in their study of aging behaviors in lithium-ion phosphate batteries charged at −10 °C. He observed that battery aging accelerates significantly when the charging rate surpasses 0.25 C.

Similarly, the capacity degradation rate k is fitted using a piecewise linear function for the charging rates of 0.2 C and 1 C at −15 °C. The fitting curve for the charging rate of 0.5 C at −15 °C has already been shown in Fig. 2.7e and will not be repeated here. The fitting parameters of each condition can be found in Tables 2.12 and 2.13. As depicted in Fig. 2.11, the fitting results demonstrate that the piecewise linear function effectively characterizes capacity degradation across various charging rates.

When exploring the relationship between the capacity degradation rate k and the charging rate I using a power function, as depicted in Fig. 2.10a, the second stage exhibits a low coefficient of determination ($R^2 = 0.44$). This suggests that the power

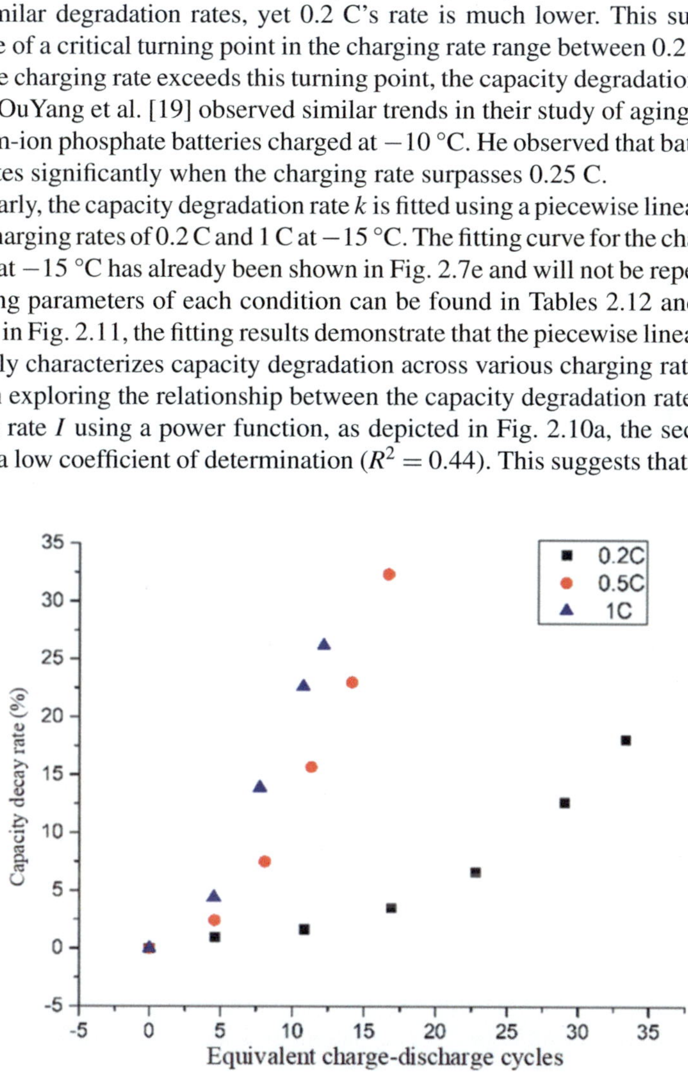

Fig. 2.10 Capacity degradation curves when charging at −15 °C at different rates up to 4.2 V

Table 2.12 Linear fitting parameters for capacity degradation rate *k* in the first stage at different charging rates

Working condition	Intercept (a)	Slope (k)	Adjusted R-square
−15 °C, 0.2 C, 4.2 V	−0.5077	0.2750	0.8929
−15 °C, 0.5 C, 4.2 V	−0.8698	1.0738	0.8889
−15 °C, 1 C, 4.2 V	−1.0061	1.7410	0.8249

Table 2.13 Linear fitting parameters for capacity degradation rate *k* in the second stage at different charging rates

Working condition	Intercept (a)	Slope (k)	Adjusted R-square
−15 °C, 0.2 C, 4.2 V	−18.0786	1.0758	0.9893
−15 °C, 0.5 C, 4.2 V	−19.6954	3.0912	0.9787
−15 °C, 1 C, 4.2 V	−10.1889	3.1001	0.9977

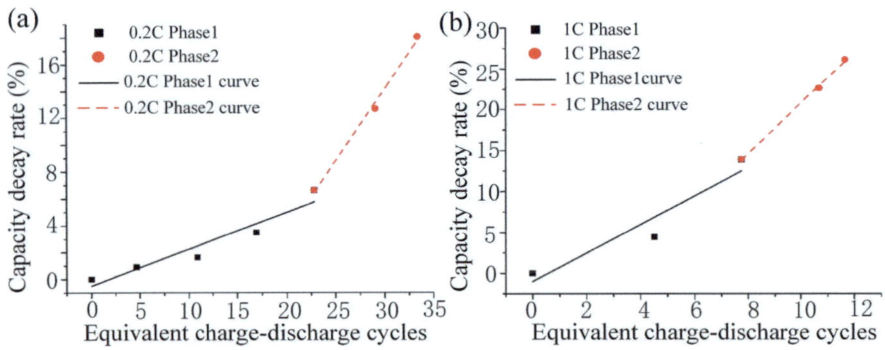

Fig. 2.11 Linear fitting curves for capacity degradation rate *k* at different rates up to 4.2 V. **a** −15 °C 0.2 C; **b** −15 °C 1 C

function is unsuitable for describing the charging rate-capacity degradation rate relationship at this stage. Observing the trend of *k* with the charging rate in Fig. 2.12a, it is found that the relationship between *k* and *I* follows an inverse exponential function, i.e.,

$$k = c \cdot \exp\left(\frac{d}{I}\right) \tag{2.12}$$

where *k* is the capacity degradation rate, Fig. 2.12 is the charging rate, and *c* and *d* are undetermined coefficients. Applying the natural logarithm to both sides of Eq. (2.12) linearizes it, resulting in:

$$\ln k = \ln c + \frac{d}{I} \tag{2.13}$$

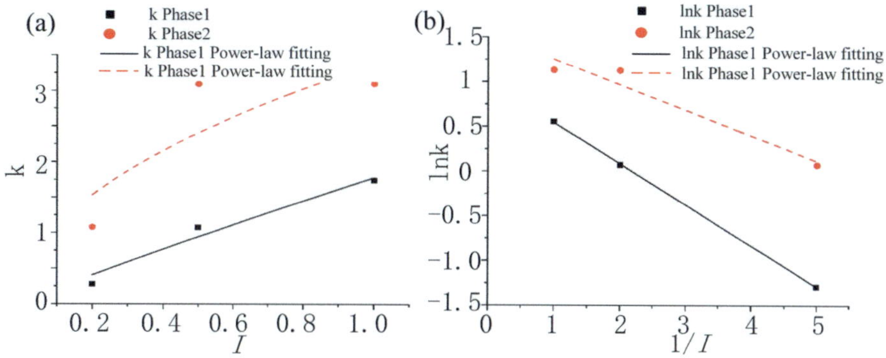

Fig. 2.12 Fitting of k versus I and $\ln k$ versus $1/I$

Fitting $\ln k$ to $1/I$ using Eq. (2.13) and is shown in Fig. 2.12b.

As shown in Fig. 2.12b, the inverse exponential function effectively captures the relationship between charging rate and capacity degradation rate. The capacity degradation rate k during the initial aging stage is characterized by:

$$\ln k_1 = 1.0041 - 0.4597 \cdot \frac{1}{I} \quad R^2 = 0.9997 \tag{2.14}$$

which can be rewritten as:

$$k_1 = \exp(1.0041) \cdot \exp\left(-0.4597 \cdot \frac{1}{I}\right) \tag{2.15}$$

The capacity degradation rate k in the second stage of aging is described by:

$$\ln k_2 = 1.5370 - 0.2847 \cdot \frac{1}{I} R^2 = 0.8868 \tag{2.16}$$

which can be rewritten as:

$$k_2 = \exp(1.5370) \cdot \exp\left(-0.2847 \cdot \frac{1}{I}\right) \tag{2.17}$$

3) Analysis of the Effect of Charging Cutoff Voltage on Capacity Degradation Rate

Figure 2.13a shows the scatter plot of capacity degradation when charging at −15 °C at a 0.5 C rate up to different cutoff voltages. The capacity degradation scatter points form two different degradation trends. In Fig. 2.13b, the x-axis range is adjusted to clearly show that the capacity degradation is faster when charging up to 4.05–4.25 V compared to 4 V.

The different capacity degradation trends shown in Fig. 2.13 suggest different aging mechanisms for the cutoff voltages of 4–4.25 V. The capacity degradation

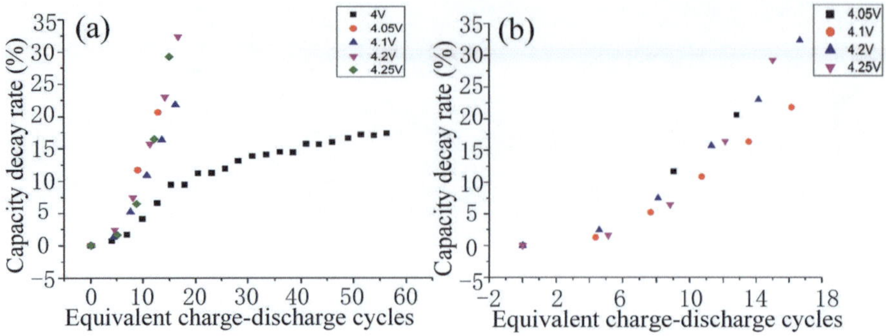

Fig. 2.13 Capacity degradation when charging at −15 °C at a 0.5 C rate up to different cutoff voltages. **a** 4–4.25 V; **b** 4.05–4.25 V

features at cutoff voltages from 4.05–4.25 V offer insights into the general impact of cutoff voltage on battery aging. Figures 2.14a–d show the piecewise linear fitting curves for capacity degradation when charging at −15 °C at a 0.5 C rate up to different cutoff voltages. The fitting parameters are listed in Tables 2.14 and 2.15.

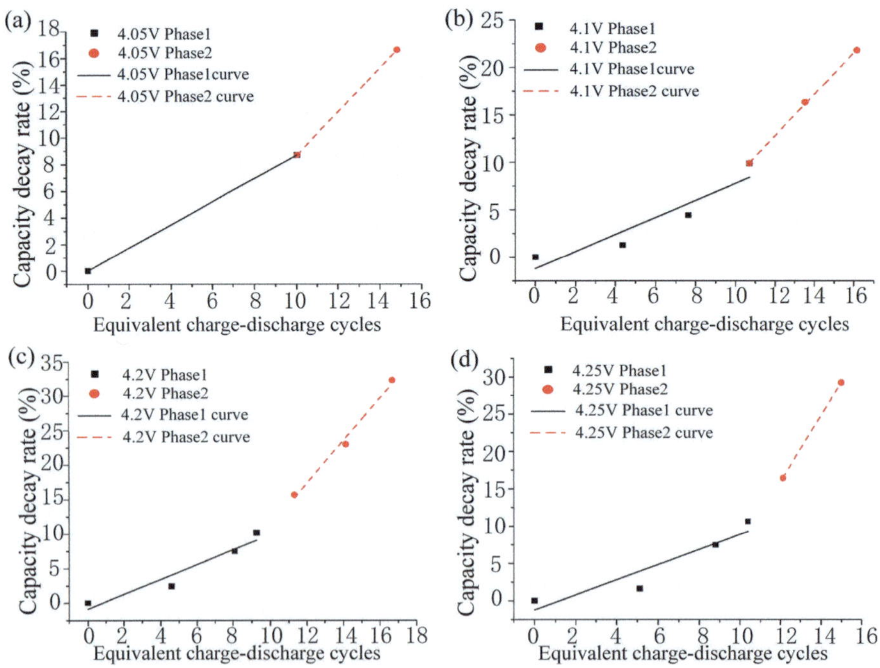

Fig. 2.14 Piecewise linear fitting curves for capacity degradation when charging at −15 °C at a 0.5 C rate up to different cutoff voltages. **a** 4.05 V; **b** 4.1 V; **c** 4.2 V; **d** 4.25 V

Table 2.14 Linear fitting parameters for capacity degradation rate k in the first stage at different cutoff voltages

Condition	Intercept (a)	Slope (k)	Adj. R-square
−15 °C 0.5 C 4.05 V	−6.2804E−16	0.8682	–
−15 °C 0.5 C 4.1 V	−1.2025	0.8970	0.8138
−15 °C 0.5 C 4.2 V	−0.8698	1.0738	0.8889
−15 °C 0.5 C 4.25 V	−1.2225	1.0112	0.8211

Table 2.15 Linear fitting parameters for capacity degradation rate k in the second stage at different cutoff voltages

Condition	Intercept (a)	Slope (k)	Adj. R-Square
15 °C 0.5 C 4.05 V	−7.8484	1.6513	–
−15 °C 0.5 C 4.1 V	−13.5197	2.1925	0.9989
−15 °C 0.5 C 4.2 V	−19.6954	3.0912	0.9787
−15 °C 0.5 C 4.25 V	−37.7885	4.4759	–

Figure 2.15 shows the relationship between the capacity degradation rate k and the charging cutoff voltage. As can be seen from Fig. 2.15, the k values for different cutoff voltages in the first stage are relatively close, varying between 0.86 and 1.01. This indicates that the effect of cutoff voltage on capacity degradation is not significant in this stage. In the latter stage of battery life, the capacity degradation rate k rises with the cutoff voltage, indicating a shift in the aging mechanism. Higher cutoff voltages intensify lithium plating, thereby accelerating capacity loss.

The relatively close k values for different cutoff voltages in the first stage can be used to obtain the k values for 4.05–4.25 V through interpolation in Table 2.14. In the second stage, k increases with increasing cutoff voltage. A power function is selected to characterize the relationship between charging stress and capacity degradation rate, that is,

$$k = a \cdot U^b \tag{2.18}$$

where k represents the capacity degradation rate, U denotes the charging cutoff voltage, and a and b are undetermined coefficients. Taking the natural logarithm of both sides of Eq. (2.18) yields:

$$\ln k = \ln a + b \cdot \ln U \tag{2.19}$$

The relationship between ln k and ln U is fitted linearly for the second stage, as shown in Fig. 2.16.

From Fig. 2.16, the capacity degradation rate k in the second stage of aging is described by:

$$\ln k_2 = -26.6174 + 19.3937 \cdot \ln U \quad R^2 = 0.9681 \tag{2.20}$$

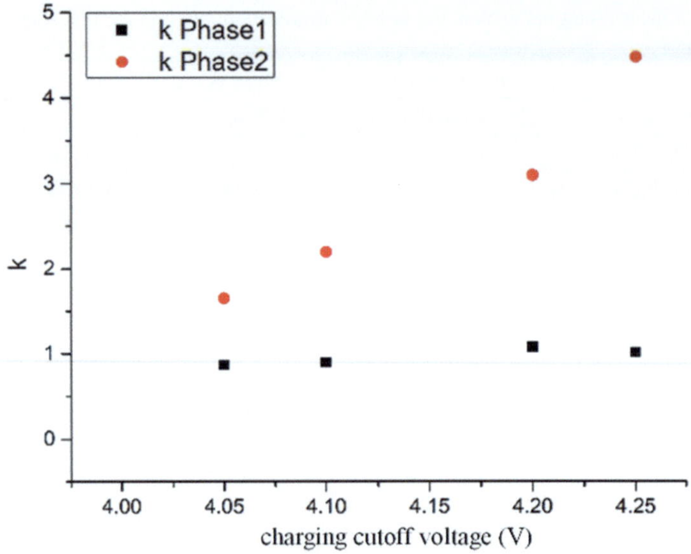

Fig. 2.15 Relationship between capacity degradation rate k and charging cutoff voltage

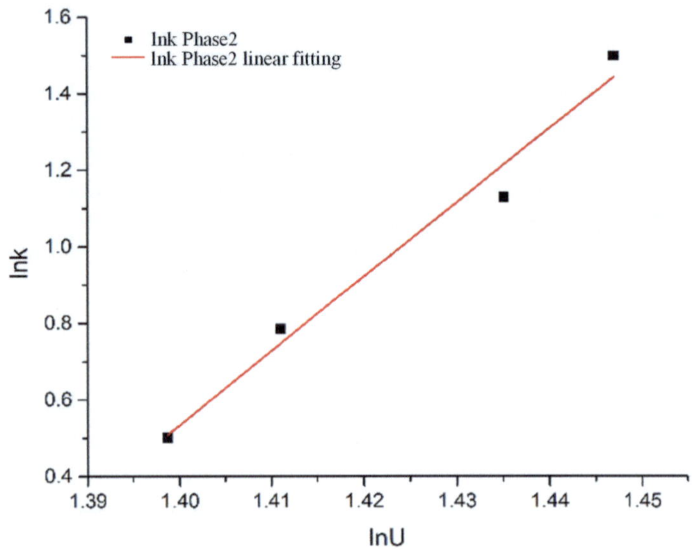

Fig. 2.16 Linear fitting of ln k versus ln U in the second stage

Which can be rewritten as:

$$k_2 = \exp(-26.6174) \cdot U^{19.3937} \tag{2.21}$$

(2) Analysis of Orthogonal Experimental Results

1) Capacity Degradation Characteristics from Orthogonal Experiments

This part focuses on examining the $L_8(2^7)$ orthogonal experiment results to explore how factors interact with each other. For analyzing battery aging, it is still segmented into two phases using a 10% capacity degradation threshold.

Figures 2.17a–d show the piecewise linear fitting curves for capacity degradation when charging at −10 °C in the $L_8(2^7)$ orthogonal experiments. Note that the orthogonal experimental group commenced later than others. To minimize the impact of the battery's initial health state on the results, the data for −10 °C, 0.5 C, 4.2 V used here are from the second repeat experiment. The fitting parameters for each condition in Fig. 2.17 are listed in Tables 2.16 and 2.17.

Similarly, the piecewise linear fitting curves for capacity degradation when charging at −15 °C in the $L_8(2^7)$ orthogonal experiments are obtained, as shown in Figs. 2.18a and b. As depicted in Fig. 2.14, the fitting curves for −15 °C, 0.5 C,

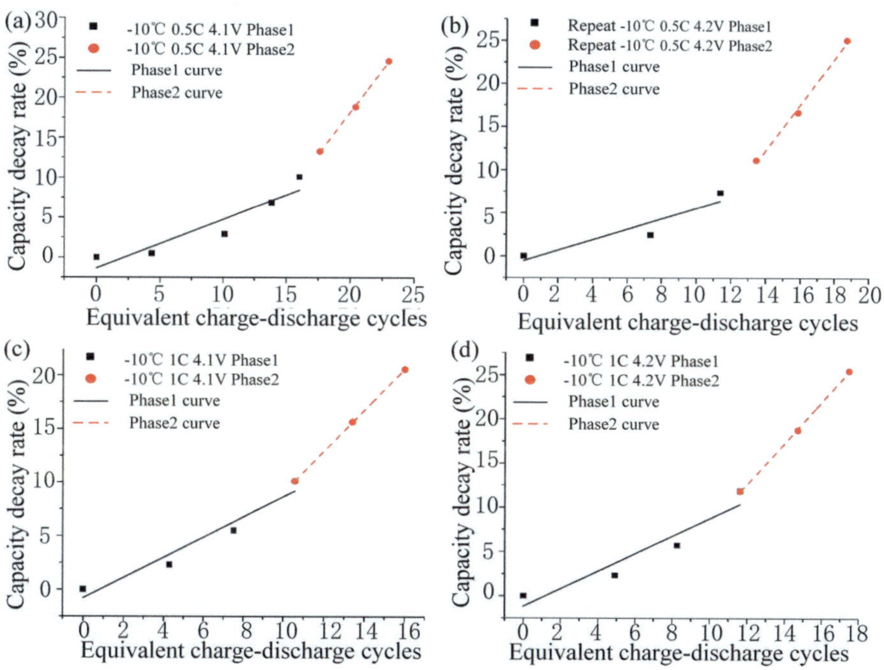

Fig. 2.17 Piecewise linear fitting curves for capacity degradation at −10 °C. **a** −10 °C 0.5 C 4.1 V; **b** Repeat −10 °C 0.5 C 4.2 V; **c** −10 °C 1 C 4.1 V; **d** −10 °C 1 C 4.2 V

Table 2.16 Linear fitting parameters for capacity degradation rate *k* in the first stage at −10 °C

Condition	Intercept (*a*)	Slope (k)	Adj. R-square
−10 °C 0.5 C 4.1 V	−1.3504	0.6113	0.8388
Re_−10 °C 0.5 C 4.2 V	−1.0961	0.8107	0.7532
−10 °C 1 C 4.1 V	−0.7928	0.9434	0.9200
−10 °C 1 C 4.2 V	−1.1683	0.9919	0.8647

Table 2.17 Linear fitting parameters for capacity degradation rate *k* in the second stage at −10 °C

Condition	Intercept (*a*)	Slope (k)	Adj. R-square
−10 °C 0.5 C 4.1 V	−24.0450	2.1153	0.9980
Re_−10 °C 0.5 C 4.2 V	−24.5866	2.6578	0.9903
−10 °C 1 C 4.1 V	−10.1940	1.9255	0.9999
−10 °C 1 C 4.2 V	−15.3615	2.3366	0.9992

4.1 V, and −15 °C, 0.5 C, 4.2 V have already been illustrated, so there is no need to repeat them here. The fitting parameters for each condition are listed in Tables 2.18 and 2.19.

Thus, the capacity degradation rate *k* for all conditions in the $L_8(2^7)$ orthogonal experimental group follows the previously summarized trends, i.e., the capacity

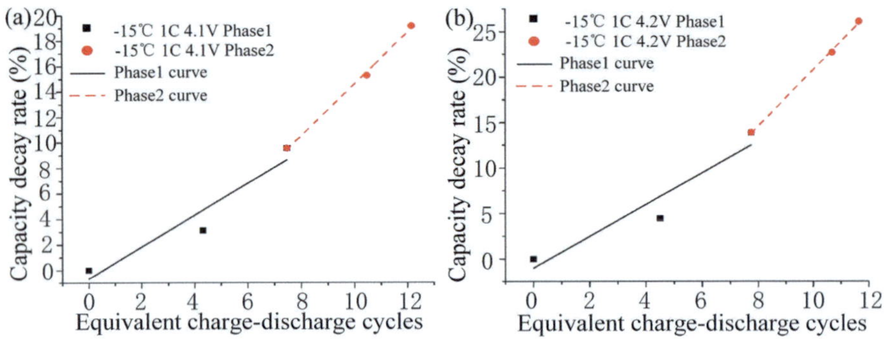

Fig. 2.18 Piecewise linear fitting curves for capacity degradation at −15 °C. **a** −15 °C 1 C 4.1 V; **b** −15 °C 1 C 4.2 V

Table 2.18 Linear fitting parameters for capacity degradation rate *k* in the first stage at −15 °C

Condition	Intercept (a)	Slope (k)	Adj. R-square
−15 °C 0.5 C 4.1 V	−0.9832	0.8795	0.8410
−15 °C 0.5 C 4.2 V	−0.8698	1.0738	0.8889
−15 °C 1 C 4.1 V	−0.6743	1.2461	0.8371
−15 °C 1 C 4.2 V	−1.0061	1.7410	0.8249

Table 2.19 Linear fitting parameters for capacity degradation rate k in the second stage at −15 °C

Condition	Intercept (a)	Slope (k)	Adj. R-square
−15 °C 0.5 C 4.1 V	−10.6831	2.0066	0.9988
−15 °C 0.5 C 4.2 V	−19.6954	3.0912	0.9787
−15 °C 1 C 4.1 V	−5.7255	2.0333	0.9951
−15 °C 1 C 4.2 V	−10.1889	3.1001	0.9977

degradation rate increases with decreasing temperature, increasing charging rate, and increasing cutoff voltage. Moreover, the capacity degradation rate is higher in the later stages of battery aging.

2) Analysis of Orthogonal Experimental Results

Orthogonal experimental results can be analyzed using two primary methods: range analysis and analysis of variance. Range analysis is a straightforward and interpretable approach that enables precise identification of the relative importance of factors influencing experimental outcomes. In contrast, the analysis of variance method is particularly effective for identifying factors with significant influence on the results [79]. Typically, these two methods are combined for analysis. Similar to the analysis of single-factor experimental results, the experimental index of interest in the orthogonal experimental results analysis is also the capacity degradation rate k.

a. Range Analysis Method

The capacity degradation rate data for each condition are used to calculate the range to determine the order of importance of factors. In this context, K_i denotes the aggregate value of experimental outcomes associated with the i-th level of a specific factor column. For instance, K_1 for column A corresponds to the cumulative sum of capacity degradation rates observed across all experimental groups where the temperature factor is set at level 1 (i.e., charging temperature of −10 °C). k_i is defined as the arithmetic mean of experimental results when the factor is assigned to level i, which is calculated as $k_i = K_i/s$, where s represents the number of times each level appears within any given column. For example, the level 1 appears four times in column A, so $s = 4$. R is called the range, and $R = \max(k_i) - \min(k_i)$. Specifically, a larger range typically indicates a more significant change in the experimental index resulting from fluctuations in factor values within the tested range. Consequently, the factor associated with the largest range in the experimental matrix is identified as the most influential on the results. This relationship highlights the utility of range analysis in quantifying the relative importance of different factors during experimental evaluation. The results of the range analysis for the capacity degradation rate k in the first stage of aging are listed in Table 2.20.

As observed from Table 2.20, the hierarchy of significance among factors influencing the capacity degradation rate during the initial aging phase is as follows: charging temperature > charging rate > charging cutoff voltage > interaction between charging temperature and cutoff voltage > interaction between charging temperature

Table 2.20 Range analysis for the first stage of the $L_8(2^7)$ orthogonal experiment

Indicator	Indicator	A	B	$A \times B$	C	$A \times C$	$B \times C$	Error column
Capacity degradation rate k	$K1$	3.3573	3.3754	4.4091	3.6803	4.4180	4.2238	3.9231
	$K2$	4.9404	4.9223	3.8886	4.6174	3.8797	4.0734	4.3747
	$k1$	0.8393	0.8438	1.1023	0.9201	1.1045	1.0559	0.9808
	$k2$	1.2351	1.2306	0.9721	1.1544	0.9699	1.0185	1.0937
	R	0.3958	0.3867	0.1301	0.2342	0.1346	0.0374	0.1129
Hierarchical importance of factors		$A > B > C > A \times C > A \times B >$ Error column $> B \times C$						

Table 2.21 Range analysis for the second stage of the $L_8(2^7)$ orthogonal experiment

Indicator	Indicator	A	B	$A \times B$	C	$A \times C$	$B \times C$	Error Column
Capacity degradation rate k	$K1$	9.0357	9.8709	9.9065	8.0807	10.6432	9.5586	9.5764
	$K2$	10.231	9.3955	9.3599	11.1857	8.6232	9.7078	9.6900
	$k1$	2.2594	2.4677	2.4766	2.0202	2.6608	2.3897	2.3941
	$k2$	2.5582	2.3489	2.3399	2.7964	2.1558	2.4269	2.4225
	R	0.2991	0.1189	0.1367	0.7762	0.5050	0.0373	0.0284
Hierarchical importance of factors		$C > A \times C > A > A \times B > B > B \times C >$ Error column						

and rate. The range of the interaction between charging rate and cutoff voltage is smaller than that of the error column, indicating that the interaction between charging rate and cutoff voltage can be neglected or its impact on capacity degradation rate can be ignored.

The range analysis results for the second stage of aging are listed in Table 2.21. During the second stage of aging, the sequence of factor significance regarding capacity degradation rate is determined as follows, based on the experimental data presented in Table 2.21, charging cutoff voltage > interaction between charging temperature and cutoff voltage > charging rate > interaction between charging temperature and rate > charging temperature > interaction between charging temperature and rate.

b. Analysis of Variance Method

The analysis of variance method can be used to test the significance of factors. Taking factor A as an example, for a given significance level α, the critical value F_α (df_A, df_e) is first looked up from the F-distribution table, where df_A and df_e are the degrees of freedom for factor A and the error column, respectively. If $F_A > F_\alpha$ (df_A, df_e), then factor A has a significant impact on the experimental results. In cases where the mean square of a factor or interaction is less than or equal to the mean square of the error column, the corresponding factor or interaction should be incorporated into the error column to form a new error term. The degrees of freedom associated with the error column must also be updated to reflect the revised configuration. For example, in the first stage analysis of variance, the mean square of B × C is less than that of the error column, so B × C is combined into the error column, and the degrees of freedom of the error column are updated to 2.

The outcomes of the analysis of variance for the first stage are presented in Table 2.22. In the first stage of aging, the interaction between charging rate and cutoff voltage has been combined into the error column and is no longer considered. The order of importance of factors affecting the capacity degradation rate is charging temperature > charging rate > charging cutoff voltage > interaction between charging temperature and cutoff voltage > interaction between charging temperature and rate, which align with the outcomes of the range analysis. The critical value $F_{0.05}(1,2) = 18.51$ is obtained from statistical tables. Following standard significance testing procedures, at the 0.05 significance threshold, charging temperature and charging rate emerge as significant factors influencing the results.

The results of the analysis of variance for the second stage are provided in Table 2.23. During this stage, the mean square values of all factors and their interactions exceed those of the error column. This finding theoretically suggests that the effects of interactions are statistically significant and cannot be disregarded in the analysis. The order of importance of factors affecting the capacity degradation rate is charging cutoff voltage > interaction between charging temperature and cutoff voltage > charging temperature > interaction between charging temperature and rate > charging rate > interaction between charging rate and cutoff voltage, which is consistent with the range analysis results. The critical value $F_{0.05}(1,1) = 161.45$ is looked up. According to the significance testing criteria, at a significance level of 0.05, the factors exhibiting significant effects on the experimental outcomes are the charging cutoff voltage and the interaction between charging temperature and cutoff voltage. In contrast, the charging temperature was considered a secondary significant factor.

Upon evaluating the outcomes of range analysis and analysis of variance across the two phases of battery aging, it is possible to conclude the following:

Table 2.22 Analysis of variance for the first stage of the $L_8(2^7)$ orthogonal experiment

Indicator	Indicator	A	B	$A \times B$	C	$A \times C$	Error column
Capacity degradation rate k	Sum of squares	0.3133	0.2991	0.0339	0.1098	0.0362	0.0283
	Degrees of freedom	1	1	1	1	1	2
	Mean square	0.3133	0.2991	0.0339	0.1098	0.0362	0.0141
	F-value	22.1430	21.1419	2.3941	7.7580	2.5600	–
Hierarchical importance of factors		$A > B > C > A \times C > A \times B$					

Table 2.23 Analysis of variance for the second stage of the $L_8(2^7)$ orthogonal experiment

Indicator	Indicator	A	B	$A \times B$	C	$A \times C$	$B \times C$	Error column
Capacity degradation rate k	Sum of squares	0.1788	0.0283	0.0374	1.2051	0.51002	0.0028	0.0016
	Degrees of freedom	1	1	1	1	1	1	1
	Mean square	0.1788	0.0283	0.0374	1.2051	0.51002	0.0028	0.0016
	F-value	110.7931	17.5075	23.142	746.651	315.991	1.7234	–
Hierarchical importance of factors		$C > A \times C > A > A \times B > B > B \times C$						

- In the first stage of aging, exert a notable influence on the capacity degradation rate of batteries. Among these factors, charging temperature is identified as the most significant contributor to capacity degradation, followed by charging rate and subsequently charging cutoff voltage.
- During the initial aging phase, the interactions among factors exhibit a limited influence on the capacity degradation rate. Specifically, the interaction between charging rate and cutoff voltage is deemed negligible in this stage.
- Compared with the first stage of aging, the impact of charging cutoff voltage on capacity degradation is strengthened in the second stage. The significant alterations in the electrode structure and the lithium plating layer on the negative electrode during the first stages of battery aging are responsible for this phenomenon. Specifically, the increase in charging cutoff voltage intensifies lithium plating, thereby accelerating the degradation of battery capacity In the second stage of aging, the charging cutoff voltage becomes a significant impact factor on the capacity degradation rate.
- Compared with the first stage of aging, the impact of factor interactions on capacity degradation is strengthened in the second stage, which may be related to changes in electrode structure and loss of active material. Therefore, when establishing the multi-stress capacity degradation model, the effects of factor interactions should be considered. Among the interactions, the interaction between charging temperature and cutoff voltage has the greatest impact, followed by the interaction between charging temperature and rate, while the interaction between charging rate and cutoff voltage still has the smallest impact.

2.3.2.2 Analysis of Sample D

(1) Cell Performance

The experimental cells were subjected to cycling at −25 °C using a 2 C CC – CV protocol. Every 10 cycles, a standard capacity test and a HPPC test were conducted.

The final discharge capacity obtained during these tests was defined as the standard capacity of the cell at the given temperature. Meanwhile, the internal resistance of the cells was determined from the HPPC test results. The pulse discharge and charge resistances were calculated using Eqs. 2.1 and 2.2, respectively.

$$R\_D = \left| \frac{U_{D\_0\_s} - U_{D\_18\_s}}{I_D} \right| \tag{2.22}$$

$$R\_C = \left| \frac{U_{C\_0\_s} - U_{C\_10\_s}}{I_C} \right| \tag{2.23}$$

The definitions of the parameters are as follows: R_D represents the pulse discharging resistance; R_C denotes the pulse charging resistance; U_{D_0s}: the initial voltage of pulse discharge; U_{D_18s}: the terminal voltage of pulse discharge; I_D: the current of pulse discharge; U_{C_0s}: the initial voltage of pulse charge; U_{C_10s}: the terminal voltage of pulse charge; I_C: the current of pulse charge.

Figure 2.19 illustrates the cell capacity and internal resistance during long-term cycling. Initially, the cell capacity experiences a rapid decline in the early stages of cycling. Following extensive cycling, the cell capacity exhibits a significant reduction. However, the degradation rate of cell capacity gradually slows down in the subsequent cycles. After 120 cycles, the cell capacity retains 54.5% of its initial value, whereas after 250 cycles, it maintains 41.2% of its initial capacity. From the 0th to the 120th cycle, the cell capacity decreases by 45.5%. In contrast, during the subsequent 100 cycles (from the 120th to the 250th cycle), the capacity loss is only 13.3%. Notably, a slight increase in cell capacity is observed after the 40th cycle, which can be attributed to the extended storage time at room temperature.

The standard capacity test results for cells undergoing low-temperature high-rate charge cycling reveal varying degrees of capacity recovery. Specifically, the cell capacity measured during the standard capacity test is consistently higher than that observed during low-temperature high-rate charge cycling. This finding contrasts with prior research [12, 81, 82], a remarkable capacity restoration occurs upon thermal transition from subzero (−25 °C) to ambient (23 °C) conditions. The recovery efficiency demonstrates progressive enhancement with extended cycling duration, reaching 80–90% capacity retention after 200 cycles. This cyclical regeneration pattern demonstrates significant reversibility of capacity degradation mechanisms under specified operating parameters.

The internal resistance exhibits an inverse relationship with cell capacity in terms of variation trends. During the early stages of cycling, the tested cell's internal resistance demonstrates a tendency to rise, following an approximately linear pattern. However, beyond 200 cycles, minimal changes in internal resistance are observed. This behavior is attributed to cell internal polarization: the application of a larger pulse current exacerbates polarization, driving an increase in internal resistance. Consequently, the pulse discharge resistance exceeds the pulse charge resistance due to these polarization effects.

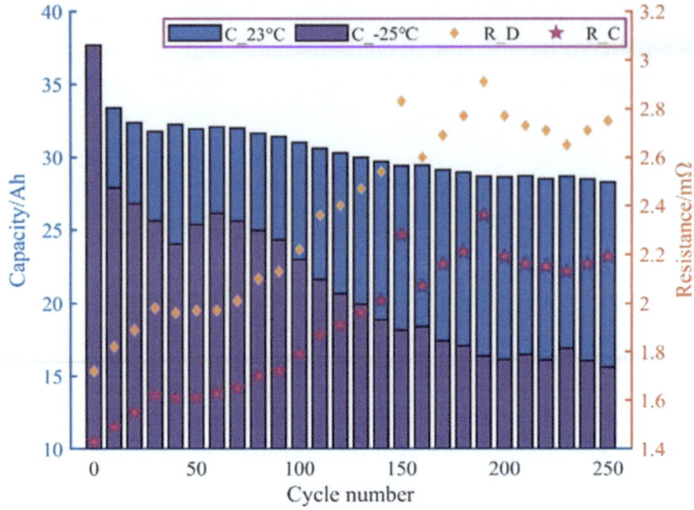

Fig. 2.19 Battery capacity and internal resistance change with cycle. (*R_D*: the pulse discharging resistance; *R_C*: the pulse charging resistance) [80]

Unlike the methods previously used by other researchers, the CC−CV charging profile is employed during the discharging process in this study. This approach is intended to reduce internal polarization and enhance the reinsertion of lithium ions into the cathode. Figure 2.20 illustrates the variation trend of the discharge capacity and its corresponding ratio to the total discharge capacity during the CC discharging process under low-temperature high-rate charge cycling conditions.

It is evident that the CC discharge capacity diminishes as cycling progresses. Contrary to the degradation pattern of total cell capacity, the CC discharge capacity exhibits a sporadic decline with cycling. A notable abrupt change occurs at the 120th cycle, where the CC discharge capacity undergoes a sharp reduction. Prior to the 120th cycle, the CC discharge capacity demonstrates an accelerating rate of decrease. Consequently, the ratio of CC discharge capacity to total discharge capacity also diminishes rapidly. Beyond the 120th cycle, although the CC discharge capacity continues to decline, the rate becomes considerably slower. Moreover, the ratio of CC discharge capacity to total discharge capacity stabilizes, remaining nearly constant at approximately 15%. This abrupt behavior in CC discharge capacity is predominantly attributed to the progressively severe internal polarization of the cell during low-temperature high-rate cycling. As internal polarization intensifies, the cell voltage experiences a rapid decline, leading to a significant reduction in discharge capacity during the CC discharge phase.

An additional notable characteristic observed during low-temperature high-rate cycling is that the discharge capacity consistently exceeds the charge capacity. Throughout the low-temperature high-rate cycling process, the cell's discharge capacity remains higher than its charge capacity. As cycling progresses, both the

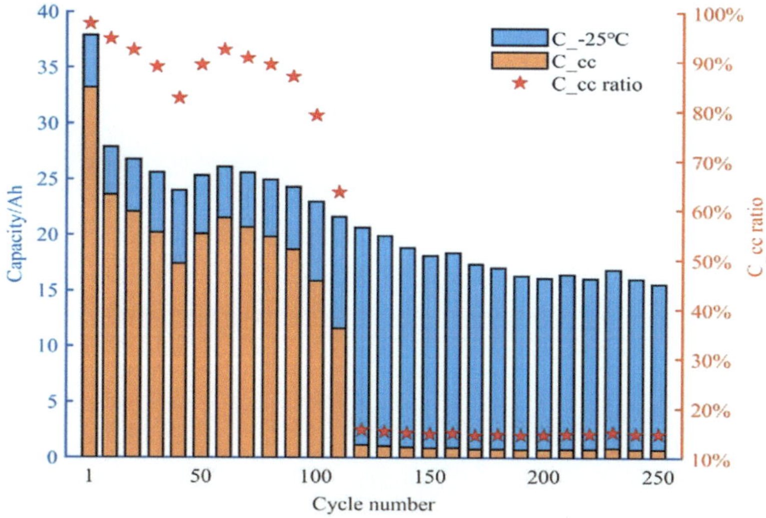

Fig. 2.20 Depicts the CC discharge capacity and its proportion relative to the total discharge capacity during the cycling process. (C_cc: CC discharge capacity; C_cc ratio: the ratio of the CC discharge capacity to the total discharge capacity) [80]

charge and discharge capacities exhibit a continuous downward trend. Furthermore, the patterns of change in charge and discharge capacities are strikingly similar, with discharge capacity consistently surpassing charge capacity. Notably, the disparity between them gradually widens as cycling continues; however, the rate of increase in this disparity remains relatively modest. By the conclusion of the low-temperature high-rate charge cycling, the discharge capacity is approximately 2 Ah greater than the charge capacity.

(2) Postmortem Results

Analysis of the cell's behavior reveals that the 120th cycle serves as a pivotal milestone. To further investigate the effects of low-temperature high-rate charging on the cell during extended cycling, three distinct cell states were examined: pristine cells, cells after 120 cycles, and cells after 250 cycles. Fully discharged cells were disassembled within an argon-filled glovebox to retrieve the corresponding anodes. Figure 2.21 presents the morphological images of anodes subjected to varying cycle counts post-disassembly. When compared to the pristine anode, the anodes from cycled cells exhibit pronounced morphological changes. Notably, the anode surfaces from cycled cells are coated with silver-white deposits, which nearly entirely obscure the graphite particles. Furthermore, inhomogeneous deposition patterns are evident, with the most severe deposition occurring at the edges. This phenomenon is likely associated with non-uniform local current distribution and uneven heat dissipation.

Figure 2.22 presents the SEM morphologies of anodes subjected to varying cycle counts. The distinct morphology of graphite particles in the pristine anode is clearly observable. Conversely, the anode subjected to 120 cycles exhibits complete coverage

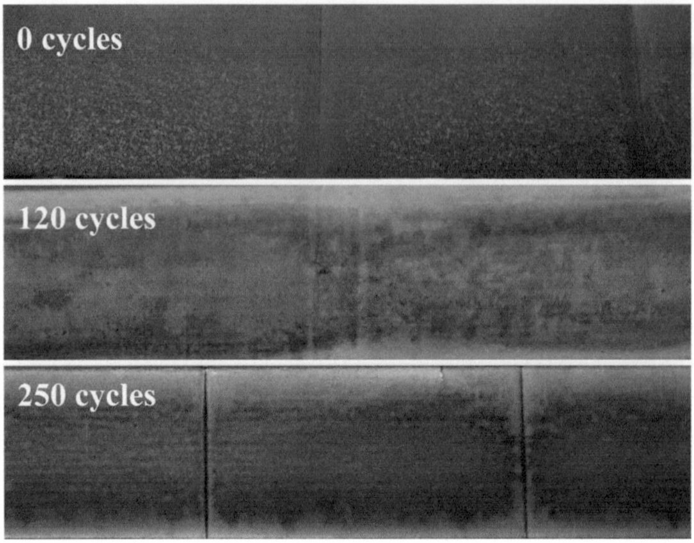

Fig. 2.21 Anode appearance after different cycles [80]

of silver-white deposits at the periphery, rendering the graphite morphology entirely obscured. In the central region, only partial coverage by silver-white substances is evident, with some graphite particles still discernible. For the anode after 250 cycles, both the edge and central regions are almost entirely coated with silver-white substances, causing the graphite morphology to be barely distinguishable. This observation preliminarily suggests that low-temperature high-rate charging exerts a pronounced detrimental effect on the cell. Furthermore, as the cycle count increases, the impact of low-temperature high-rate charging on the anode morphology becomes progressively more severe.

Figure 2.22a–f provides an enlarged view of the anode surface morphology, revealing the presence of numerous laterally growing lithium dendrites that become increasingly intertwined with cycling. Observations indicate a dense population of lithium dendrites at the anode edge after 120 cycles. Despite variations in width, length, and shape, these dendrites are distinctly identifiable. Measurements reveal that their widths typically fall within the range of 0.4–0.8 μm, while their lengths exhibit a broader distribution. Longer dendrites exceed 5 μm in length, whereas shorter ones are less than 0.7 μm. Some elongated dendrites exhibit deformation, likely due to spatial constraints causing compression. In contrast, the central region of the anode contains fewer lithium dendrites, and their dimensions are relatively smaller.

After 250 cycles, the lithium dendrites at the anode edge become interwoven, making individual dendrites less distinguishable compared to those after 120 cycles. The visible lithium dendrites also exhibit a reduction in size. Conversely, the central region of the anode undergoes significant morphological changes. There is a marked

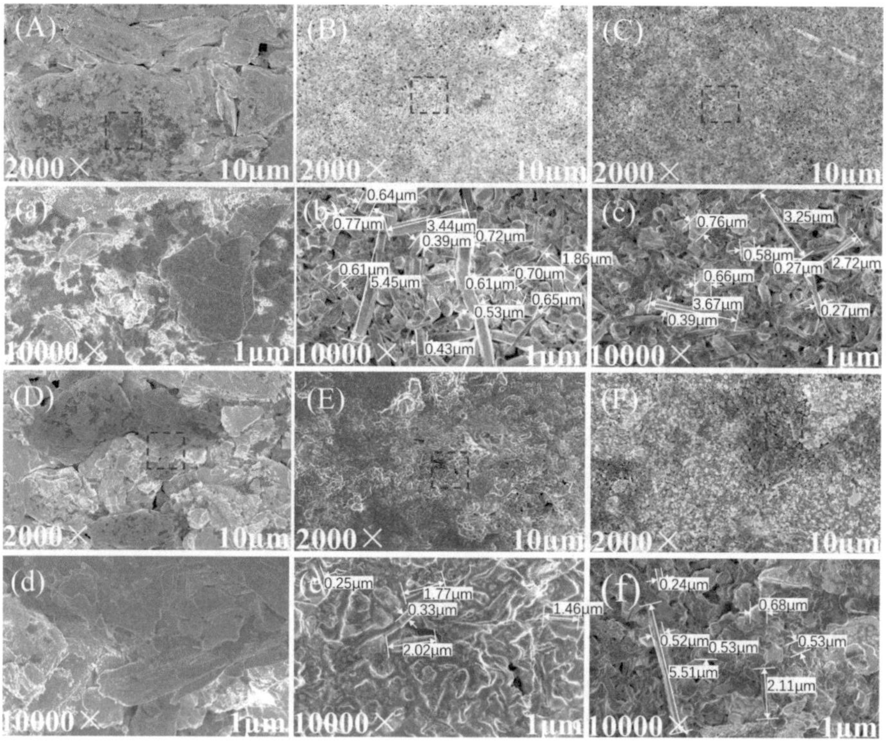

Fig. 2.22 SEM images of the anode surface after different cycles. **A**, **a**; **D**, **d**: 0 cycle; **B**, **b**; **E**, **e**: 120 cycles; **C**, **c**; **F**, **f**: 250 cycles. **A**–**C**: Edge; **D**–**F**: Center. **a**–**f**: Enlarged view of (**A**–**F**), respectively [80]

increase in the number of lithium dendrites, and their dimensions are notably larger. Upon closer examination, the longer lithium dendrites in this region exceed 5 μm in length, which is considerably greater than those observed in the central part of the anode after 120 cycles.

SEM analysis of the anode surface reveals that lithium plating is the dominant factor contributing to the accelerated capacity degradation of LIBs under low-temperature high-rate charging conditions. This lithium plating exhibits a highly non-uniform distribution, initially forming at the anode edge and progressively propagating toward the central region as cycling continues. The deposited metallic lithium manifests as numerous lithium dendrites on the anode surface. Contrary to conventional assumptions, these dendrites grow laterally rather than vertically, likely constrained by the limited internal space within the cell. As cycling progresses, the lithium dendrites become increasingly intertwined. Eventually, they coalesce into a continuous layer, obscuring individual dendrite morphologies and rendering them indistinct.

To investigate the evolution of lithium plating during low-temperature high-rate charge cycling, the cross-sectional morphology of the anode was analyzed. As shown in Fig. 2.23, the dense layer observed above the cross-section corresponds to the plated lithium. Variations in the thickness of this dense layer indicate non-uniform lithium plating across the anode. Measurements reveal that the thickness of lithium deposition is significantly greater at the edge compared to the central region. Specifically, after 120 cycles, the lithium deposition thickness reaches approximately 15 μm at the edge, whereas it is about 7 μm at the central region.

Upon further magnification of the cross-sectional morphology, a thick film covering the anode surface is evident. This film is speculated to be the SEI layer, beneath which the plated lithium resides. Unlike the lithium deposition, the SEI film exhibits a thinner profile at the edge and a thicker profile at the central region. The SEI film thickness is approximately 0.2 μm at the edge and about 0.4 μm at the central region. This discrepancy may be attributed to the central region's poorer heat dissipation compared to the edge, resulting in higher local temperatures. Elevated temperatures accelerate the reaction between plated lithium and the electrolyte, thereby thickening the SEI film in the central region.

After 250 cycles, the thickness of lithium deposition has significantly increased in both the edge and central regions of the anode, indicating more severe lithium plating compared to the anode after 120 cycles. Additionally, lithium deposition is observed within some graphite cracks located far from the anode surface. Furthermore, increased stress from lithium plating has caused some graphite particles to fracture. Meanwhile, the thickness of the SEI film in the edge region remains relatively unchanged, whereas in the central region, the SEI film thickness has increased substantially. This suggests that the central region experiences more pronounced

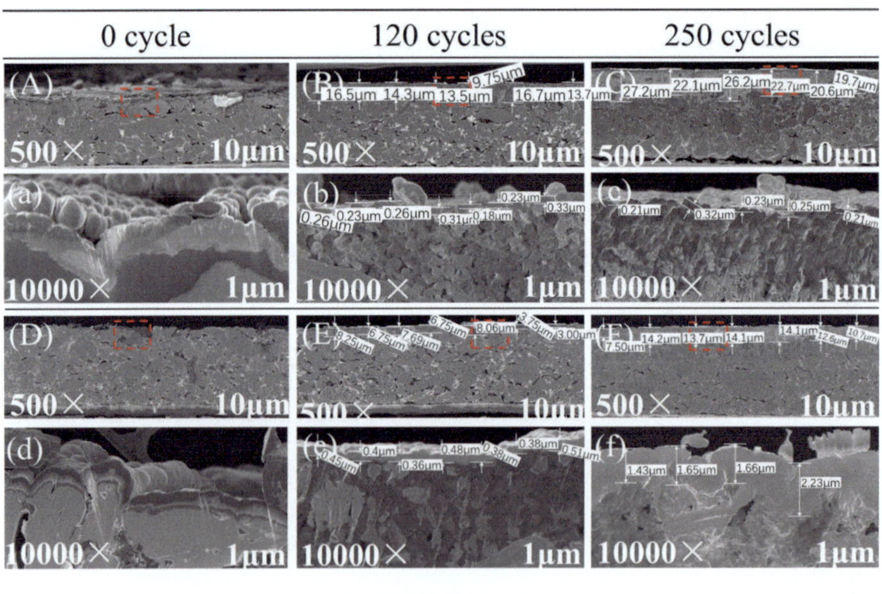

Fig. 2.23 Cross-sectional SEM images of the anode after different cycles. **A–C**: Edge; **D–F**: Center. **a–f**: Enlarged view of **A–F**, respectively [80]

SEI film growth, likely due to elevated local temperatures and increased reactivity between plated lithium and the electrolyte.

To further investigate the effects of lithium plating, XPS analysis was conducted to examine the compositional changes in the anode. XPS measurements were performed on both the edge and central regions of anodes subjected to different cycle counts. In addition to analyzing the surface composition of the anode, a 3000 eV argon ion beam was used to etch the anode for 120 s, allowing for a more in-depth examination of lithium deposition. The XPS spectra are presented in Figs. 2.24 and 2.25.

In the C 1s spectrum, four peaks are positioned at approximately 282.5, 284.8, 286.8, and 290.1 eV, corresponding to carbide, carbon–carbon single bond (C–C), carbon–oxygen single bond (C–O), and carbonate (CO_3^{2-}), respectively. In the O 1s spectrum, the peaks for metal oxide oxygen bond (M–O), CO_3^{2-}, and C–O are found at approximately 528.5, 532.0, and 533.4 eV, respectively. The F 1s spectrum exhibits two peaks associated with lithium fluoride (LiF) at 685.2 eV and phosphorus–fluorine single bond (P–F) at 687.1 eV. In the Li 1s spectrum, the peaks corresponding to lithium (Li), lithium oxide (Li_2O), and LiF/Li_2CO_3 are observed at approximately 52.5, 54.1, and 55.6 eV, respectively.

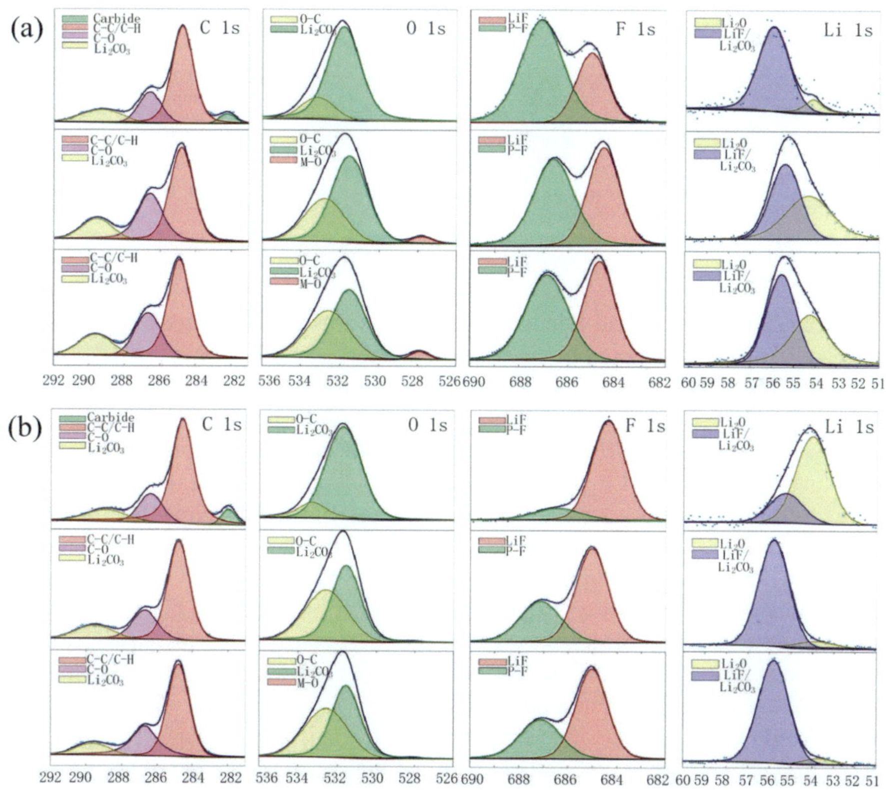

Fig. 2.24 XPS patterns of the anode surface after different cycles. **a** Edge; **b** Center [80]

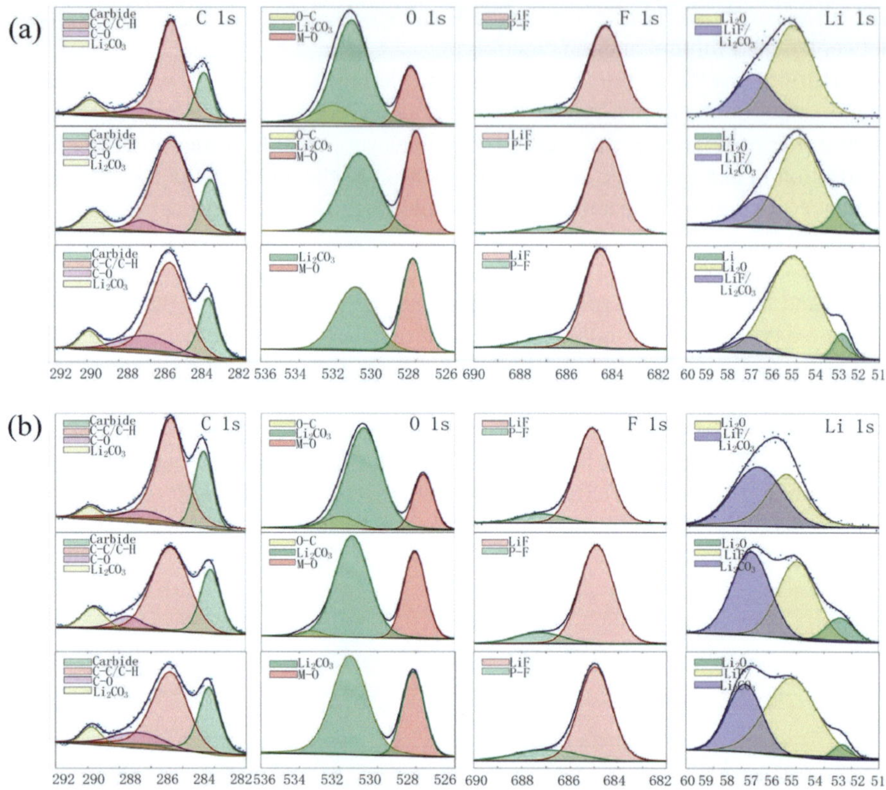

Fig. 2.25 XPS patterns of the anode etching after different cycles. **a** Edge; **b** Center [80]

The analysis of chemical bonds and groups on the anode surface allows for the inference of its composition. The surface components primarily consist of organic and inorganic substances. Organic substances are predominantly organic lithium salts, particularly alkyl lithium compounds, while inorganic substances mainly include Li_2CO_3, LiF, and Li_2O. Notably, elemental lithium is absent on the anode surface, indicating that the plated metallic lithium has undergone side reactions. In the C 1s spectrum, the disappearance of the carbide peak further confirms that the anode surface is coated with reaction products. This aligns with the SEM observations, where a layer of material-composed primarily of Li_2CO_3, LiF, Li_2O, and organic lithium salts-covers the anode surface. This layer is attributed to the reaction between plated lithium and the electrolyte. In the O 1s spectrum, the increased content of organic substances is attributed to the formation of additional organic lithium salts resulting from the interaction between plated lithium and the electrolyte. The rise in M−O content, indicative of increased Li_2O, is corroborated by the Li 1s spectrum. Furthermore, a comparison between the central and edge regions reveals minimal changes in the central region, reinforcing the conclusion that lithium plating is spatially inhomogeneous.

Upon anode etching, the following observations are made: the signal intensities of C–C, C–O, and CO_3^{2-} functional groups diminish, whereas those of LiF and Li_2O increase. This suggests a heterogeneous composition within the SEI film. Specifically, the outer SEI layer predominantly consists of Li_2CO_3 and lithium alkyl, while the inner layer is rich in LiF and Li_2O. Notably, the emergence of an elemental lithium peak after etching indicates that the plated lithium is encapsulated by the SEI film, corroborating the SEM findings. The plated metallic lithium is present on the surface of layered graphite and its interstices, covered by an SEI film formed through the reaction of lithium with the electrolyte. However, as aging progresses, although the rate of lithium plating diminishes, the continuous consumption of plated lithium due to its reaction with the electrolyte reduces the residual lithium, as evidenced by the Li 1s spectrum.

(3) Discussion

The primary degradation mechanisms during cycling under low-temperature high-rate charging conditions are depicted in Fig. 2.26. Unlike the typical accelerated degradation behavior of LIBs at low temperatures, the cell capacity exhibits a decelerated degradation trend. Under these conditions, lithium plating is identified as the dominant degradation mechanism. In the early stages of cyclic aging, the highly reactive plated lithium continuously undergoes side reactions with the electrolyte, leading to the consumption of elemental lithium. Additionally, the side reaction products thicken the SEI film, further reducing the available recyclable lithium. This significant loss of recyclable lithium results in a rapid decline in cell capacity. Simultaneously, lithium plating causes a shift in the SoC window, rendering the high SoC region of the anode electrochemically inaccessible. As a consequence, the rate of lithium plating gradually diminishes, indicating a slowdown in the intercalation of lithium into graphite. This deceleration in lithium plating contributes to the observed decelerated degradation of cell capacity. Furthermore, the cracking of graphite particles also contributes to capacity loss. As the rate of lithium plating decreases, the side reactions between plated lithium and the electrolyte are reduced, leading to a stabilization of the cell's internal resistance in the later stages of cyclic aging. This behavior highlights the complex interplay between lithium plating, SEI film evolution, and structural degradation in determining the overall capacity retention of LIBs under low-temperature high-rate charging conditions.

This study employs the CC–CV discharge method, differing from the conventional CC discharge approach. As the number of cycles increases, the discharge capacity during the CC discharge phase exhibits a downward trend. A sharp decline in CC discharge capacity is observed at approximately 120 cycles. At low temperatures, the SEI film thickens with cycling due to the reaction between plated lithium and the electrolyte. These side reactions involving plated lithium lead to an increase in cell internal resistance. Moreover, the reduced electrolyte conductivity, slower electrochemical reaction rate at the anode-electrolyte interface, and decreased solid-phase diffusion rate of lithium ions in graphite, which all attributable to lower temperatures contribute to heightened internal polarization. As low-temperature cyclic aging progresses, the escalating internal polarization accelerates the decline in CC

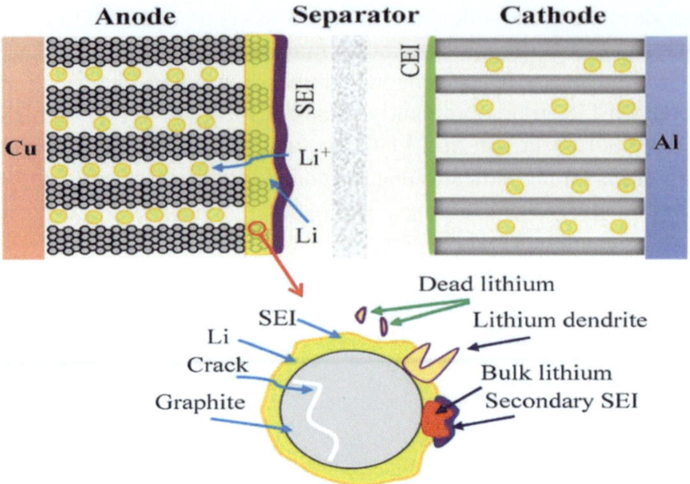

Fig. 2.26 Degradation mechanisms of LIBs during cycling under high-rate charging at low temperature [80]

discharge capacity during discharge. Consequently, the CC discharge capacity experiences a significant drop around the 120th cycle and remains largely unchanged in subsequent cyclic aging.

Under low-temperature high-rate charging conditions, the discharge capacity during the discharge process exceeds the charge capacity during the charge process for the tested cell. As cycling progresses, substantial metallic lithium deposition occurs on the anode surface. However, only a small portion of this plated lithium participates in side reactions, such as reactions with the electrolyte at low temperatures, forming irreversible lithium. Most of the plated lithium remains highly reactive and actively participates in intercalation and deintercalation reactions at both the cathode and anode. The potential of the plated metallic lithium on the anode surface is lower than that of Li_xC_6 in the graphite layer, causing the plated lithium to preferentially dissolve and migrate to the cathode during discharge. Consequently, during discharge, lithium ions intercalated into the graphite and plated lithium migrate through the electrolyte to the cathode, where they intercalate. In contrast, during charging, some active lithium is plated onto the anode surface. This results in a greater number of lithium ions intercalating into the cathode during discharge compared to those intercalating into the anode during charging. This mechanism explains the observed phenomenon of discharge capacity exceeding charge capacity under low-temperature high-rate charging conditions.

At low temperatures, the rate of side reactions is reduced, allowing most of the plated lithium on the anode surface to retain its reactivity. When the ambient temperature rises, this plated lithium can be reactivated and reintercalated into the graphite. Consequently, when cells cycled under low-temperature high-rate charging conditions undergo a standard capacity test at 23 °C, the plated metallic lithium becomes active again, participating in normal intercalation and deintercalation processes. This results in the recovery of cell capacity. In the initial stage of low-temperature high-rate cycling, the primary loss of cell capacity is due to the consumption of plated lithium in side reactions, leading to the formation of irreversible lithium. This results in a relatively low cell capacity recovery rate. However, in the later stages of cycling, the low temperature significantly reduces the rate of side reactions. Most of the plated lithium remains uninvolved in these side reactions and retains its reactivity. When the ambient temperature increases, this plated lithium can be reactivated and participate in normal intercalation and deintercalation processes. As a result, the cell capacity recovery rate becomes very high, reaching 80–90%.

2.3.3 Summary of Low-Temperature Cycling Degradation Mechanism

The experimental investigation of sample C forms the basis of this chapter, which primarily focuses on analyzing the results of single-factor reference experiments and $L_8(2^7)$ orthogonal experiments to determine the capacity degradation trends associated with variations in charging parameters. The analysis begins by examining the capacity degradation characteristics under diverse charging conditions. Subsequently, a piecewise linear function is employed to quantify the capacity degradation rate k for each specific charging condition. Further exploration delves into the effects of charging temperature T, charging rate I, and charging cutoff voltage U on the capacity degradation rate k. It is revealed that the capacity degradation rate k can be represented by the Arrhenius equation concerning T, an inverse exponential function concerning I, and a power function concerning U. The outcomes of the $L_8(2^7)$ orthogonal experiments, subjected to range analysis and analysis of variance, not only confirm the presence of interactions among the factors but also establish the hierarchy of factor importance. In the first stage of aging (capacity degradation rate not exceeding 10%), the order of importance of factors is charging temperature > charging rate > charging cutoff voltage > interaction between charging temperature and cutoff voltage > interaction between charging temperature and rate, with charging temperature and rate being significant impact factors. In the second stage of aging, where the capacity degradation rate surpasses 10%, the hierarchy of factor significance is as follows: charging cutoff voltage > interaction between charging temperature and cutoff voltage > charging temperature > interaction between charging temperature and rate > charging rate > interaction between charging rate and cutoff voltage, with charging cutoff voltage, interaction between charging temperature and cutoff voltage, and charging temperature being significant impact factors.

Based on the experimental investigation of sample D, this chapter investigates the degradation mechanisms of LIBs under low-temperature high-rate charging conditions, focusing on a large-capacity commercial prismatic battery (37 Ah, NMC523 cathode, graphite anode). The experiments involved three stages: standard capacity testing, hybrid pulse power characteristic (HPPC) testing, and low-temperature high-rate charging cycles at −25 °C with a 2 C rate. The results show that cell capacity decreases significantly during initial cycles but exhibits a decelerating degradation trend over prolonged cycling. Notably, capacity recovery is observed when transitioning from low to room temperature, with recovery rates reaching 80–90% after 200 cycles. Post-mortem analysis indicates pronounced lithium deposition on the anode, particularly at the periphery, accompanied by extensive lithium dendrite growth and an increase in SEI film thickness. The findings suggest that lithium plating and SEI film formation are the dominant degradation mechanisms under these conditions. These insights provide critical guidance for optimizing battery design and management strategies to reduce capacity loss in cold environments.

2.4 Degradation Mechanism During AC Heating

The heating methods for lithium batteries can be broadly categorized into internal and external heating approaches. Among these, the internal heating method, particularly mutual pulse heating, has been identified as a superior technique due to its ability to achieve uniform temperature distribution within the cell while consuming minimal battery energy. Compared to external heating methods, internal heating offers several distinct advantages. First, internal heating generates heat directly within the battery, eliminating the need for lengthy heat conduction paths associated with external heating. This approach reduces the risk of localized hotspots near heating devices and enhances energy utilization efficiency, promoting more uniform temperature distribution across the battery. Second, while external heating efficiency can be improved with optimized heaters and thermal designs, internal heating provides greater design flexibility. Unlike external methods, internal heating is not restricted by battery geometry (e.g., cylindrical or prismatic cells) or the structural design of battery modules. These characteristics have made internal heating, particularly AC heating methods, a subject of growing interest among researchers in recent years.

A systematic investigation into the battery SoH performance following hundreds of AC heating cycles with varying excitation parameters is of paramount importance. AC heating involves both discharging and charging processes, which may lead to performance degradation due to lithium plating, particularly during low-temperature charging at low frequencies. To accelerate the implementation of AC heating methods, experimental validation is crucial to ensure their reliability and effectiveness in mitigating such detrimental effects. This underscores the necessity of rigorous testing to confirm the method's practical applicability and long-term viability.

2.4.1 Experimental Design

The experimental study employs 30 Ah commercial lithium-ion cells featuring a $LiFePO_4$ cathode and a graphite anode, as depicted in Fig. 2.27 [83]. The battery is sample E. The manufacturer specifies these cells for potential use in electric vehicles, with detailed specifications provided in Table 2.24. The experimental protocol includes AC pulse heating cycle tests and calibration tests. To monitor the internal temperature during AC heating operations, a T-type thermocouple (Omega, USA) with an external diameter of 0.28 mm and insulated coating is inserted into the geometric center of Cell 2. The thermocouple is embedded between electrode layers during the manufacturing process, prior to electrolyte injection, ensuring minimal structural deviation from a standard cell. As illustrated in Fig. 2.27, the thermocouple is secured to a separator layer using heat-resistant tape. All procedures are conducted within an argon-filled glove box to maintain an inert atmosphere, and a resin is applied to seal the thermocouple lead wire entry point, preventing leakage. Prior to conducting AC thermal tests, the thermocouple-embedded cell undergoes recharacterization via EIS and discharge tests. The results, presented in Figs. 2.28 and 2.29, demonstrate negligible changes in cell impedance and capacity, confirming that the embedded thermocouple does not significantly alter the cell's electrical characteristics. This validation ensures the cell retains its charge storage capability with minimal performance degradation, thereby validating the experimental setup for subsequent thermal analyses.

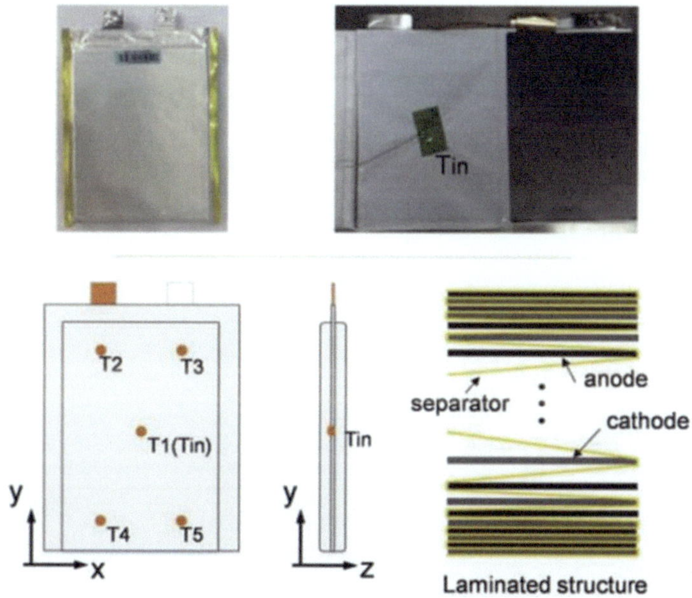

Fig. 2.27 The battery sample and locations of the thermocouples [83]

Table 2.24 Specifications of the sample E [83]

Nominal voltage	3.2 V
Nominal voltage	30 Ah
Electrode chemistry	$LiFePO_4$/C (graphite)
Internal resistance	≤4 mΩ
Core size	13 mm × 132 mm × 184 mm
Storage temperature	−20 °C – 45°C
Normal charge voltage	3.7 V
Discharge ending voltage	2.5 V
Weight	0.675 kg
Energy density	144 Wh/kg
Manufacturer	SAPT (China)*

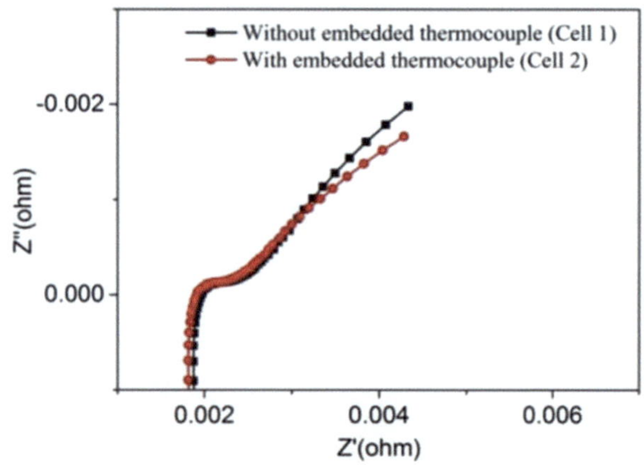

Fig. 2.28 Impedance spectra obtained at 50% SoC using various current amplitudes and the testing frequency was varied from 100 kHz down to 0.01 Hz [83]

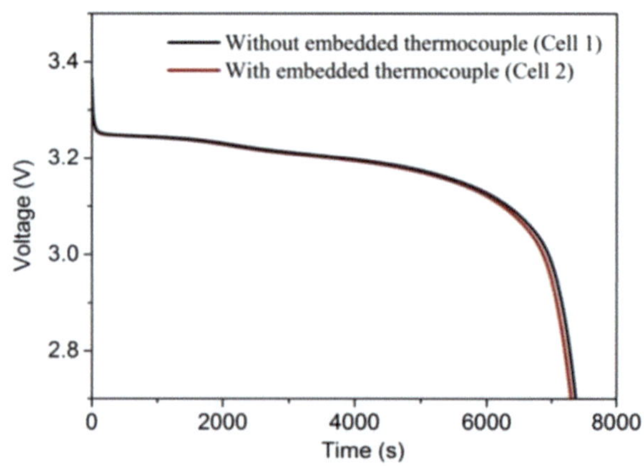

Fig. 2.29 The discharge curves with and without embedded thermocouple under 0.5 C at 25 °C [83]

2.4.1.1 AC Heating Cycle Tests

The AC pulse heating cycle tests are designed to explore the impact of the AC heating method on the battery's SoH. The excitation parameters matrix is detailed in Table 2.25. This test matrix (Cells 3–8) primarily examines how excitation frequency influences battery aging under normal voltage constraints. The rationale behind the selection of current amplitudes and voltage limitations will be elaborated in Sect. 2.4.2. Cell 9, serving as a fresh cell, acts as the control group in this study. A total of 240 pulse heating cycles are executed to assess cell aging, with each cycle comprising 1800s of continuous heating at low temperature, followed by a sufficient equilibration period (e.g., 3 h) for the cell temperature to align with the chamber temperature (−25 °C). Each cell undergoes approximately 120 h of total heating time, and the data collected over one year (including rest and calibration periods) provide valuable insights into the aging behavior of the tested cells. All rectangular pulse excitations are generated using a bipolar power supply (KIKUSUI PBZ20-80), while a digital oscilloscope (Tektronix DPO 3054) monitors the current waveforms. Prior to the AC heating experiments, the SoC is adjusted to 50%. The rationale for selecting a 50% SoC is based on the following considerations:

Table 2.25 Matrix of excitation parameters employed in the AC pulse heating tests [83]

Cell name	Frequency (Hz)						Current (A)			Voltage limitations (V/V)			Heating time (s) (one cycle)	Cycle
Cell 1	300	30	10	2	1	0.5	20	40	60	3.7/ 2.5	3.8/ 2.4	3.9/ 2.3	1800	
Cell 2														
Cell 3	300							40		3.7/ 2.5			1800	240
Cell 4		30						40		3.7/ 2.5			1800	240
Cell 5			10					40		3.7/ 2.5			1800	240
Cell 6				2				40		3.7/ 2.5			1800	240
Cell 7					1			40		3.7/ 2.5			1800	240
Cell 8						0.5		40		3.7/ 2.5			1800	240
Cell 9														
Cell 10	CC–CV charge (0.5 C, 3.7 V, 0.01 C) at −20 °C; CC discharge (0.5 C, 2.5 V) at 25 °C													36

(1) Minimizing SoC Fluctuations: At 50% SoC, the charge and discharge half-cycles during AC heating are unlikely to cause significant deviations in SoC. This stability facilitates comparative analysis of other heating parameters (frequency, current amplitude, and voltage limitations) by isolating their effects from SoC variations.
(2) Voltage Regulation: Higher SoC levels can lead to accelerated voltage increases during the charging half-cycle. However, by incorporating voltage limitations, the current amplitudes can be effectively controlled to maintain the battery within predefined over-charge and over-discharge conditions. As previously established, the battery reaches voltage limitations at 50% SoC at the onset of the heating test, making this SoC level representative of higher SoC conditions.
(3) Efficiency and Representativeness: Choosing 50% SoC reduces the total runtime of the tests while maintaining representativeness. Drawing from prior research, this SoC level is considered indicative of observations across other SoC levels in AC heating studies.

Each cell is enclosed in a sealed box, and five thermocouples are affixed to the surface of each cell, as depicted in Fig. 2.27. The temperature data from these thermocouples are measured and recorded using HIOKI temperature instruments (LR8510, LR8410-30).

2.4.1.2 Calibration Tests

An Arbin BT2000 instrument is utilized to calibrate the battery capacity and DC pulse resistance following every 40 pulse heating cycles. EIS calibration is performed using a Solartron SI 1287/1255B workstation with a PBI250-10 power booster, where the impedance measurement frequency spans from 10 kHz to 0.01 Hz with a 2 A perturbation current. All calibration tests are conducted at ambient temperature. To investigate the internal dynamics of the electrodes post-AC heating, cell disassembly is performed in an argon-filled LABstar MBRAUN Glovebox, maintaining an atmosphere with water content < 0.1 ppm and oxygen content < 5 ppm. The $LiFePO_4$ cathode, characterized by an ordered-olivine structure, exhibits reduced ion diffusivity and active volume availability due to potential blockages in its diffusion channels by impurities or defects [84]. The sluggish kinetics of electron and lithium-ion transport in $LiFePO_4$-based cathodes limit the energy and power output of $LiFePO_4$/C cells at low temperatures [85, 86]. However, the primary aging mechanism under cold conditions is the plating of metallic lithium on the anode, leading to active material loss and SEI film growth [19]. Anode samples near the battery tabs are prepared for SEM and EDS analysis, as this region is more prone to lithium metal deposition [81]. SEM and EDS analyses are conducted using a JEM 2010 EX microscope equipped with an energy dispersive X-ray spectrometer to determine bulk atomic concentrations.

2.4.2 Results and Analysis

This study examines the temperature dynamics of LIBs during AC heating, with a focus on the influence of current amplitude, excitation frequency, and voltage limitations. Experimental results indicate that higher current amplitudes (20, 40, and 60 A) significantly enhance temperature increases, with faster temperature rises observed at lower frequencies (300–0.5 Hz). The charge transfer resistance decreases as current increases and stabilizes at higher current levels. However, impedance spectra continue to exhibit frequency-dependent variations, particularly at low temperatures. This suggests that lower frequencies prolong the charge/discharge duration and intensify the electrochemical reaction process.

In the design of AC heating experiments, voltage limit protection is a critical consideration. Over-voltage and under-voltage can cause SEI layer decomposition, potentially triggering exothermic reactions, gassing, and thermal runaway [87, 88]. Despite varying voltage limitations, no significant differences in temperature evolution were observed within 1800 s of heating. To prevent irreversible damage, the experiments avoid over-voltage and under-voltage. Based on empirical data, the upper voltage limit is set to 3.7 V, and the lower limit to 2.5 V. From Fig. 2.31, it is evident that impedance decreases with increasing current amplitude and stabilizes at higher currents. Considering both battery impedance characteristics and cycling safety, a pulse current amplitude of 40 A is selected for the AC heating cycle tests. As depicted in Fig. 2.30, impedance spectra appear as semicircular arcs, which vary significantly with excitation frequency, particularly at low temperatures. The excitation frequency correlates with the depth of the electrochemical reaction process [25], with lower frequencies indicating longer charge/discharge durations. Cell 3–Cell 8 are used to investigate how different excitation frequencies affect battery SoH. Low-temperature cycle tests help assess battery performance during low-temperature charging. Battery capacity, DC resistance, EIS, SEM, and EDS are employed to evaluate SoH. Due to the large number of tests, only representative calibration curves are presented.

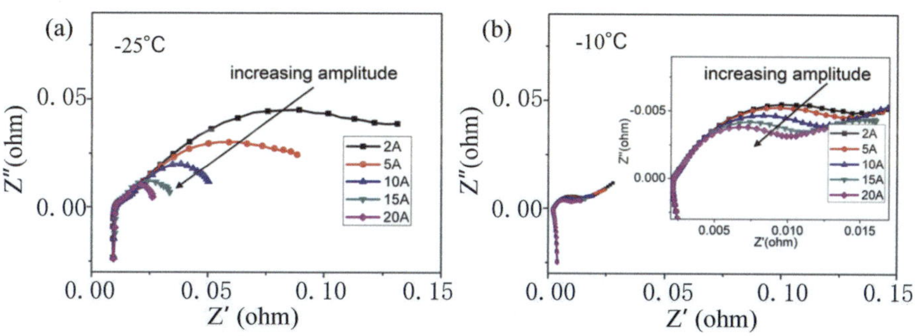

Fig. 2.30 The impedance spectra at 50% SoC under various current amplitudes, including 2, 5, 10, 15, and 20 A [83]

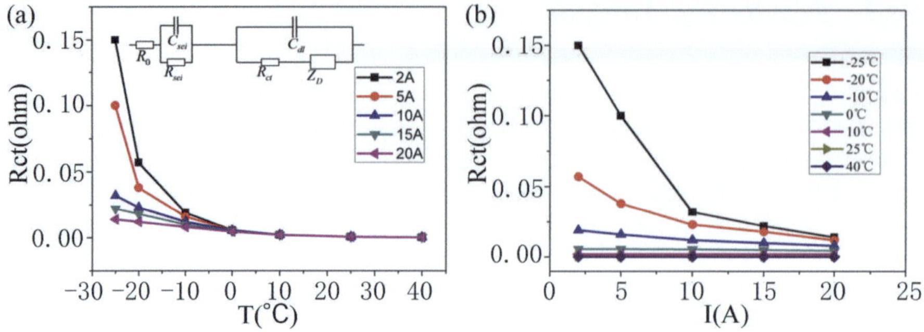

Fig. 2.31 Effects of current amplitudes and temperature on R_{ct} [83]

The impact of AC heating method under different experimental conditions on the cell capacity fade is compared in Fig. 2.32. The cell capacity resulting from different excitation frequencies is very similar for all cells in Fig. 2.32a. The capacity peaks at the first 80 heating cycles due to the battery activation. Then the capacity stays on a higher level even after the following several hundred heating cycles. The capacity evolution of CC–CV charging at −20 °C is illustrated in Fig. 2.32b, indicating that the capacity degradation accelerates after 5 charging cycles. The alternative reason is related to the behavior of lithium plating at low temperature, and the details will be discussed in the following pages.

Figure 2.32 compares the impact of the AC heating method under various experimental conditions on cell capacity degradation. In Fig. 2.32a, cells subjected to different excitation frequencies exhibit similar capacity trends. Capacity peaks during the first 80 heating cycles due to battery activation and remains high through subsequent cycles, indicating stability. In contrast, Fig. 2.32b shows capacity evolution for CC–CV charging at −20 °C. Here, capacity degradation accelerates after five cycles. This rapid decline is attributed to lithium plating behavior at low temperatures, which will be detailed in subsequent sections.

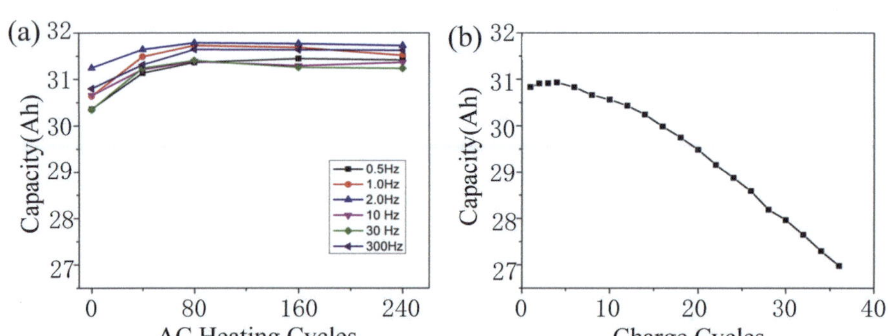

Fig. 2.32 The capacity calibrations for AC pulse heating cycle tests (**a**) and LT_C tests (**b**). **a** Cell 3-Cell 8; **b** Cell 10 [83]

After 240 cycles of heating (one cycle = continuous heating for 1800 s at low temperature), the DC resistance and EIS spectra at 0.5 Hz show no significant changes, as depicted in Fig. 2.33a and c, respectively. This is consistent with the capacity remaining relatively stable. It can be inferred that there is almost no degradation in the cells during the AC heating process. From the DC resistance results in Fig. 2.33b, there is only a slight increase after 36 low-temperature charges. The cell exhibits a shift on the real axis, indicating an increase in ohmic resistance R_0, as shown in Fig. 2.33d. The observed increase in resistance is attributed to a reduction in the ionic conductivity of the electrolyte. This degradation of the electrolyte is caused by the formation of surface films on the plated lithium. The contact of lithium metal with the electrolyte leads to the reduction of electrolyte solvents and thus irreversible consumption, resulting in reduced ionic conductivity [82]. SEM and EDS tests on disassembled cells are performed to further confirm these conclusions.

Cells 3–10 are disassembled to examine their internal morphology, while Cell 9, serving as the fresh cell, undergoes only essential activation tests. Prior to disassembly, all cells are fully discharged using a CC procedure at a 0.5 C current rate, followed by a CV procedure with a 0.01 C cut-off current rate. SEM images of the anode surfaces are presented in Fig. 2.34.

The fresh and AC-heated cells exhibit comparable anode morphologies, with no evidence of dendritic or mossy structures. In contrast, the LT_C graphite particles show distinct morphological changes, as seen in Fig. 2.34d, where needle-like

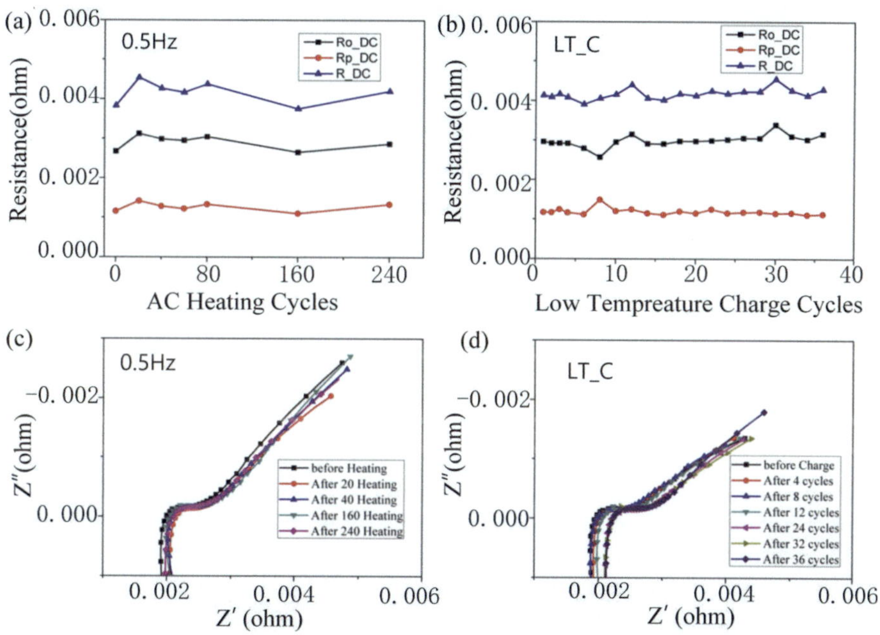

Fig. 2.33 The DC resistance and EIS results of AC pulse heating cycle tests and LT_C tests. **a c** Cell 8; **b d** Cell 10 [83]

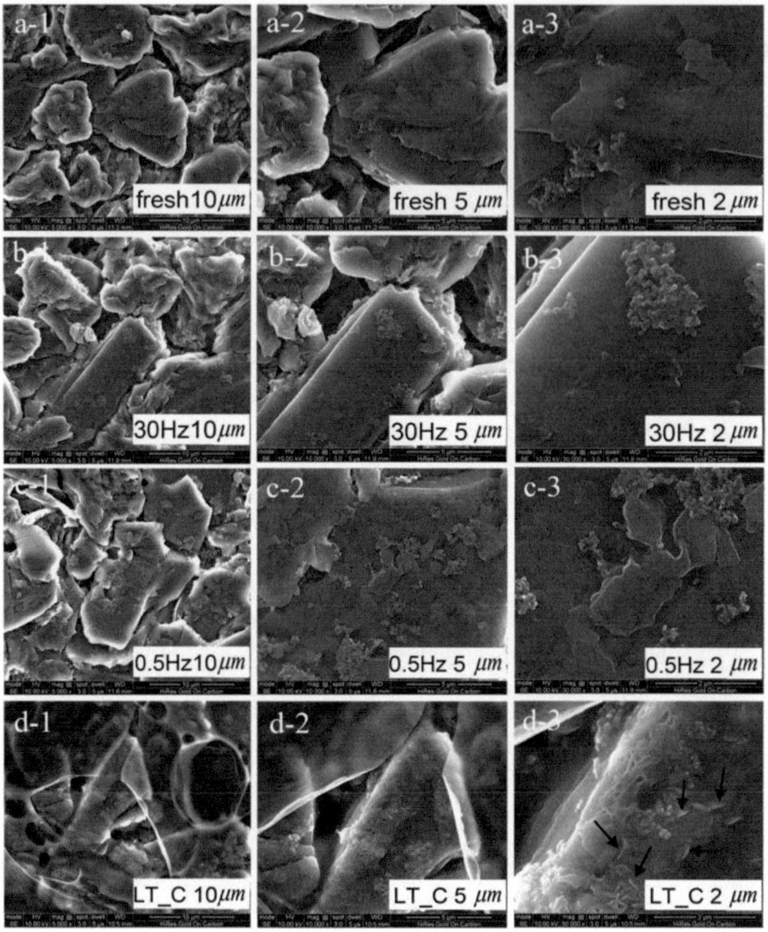

Fig. 2.34 The SEM results of AC pulse heating cycle tests and LT_C tests. **a** Cell 9; **b** Cell 4; **c** Cell 8; **d** Cell 10 [83]

features—potentially indicative of metallic lithium deposition that are visible at higher magnification. These features, lying on the surface, may penetrate the separator, potentially causing internal short-circuiting [19]. EDS results (Fig. 2.35) reveal C (carbon) and F (fluorine) peaks in the fresh cell and AC-heated samples (Fig. 2.35a, c), where C corresponds to the graphite matrix and F to the SEI and electrolyte (lithium hexafluorophosphate solution). In Fig. 2.35d, the presence of O (oxygen) likely represents oxidized lithium metal formed during sample transfer to the electron microscope vacuum chamber. This suggests no lithium plating occurred during AC heating under the specified conditions, as indirectly indicated by the absence of lithium peaks.

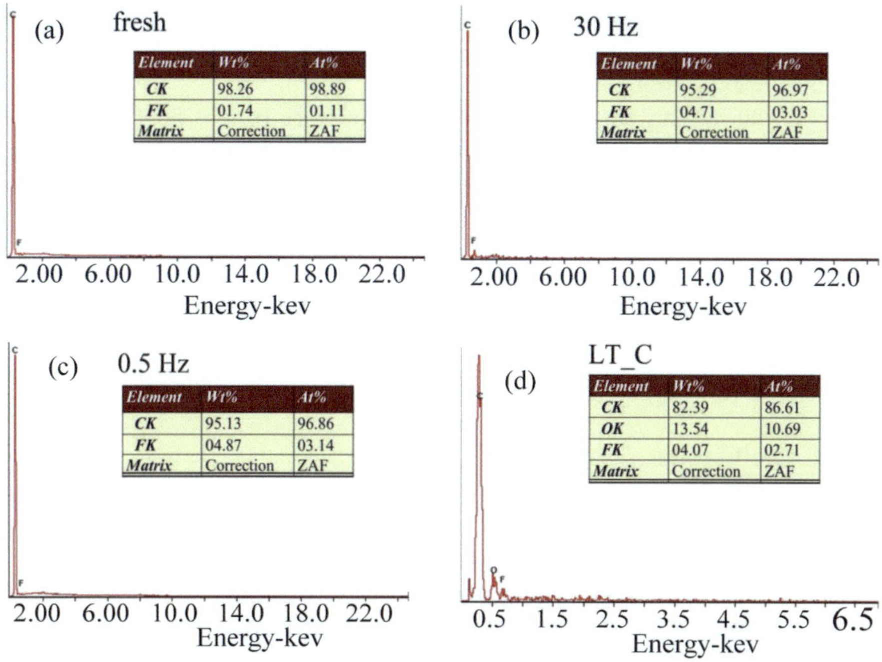

Fig. 2.35 The EDS results of AC pulse heating cycle tests and LT_C tests. **a** Cell 9; **b** Cell 4; **c** Cell 8; **d** Cell 10 [83]

Lithium plating is observed on carbon-based negative electrodes during overcharging of lithium-ion cells, but limiting the maximum charging voltage can prevent this deposition [26, 89]. Thus, one explanation for the absence of lithium plating in this study is the voltage limitations set during AC pulse heating, which prevent the negative electrode potential from reaching zero. In AC excitation, lithium-ions and electrons alternate between the electrolyte, separator, and electrodes, with electrochemical reactions and diffusion processes occurring intermittently. Previous research [90] noted capacity attenuation in 18650 cells at excitation frequencies below 10 Hz with 10 A (4.3 C), but short-term pulses, even at 0.5 Hz, caused no irreversible damage in $LiFePO_4$/C high-power batteries under normal voltage protection. Lithium deposition is reversible and can fully dissolve during discharge with sufficient relaxation time [26, 91].

Another factor is the low charging current (40 A/1.3 C), which allows delithiation during discharge. Lithium plating depends not only on temperature and charging rate but also on electrolyte composition, electrode materials, and the anode-to-cathode capacity ratio [91]. The rate-limiting steps for low-temperature performance—charge transfer and diffusion—vary with battery chemistry and structure (e.g., particle size, porosity). The superior AC performance of $LiFePO_4$/C cells may stem from their uniform carbon network and fine primary particles (Fig. 2.34a–c), enhancing lithium-ion migration and reducing diffusion time in electrodes with low diffusion kinetics

[92]. The sluggish ion kinetics in $LiFePO_4$-based cathodes limit low-temperature battery response, and cell life reduction primarily results from anode lithium plating rather than intercalation.

Overall, no lithium plating or detrimental effects on battery SoH occur when cathode lithium kinetics exceed anode kinetics during AC heating. Future work will use in situ neutron diffraction to investigate lithium-ion dynamics in electrodes [93, 94]. For practical implementation, both heating parameters and battery material characteristics must be considered. Ongoing AC heating cycle tests with high-energy pouch cells ($LiNi_{1/3}Co_{1/3}Al_{1/3}O_2$ cathode and carbon anode) include comprehensive calibration tests (capacity, EIS, SEM, EDS). Results will be reported in a subsequent publication.

2.5 Conclusions

This chapter presents a comprehensive analysis of the aging mechanisms of lithium-ion batteries under low-temperature conditions.

Section 2.1 examines the aging phenomena by focusing on the interactions between the cathode, anode, and electrolyte components. Under low-temperature conditions, the charge transfer impedance, CEI impedance, and solid-phase diffusion impedance of the cathode materials are observed to increase concurrently, accompanied by a rise in grain boundary impedance. During low-temperature charging, substantial lithium-ion accumulation occurs on the surface of the graphite anode, leading to lithium metal deposition, commonly referred to as lithium plating. The deposited lithium undergoes irreversible reactions with the electrolyte, forming a new SEI film, which contributes to an increase in SEI impedance. As the temperature decreases, the ionic conductivity of the electrolyte diminishes while its viscosity rises, resulting in reduced ion diffusion rates within the electrolyte. Additionally, these changes elevate the energy barrier required for lithium-ion desolvation, thereby impacting the embedding kinetics of lithium ions.

Section 2.2 elaborates on the conditions under which lithium plating occurs, specifically when the charging current surpasses the intercalation kinetics of the anode material, resulting in lithium deposition on the anode surface. Several factors contribute to lithium plating, including hazardous operating conditions such as low temperatures, elevated C-rates, and high SoC, as well as cell-related issues like anode/cathode imbalance (imbalanced N/P ratio), separator deformation, and aging phenomena such as active material loss (LLI) and structural degradation (LAM). Lithium plating can manifest in diverse morphologies, including dendritic, granular, and mossy structures. Among these, dendritic growth is identified as the most hazardous due to its potential to penetrate the separator and induce internal short circuits, thereby posing significant safety risks.

Section 2.3 provides an in-depth experimental analysis of the degradation behavior of LIBs during low-temperature cycling. The study employs two distinct battery types to investigate the influence of charging temperature, rate, and cutoff voltage on capacity loss: sample C and sample D. The experimental design aims to elucidate how these operational parameters affect capacity degradation under low-temperature conditions, offering insights into the mechanisms governing battery performance and longevity in such environments. The results reveal that capacity degradation is accelerated at lower temperatures and higher charging rates, with lithium plating identified as the dominant degradation mechanism. This phenomenon is particularly severe at −15 °C and 2 C charging rate, leading to rapid capacity loss and increased internal resistance. The study also finds that capacity degradation exhibits a decelerating trend over prolonged cycling, with capacity recovery observed when transitioning from low to room temperature. Post-mortem analysis shows significant lithium dendrite formation and SEI film thickening on the anode, especially after 250 cycles. These findings highlight the critical role of lithium plating in capacity fade and suggest that optimizing charging protocols and managing battery temperature are essential strategies to mitigate degradation in low-temperature environments.

Section 2.4 evaluates the efficacy of the AC heating method as a potential thermal management strategy for large laminated lithium-ion batteries, particularly under low-temperature conditions. Detailed experimental protocols, including AC pulse heating cycle tests aligned with battery SoH assessments and calibration procedures, were employed to validate the feasibility of this approach. Initial experiments demonstrated that cells equipped with embedded thermocouples exhibited uniform temperature distribution between the surface and interior during AC heating, indicating effective thermal homogenization. Hundreds of AC heating cycles were performed to assess the long-term impact on cell SoH. Calibration results based on battery capacity retention, DC resistance measurements, and EIS revealed no significant degradation in performance. Additionally, post-experiment disassembly and analysis using SEM and EDS confirmed that low-temperature AC excitation, even at low frequencies (0.5 Hz), does not induce irreversible structural or chemical damage to the battery under normal voltage protection constraints for high-power $LiFePO_4$/C cells. These findings highlight the method's suitability for low-temperature thermal management in EVs and HEVs, suggesting promising applications in enhancing battery performance and longevity under challenging thermal conditions.

References

1. Xu J, Wang X, Yuan N Y, et al. Extending the low temperature operational limit of Li-ion battery to −80 °C [J]. Energy Storage Materials, 2019, 23: 38–39.
2. Liang J f, Li Y M, Yuan H, et al. Research progress of low-temperature lithium-ion battery [J]. Journal of Beijing University of Aeronautics and Astronautics, 2020.
3. Zhao S X, Guo S T, Zhao J W, et al. Development on Low-temperature Performance of Lithium Ion Batteries [J]. Journal of the Chinese Ceramic Society, 2016, 44(1): 19–28.
4. Gao F, Tang Z Y. Kinetic behavior of LiFePO4/C cathode material for lithium-ion batteries [J]. Electrochimica Acta, 2008, 53(15): 5071–5075.
5. Liao L X, Zuo P J, Ma Y L, et al. Effects of temperature on charge/discharge behaviors of LiFePO4 cathode for Li-ion batteries [J]. Electrochimica Acta, 2012, 60: 269–273.

6. Ogihara N, Itou Y, Kawauchi S. Ion Transport in Porous Electrodes Obtained by Impedance Using a Symmetric Cell with Predictable Low-Temperature Battery Performance [J]. J Phys Chem Lett, 2019, 10(17): 5013–5018.
7. Hu X S, Zheng Y S, Howey D A, et al. Battery warm-up methodologies at subzero temperatures for automotive applications: Recent advances and perspectives [J]. Progress in Energy and Combustion Science, 2020, 77: 100806.
8. Wu W X, Ma R X, Liu J Z, et al. Impact of low temperature and charge profile on the aging of lithium-ion battery: Non-invasive and post-mortem analysis [J]. International Journal of Heat and Mass Transfer, 2021, 170.
9. Hu D Z, Chen G, Tian J, et al. Unrevealing the effects of low temperature on cycling life of 21700-type cylindrical Li-ion batteries [J]. Journal of Energy Chemistry, 2021, 60: 104–110.
10. Li J Z, Li S F, Zhang Y X, et al. Multiphase, Multiscale Chemomechanics at Extreme Low Temperatures: Battery Electrodes for Operation in a Wide Temperature Range [J]. Advanced Energy Materials, 2021, 11(37).
11. Zhang G X, Wei X Z, Han G S, et al. Lithium plating on the anode for lithium-ion batteries during long-term low temperature cycling [J]. Journal of Power Sources, 2021, 484.
12. Chen X R, Yan C, Ding J F, et al. New insights into "dead lithium" during stripping in lithium metal batteries [J]. Journal of Energy Chemistry, 2021, 62: 289–294.
13. Zhu G L, Wen K C, Lv W Q, et al. Materials insights into low-temperature performances of lithium-ion batteries [J]. Journal of Power Sources, 2015, 300: 29–40.
14. Waldmann T, Hogg B-I, Wohlfahrt3Mehrens M. Li plating as unwanted side reaction in commercial Li-ion cells – A review [J]. Journal of Power Sources, 2018, 384: 107–124.
15. Dong X, Wang Y G, Xia Y. Promoting Rechargeable Batteries Operated at Low Temperature [J]. Acc Chem Res, 2021, 54(20): 3883–894.
16. Liu C, Neale Z G, Cao G. Understanding electrochemical potentials of cathode materials in rechargeable batteries [J]. Materials Today, 2016, 19(2): 109–123.
17. Piao N, Gao X N, Yang H C, et al. Challenges and development of lithium-ion batteries for low temperature environments [J]. eTransportation, 2022, 11: 100145.
18. Hou J B, Yang M, Wang D Y, et al. Fundamentals and Challenges of Lithium Ion Batteries at Temperatures between −40 and 60 °C [J]. Advanced Energy Materials, 2020, 10(18).
19. Ouyang M G, Chu Z Y, Lu L G, et al. Low temperature aging mechanism identification and lithium deposition in a large format lithium iron phosphate battery for different charge profiles [J]. Journal of Power Sources, 2015, 286: 309–320.
20. Chen S Q, Wei X Z, Zhang G X, et al. Active and passive safety enhancement for batteries from force perspective [J]. Renewable & Sustainable Energy Reviews, 2023, 187.
21. Jiang B, Zhu Y L, Zhu J G, et al. An adaptive capacity estimation approach for lithium-ion battery using 10-min relaxation voltage within high state of charge range [J]. Energy, 2023, 263.
22. Yuan H, Du R B, Wang X Y, et al. Advanced Online Broadband Impedance Spectrum Acquisition of Fuel Cells by S-Transform [J]. Ieee Transactions on Industrial Electronics, 2023, 70(4): 3740–3750.
23. GUAN W. Analasis on key technologies and development trend of LTE [J]. Guangxi Communication Technology, 2009, (01): 5–8+38.
24. Ahmed S, Bloom I, Jansen A N, et al. Enabling fast charging – A battery technology gap assessment [J]. Journal of Power Sources, 2017, 367: 250–262.
25. Legrand N, Knosp B, Desprez P, et al. Physical characterization of the charging process of a Li-ion battery and prediction of Li plating by electrochemical modelling [J]. Journal of Power Sources, 2014, 245: 208–216.
26. Arora P, Doyle M, White R E. Mathematical modeling of the lithium deposition overcharge reaction in lithium-ion batteries using carbon-based negative electrodes [J]. Journal of the Electrochemical Society, 1999, 146(10): 3543–3553.
27. Liu Q Q, Du C Y, Shen B, et al. Understanding undesirable anode lithium plating issues in lithium-ion batteries [J]. RSC Advances, 2016, 6(91): 88683–88700.

28. Chandrasekaran R. Quantification of bottlenecks to fast charging of lithium-ion-insertion cells for electric vehicles [J]. Journal of Power Sources, 2014, 271: 622–632.
29. Li Z, Huang J, Yann Liaw B, et al. A review of lithium deposition in lithium-ion and lithium metal secondary batteries [J]. Journal of Power Sources, 2014, 254: 168–182.
30. Ecker M, Shafiei Sabet P, Sauer D U. Influence of operational condition on lithium plating for commercial lithium-ion batteries—Electrochemical experiments and post-mortem-analysis [J]. Applied Energy, 2017, 206: 934–946.
31. Hein S, Latz A. Influence of local lithium metal deposition in 3D microstructures on local and global behavior of Lithium-ion batteries [J]. Electrochimica Acta, 2016, 201: 354–365.
32. Hein S, Danner T, Latz A. An Electrochemical Model of Lithium Plating and Stripping in Lithium Ion Batteries [J]. ACS Applied Energy Materials, 2020, 3(9): 8519–8531.
33. Wang H S, Zhu Y Y, Kim S C, et al. Underpotential lithium plating on graphite anodes caused by temperature heterogeneity [J]. Proceedings of the National Academy of Sciences, 2020, 117(47): 29453–29461.
34. Harris S J, Timmons A, Baker D R, et al. Direct in situ measurements of Li transport in Li-ion battery negative electrodes [J]. Chemical Physics Letters, 2010, 485(4): 265–274.
35. Zhao X C, Yin Y, Hu Y, et al. Electrochemical-thermal modeling of lithium plating/stripping of Li(Ni0.6Mn0.2Co0.2)O2/Carbon lithium-ion batteries at subzero ambient temperatures [J]. Journal of Power Sources, 2019, 418: 61–73.
36. Tomaszewska A, Chu Z, Feng X, et al. Lithium-ion battery fast charging: A review [J]. eTransportation, 2019, 1: 100011.
37. Ren D S, Smith K, Guo D X, et al. Investigation of Lithium Plating-Stripping Process in Li-Ion Batteries at Low Temperature Using an Electrochemical Model [J]. Journal of The Electrochemical Society, 2018, 165(10): A2167.
38. Yang X G, Ge S H, Liu T, et al. A look into the voltage plateau signal for detection and quantification of lithium plating in lithium-ion cells [J]. Journal of Power Sources, 2018, 395: 251–261.
39. Cheng X B, Zhang Q. Growth Mechanisms and Suppression Strategies of Lithium Metal Dendrites [J]. Progress in Chemistry, 2018, 30(01): 51–72.
40. Waldmann T, Hogg B-I, Kasper M, et al. Interplay of Operational Parameters on Lithium Deposition in Lithium-Ion Cells: Systematic Measurements with Reconstructed 3-Electrode Pouch Full Cells [J]. Journal of The Electrochemical Society, 2016, 163(7): A1232.
41. Liu G Y, Lu W. A Model of Concurrent Lithium Dendrite Growth, SEI Growth, SEI Penetration and Regrowth [J]. Journal of The Electrochemical Society, 2017, 164(9): A1826.
42. Sahraei E, Meier J, Wierzbicki T. Characterizing and modeling mechanical properties and onset of short circuit for three types of lithium-ion pouch cells [J]. Journal of Power Sources, 2014, 247: 503–516.
43. Orendorff C J, Roth E P, Nagasubramanian G. Experimental triggers for internal short circuits in lithium-ion cells [J]. Journal of Power Sources, 2011, 196(15): 6554–6558.
44. Tanim T R, Paul P P, Thampy V, et al. Heterogeneous Behavior of Lithium Plating during Extreme Fast Charging [J]. Cell Reports Physical Science, 2020, 1(7): 100114.
45. Lin X K, Khosravinia K, Hu X S, et al. Lithium Plating Mechanism, Detection, and Mitigation in Lithium-Ion Batteries [J]. Progress in Energy and Combustion Science, 2021, 87: 100953.
46. Guan T, Sun S, Yu F B, et al. The degradation of LiCoO2/graphite batteries at different rates [J]. Electrochimica Acta, 2018, 279: 204–212.
47. Ge H, Aoki T, Ikeda N, et al. Investigating Lithium Plating in Lithium-Ion Batteries at Low Temperatures Using Electrochemical Model with NMR Assisted Parameterization [J]. Journal of The Electrochemical Society, 2017, 164(6): A1050.
48. Campbell I D, Marzook M, Marinescu M, et al. How Observable Is Lithium Plating? Differential Voltage Analysis to Identify and Quantify Lithium Plating Following Fast Charging of Cold Lithium-Ion Batteries [J]. Journal of The Electrochemical Society, 2019, 166(4): A725.
49. Fear C, Adhikary T, Carter R, et al. In Operando Detection of the Onset and Mapping of Lithium Plating Regimes during Fast Charging of Lithium-Ion Batteries [J]. ACS Applied Materials & Interfaces, 2020, 12(27): 30438–30448.

50. Ringbeck F, Rahe C, Fuchs G, et al. Identification of Lithium Plating in Lithium-Ion Batteries by Electrical and Optical Methods [J]. Journal of The Electrochemical Society, 2020, 167(9): 090536.
51. Ouyang D X, He Y P, Weng J W, et al. Influence of low temperature conditions on lithium-ion batteries and the application of an insulation material [J]. RSC Advances, 2019, 9(16): 9053–9066.
52. Ma S, Jiang M, Tao P, et al. Temperature effect and thermal impact in lithium-ion batteries: A review [J]. Progress in Natural Science: Materials International, 2018, 28(6): 653-666.
53. Börner M, Horsthemke F, Kollmer F, et al. Degradation effects on the surface of commercial LiNi0.5Co0.2Mn0.3O2 electrodes [J]. Journal of Power Sources, 2016, 335: 45–55.
54. Yang X G, Zhang G S, Ge S H, et al. Fast charging of lithium-ion batteries at all temperatures [J]. Proceedings of the National Academy of Sciences, 2018, 115(28): 7266-7271.
55. Yang X G, Liu T, Gao Y, et al. Asymmetric Temperature Modulation for Extreme Fast Charging of Lithium-Ion Batteries [J]. Joule, 2019, 3(12): 3002-3019.
56. Tanim T R, Dufek E J, Dickerson C C, et al. Electrochemical Quantification of Lithium Plating: Challenges and Considerations [J]. Journal of The Electrochemical Society, 2019, 166(12): A2689.
57. Rangarajan S P, Barsukov Y, Mukherjee P P. In operando signature and quantification of lithium plating [J]. Journal of Materials Chemistry A, 2019, 7(36): 20683-20695.
58. Logan E R, Dahn J R. Electrolyte Design for Fast-Charging Li-Ion Batteries [J]. Trends in Chemistry, 2020, 2(4): 354–366.
59. Keil P, Jossen A. Charging protocols for lithium-ion batteries and their impact on cycle life—An experimental study with different 18650 high-power cells [J]. Journal of Energy Storage, 2016, 6: 125–141.
60. Zhang L, Zhang Z C, Redfern P C, et al. Molecular engineering towards safer lithium-ion batteries: a highly stable and compatible redox shuttle for overcharge protection [J]. Energy & Environmental Science, 2012, 5(8): 8204–8207.
61. Juarez-Robles D, Vyas A A, Fear C, et al. Overcharge and Aging Analytics of Li-Ion Cells [J]. Journal of The Electrochemical Society, 2020, 167(9): 090547.
62. Liu S, Xiong L, He C. Long cycle life lithium ion battery with lithium nickel cobalt manganese oxide (NCM) cathode [J]. Journal of Power Sources, 2014, 261: 285–291.
63. Cannarella J, Arnold C B. The Effects of Defects on Localized Plating in Lithium-Ion Batteries [J]. Journal of The Electrochemical Society, 2015, 162(7): A1365.
64. Abe T, Fukuda H, Iriyama Y, et al. Solvated Li-Ion Transfer at Interface Between Graphite and Electrolyte [J]. Journal of The Electrochemical Society, 2004, 151(8): A1120.
65. Smart M C, Ratnakumar B V. Effects of Electrolyte Composition on Lithium Plating in Lithium-Ion Cells [J]. Journal of The Electrochemical Society, 2011, 158(4): A379.
66. Xu K, von Cresce A, Lee U. Differentiating Contributions to "Ion Transfer" Barrier from Interphasial Resistance and Li+ Desolvation at Electrolyte/Graphite Interface [J]. Langmuir, 2010, 26(13): 11538-43.
67. Love C T, Baturina O A, Swider-Lyons K E. Observation of Lithium Dendrites at Ambient Temperature and Below [J]. ECS Electrochemistry Letters, 2015, 4(2): A24.
68. Yao F, Güneş F, Ta H Q, et al. Diffusion Mechanism of Lithium Ion through Basal Plane of Layered Graphene [J]. Journal of the American Chemical Society, 2012, 134(20): 8646–8654.
69. Dubarry M, Truchot C, Liaw B Y. Synthesize battery degradation modes via a diagnostic and prognostic model [J]. Journal of Power Sources, 2012, 219: 204–216.
70. Anseán D, Dubarry M, Devie A, et al. Operando lithium plating quantification and early detection of a commercial LiFePO4 cell cycled under dynamic driving schedule [J]. Journal of Power Sources, 2017, 356: 36–46.
71. Dubarry M, Baure G, Anseán D. Perspective on State-of-Health Determination in Lithium-Ion Batteries [J]. Journal of Electrochemical Energy Conversion and Storage, 2020, 17(4).
72. Baure G, Dubarry M. Synthetic vs. Real Driving Cycles: A Comparison of Electric Vehicle Battery Degradation [J/OL] 2019, 5(2): https://doi.org/10.3390/batteries5020042.

73. Schuster S F, Bach T, Fleder E, et al. Nonlinear aging characteristics of lithium-ion cells under different operational conditions [J]. Journal of Energy Storage, 2015, 1: 44–53.
74. Waldmann T, Wilka M, Kasper M, et al. Temperature dependent ageing mechanisms in Lithium-ion batteries – A Post-Mortem study [J]. Journal of Power Sources, 2014, 262: 129–335.
75. Dai H F, Zhu J G. Battery life test [M]. Elsevier, 2024.
76. Chen k. Design and Analysis of Experimental [M]. Tsinghua University Press, 2005.
77. You H Z, Dai H F, Li L Z, et al. Charging Strategy Optimization at Low Temperatures for Li-Ion Batteries Based on Multi-Factor Coupling Aging Model [J]. Ieee Transactions on Vehicular Technology, 2021, 70(11): 11433–11445.
78. Waldmann T, Kasper M, Wohlfahrt-Mehrens M. Optimization of Charging Strategy by Prevention of Lithium Deposition on Anodes in high-energy Lithium-ion Batteries—Electrochemical Experiments [J]. Electrochimica Acta, 2015, 178: 525–532.
79. Li Y Y, Hu C R. Experiment Design and Data Processing [M]. Chemical Industry Press, 2017.
80. Zhang G, Wei X, Chen S, et al. Investigation the Degradation Mechanisms of Lithium-Ion Batteries under Low-Temperature High-Rate Cycling [J]. ACS Applied Energy Materials, 2022, 5(5): 6462–6471.
81. Burow D, Sergeeva K, Calles S, et al. Inhomogeneous degradation of graphite anodes in automotive lithium ion batteries under low-temperature pulse cycling conditions [J]. Journal of Power Sources, 2016, 307: 806–814.
82. Petzl M, Kasper M, Danzer M A. Lithium plating in a commercial lithium-ion battery – A low-temperature aging study [J]. Journal of Power Sources, 2015, 275: 799–807.
83. Zhu J, Sun Z, Wei X, et al. Experimental investigations of an AC pulse heating method for vehicular high power lithium-ion batteries at subzero temperatures [J]. Journal of Power Sources, 2017, 367: 145–157.
84. Morgan D, Van der Ven A, Ceder G. Li Conductivity in LixMPO4 (M = Mn, Fe, Co, Ni) Olivine Materials [J]. Electrochemical and Solid-State Letters, 2004, 7(2): A30–A2.
85. Park O K, Cho Y, Lee S, et al. Who will drive electric vehicles, olivine or spinel? [J]. Energy & Environmental Science, 2011, 4(5): 1621–1633.
86. Kim S-W, Seo D-H, Kim H, et al. A comparative study on Na2MnPO4F and Li2MnPO4F for rechargeable battery cathodes [J]. Physical Chemistry Chemical Physics, 2012, 14(10): 3299–3–303.
87. Wu B, Yufit V, Marinescu M, et al. Coupled thermal–electrochemical modelling of uneven heat generation in lithium-ion battery packs [J]. Journal of Power Sources, 2013, 243: 544–554.
88. Spotnitz R, Franklin J. Abuse behavior of high-power, lithium-ion cells [J]. Journal of Power Sources, 2003, 113(1): 81–100.
89. Smith K, Wang C-Y. Power and thermal characterization of a lithium-ion battery pack for hybrid-electric vehicles [J]. Journal of Power Sources, 2006, 160(1): 662–673.
90. Zhu J G, Sun Z C, Wei X Z, et al. An alternating current heating method for lithium-ion batteries from subzero temperatures [J]. International Journal of Energy Research, 2016, 40(13): 1869–1883.
91. Uhlmann C, Illig J, Ender M, et al. In situ detection of lithium metal plating on graphite in experimental cells [J]. Journal of Power Sources, 2015, 279: 428–438.
92. Zhao N N, Zhi X K, Wang L, et al. Effect of microstructure on low temperature electrochemical properties of LiFePO4/C cathode material [J]. Journal of Alloys and Compounds, 2015, 645: 301–308.
93. Hoelzel M, Senyshyn A, Juenke N, et al. High-resolution neutron powder diffractometer SPODI at research reactor FRM II [J]. Nuclear Instruments and Methods in Physics Research Section A: Accelerators, Spectrometers, Detectors and Associated Equipment, 2012, 667: 32–37.
94. Senyshyn A, Mühlbauer M J, Nikolowski K, et al. "In-operando" neutron scattering studies on Li-ion batteries [J]. Journal of Power Sources, 2012, 203: 126–129.

Chapter 3
AC Heating Theory and Principles

Abstract At low temperatures, the power and capacity characteristics of lithium-ion batteries (LIBs) deteriorate significantly. Furthermore, low-temperature charging can induce lithium plating on the surface of graphite particles in the battery anode, resulting in irreversible damage. To enhance the low-temperature performance of LIBs, two approaches can be adopted. The first focuses on optimizing the electrolyte and the cathode and anode materials to enhance the intrinsic low-temperature performance of the batteries. The second approach involves designing thermal management solutions tailored to meet automotive requirements. While advancements in electrolyte composition and electrode materials can improve the low-temperature performance of LIBs, such developments are typically experimental and unlikely to achieve significant breakthroughs in the short term, making them insufficient to meet current automotive demands. One effective strategy to address this challenge is the low-temperature alternating current (AC) heating, which uses AC excitation to leverage the internal impedance of the battery for rapid heating. This method offers advantages such as low energy consumption, uniform temperature distribution, low cost, and high heating efficiency. This chapter focuses on the AC heating theory and principles for automotive LIBs. It begins by introducing the thermal characteristics used for AC heating temperature calculations, emphasizing the mechanisms of heat generation, transfer, and dissipation. Among these, the heat generation mechanism plays a fundamental role in temperature rise during AC heating and is primarily influenced by battery impedance. However, at low temperatures, the electrode processes in LIBs exhibit strong nonlinear behavior, as battery impedance is affected by multiple factors, including current amplitude and temperature. To address this complexity, this chapter investigates the nonlinear relationship between battery impedance, current, and temperature. The relationships between impedance, temperature, and current amplitude are modeled using the Arrhenius equation and the Butler–Volmer equation. Finally, the chapter examines the effects of two commonly used AC heating waveforms—sinusoidal AC heating and bidirectional pulse current (BPC) heating—on battery temperature rise.

Keywords Low-temperature · AC heating · Impedance · Thermal management · Excitation parameters

J. Zhu et al., *Alternating Current (AC) Heating for Lithium-Ion Batteries in Electric Vehicles*, https://doi.org/10.1007/978-981-96-9071-8_3

3.1 Battery Heat Generation Mechanism

3.1.1 Fundamental Theory of Thermal Analysis

Under automotive operating conditions, the power battery pack serves as a heat source. The internal heat generation rate of the power battery is influenced by factors such as operating current, internal resistance, and state of charge (SOC), while its heat dissipation environment is determined by the thermal management system of the battery pack. Due to the uncertain variation patterns of operating current in electric vehicle battery packs and the time-dependent nature of the internal resistance, the heat generation and dissipation process of power batteries represents a typical non-steady-state heat conduction process with a time-varying internal heat source. This process adheres to the energy conservation equation within the system domain, expressed in its most general form as the heat diffusion equation, shown in Eq. (3.1). The heat transfer at the system boundary complies with the formula shown in Eq. (3.2).

$$\rho_k C_{\rho,k} \frac{\partial T}{\partial t} = \nabla(\lambda_k \nabla T) + Q \tag{3.1}$$

$$-\lambda_k \frac{\partial T}{\partial n} = h(T - T_{Am}) \tag{3.2}$$

In the equations, Q represents the volumetric heat generation rate of the power battery ($\mathrm{W} \cdot \mathrm{m}^{-3}$), ρ_k is the density of the power battery ($\mathrm{Kg} \cdot \mathrm{m}^{-3}$), $C_{\rho,k}$ denotes the specific heat capacity of the power battery ($\mathrm{J} \cdot \mathrm{Kg}^{-1} \cdot \mathrm{K}^{-1}$), λ_k is the thermal conductivity of the battery cell in a specific direction ($\mathrm{W} \cdot \mathrm{m}^{-1} \cdot \mathrm{K}^{-1}$), h is the convective heat transfer coefficient at the system boundary ($\mathrm{W} \cdot \mathrm{m}^{-2} \cdot \mathrm{K}^{-1}$), and $\partial T / \partial n$ is the temperature gradient in the direction normal to the isothermal surface ($\mathrm{K} \cdot \mathrm{m}^{-1}$). The theoretical application of the thermal characteristics encompasses any infinitesimal element within the power battery.

The left-hand side of Eq. (3.1) represents the increase in thermodynamic energy of an infinitesimal element of the power battery over a unit time. The first term on the right-hand side signifies the net energy transfer through the interfaces of the infinitesimal element within the domain per unit time, i.e., the difference between inflow and outflow energy. Solving the time-dependent temperature field within the domain with high precision is typically challenging. However, when the thermal resistance to conduction within the domain is negligible compared to the convective thermal resistance at the boundary, the conduction thermal resistance within the domain can be ignored. This simplification is known as the lumped thermal capacity method in heat transfer studies. It assumes that during transient processes, the temperature within the domain is uniformly distributed spatially. This method is widely employed in thermal studies of lithium-ion batteries. The second term on the right-hand side represents the heat generation rate within the battery. In Eq. (3.2), the left-hand side corresponds to the energy increase per unit time on the boundary surface of the system, satisfying the third-type boundary condition through heat transfer. The right-hand side represents the energy transferred from the system boundary surface to the environment per unit time.

The thermal characteristics for a single power battery can thus be categorized into the following components:

(1) Calculation of the internal heat generation rate Q within the battery domain.
(2) Determination of internal heat conduction and boundary heat dissipation mechanisms, as well as calculation of the relevant thermal properties of the battery, such as thermal conductivity λ_k, specific heat capacity $C_{\rho,k}$, and heat transfer coefficient h.
(3) Specification of the boundary and initial conditions for the entire model system, including parameters such as initial temperature.

3.1.2 Battery Heat Generation Mechanism

In existing thermal models of lithium-ion power batteries, the heat source is predominantly based on the heat generation rate model proposed by Bernardi et al. [1]. This model assumes that the power battery acts as a uniform heat source, and the temperature variation over time is related to internal reactions, phase transitions, mixing processes, and heat transfer characteristics of the battery. The specific heat generation rate equation by Bernardi is expressed as shown in Eq. (3.3):

$$q = \sum_j I_j (U_{j,avg} - T\frac{\partial U_{j,avg}}{\partial T}) - IU + q_{mixing} + q_{phase-change} \tag{3.3}$$

where, q represents the total heat generation rate (W), and I_j denotes the reaction current of the infinitesimal element within the battery, $T\partial U_{j,avg}/\partial T$ is the reversible heat per unit of the battery's infinitesimal element, while $U_{j,avg}$ corresponds to the average open-circuit potential of the electrode against lithium (vs. Li/Li^+) for the infinitesimal element. I is the total current of the power battery, and U is the terminal voltage of the power battery.

The first term on the right of Eq. (3.3) represents the enthalpy of the electrochemical reaction. The second one corresponds to the electrical work performed by the power battery. The third accounts for the heat generated due to concentration gradients arising from uneven mixing of materials within the battery, which occurs because of non-uniform reaction rates in different regions of the battery. When the relationship between the enthalpy of the mixture and its concentration is nonlinear, this heat cannot be ignored. The final term represents the heat generated due to material phase transitions during the chemical reaction.

Under normal charging and discharging conditions of the power battery, the following assumptions can be made. When the research object is a battery electrode with negligible thickness in the through-thickness direction, or the temperature difference across the thickness of the battery is negligible, the effects of side reactions leading to battery aging, such as phase transition heat generation, can be ignored. The electrochemical system in the model is assumed to have excellent transport properties, and any concentration gradients are considered negligible. Therefore, q_{mixing} and $q_{phase-change}$ can be omitted. Consequently, the heat source model simplifies to:

$$q = \sum_j I_j U_{j,avg} - \sum_j I_j T \frac{\partial U_{j,avg}}{\partial T} - IU \tag{3.4}$$

During charging/discharging processes, the reversible and irreversible heat generation accounts for the majority of total heat generation. In literature, Eq. (3.4) is commonly used as the heat source for thermal models of batteries. Temperature results derived from this model have been found to be within an acceptable error range when compared to actual temperature conditions.

According to the research on electrochemical porous electrode theory [2, 3], for a homogeneously mixed cell, the heat generation rate primarily consists of the following three components: Heat generated from chemical reactions inside the battery: This heat is negative during charging, indicating that the chemical reaction is endothermic, and positive during discharging, indicating that the reaction is exothermic. Polarization heat: This is caused by charge imbalance at the solid–liquid interface of the reaction medium. Ohmic heat: This is generated due to the Joule heating effect of charged particles. Among these, Joule and polarization heat are classified as irreversible heat, while chemical reaction heat is classified as reversible heat. By analyzing an infinitesimal reaction unit, the heat generation rate can be expressed as:

$$Q = Q_{act} + Q_{rev} + Q_{ohm} \tag{3.5}$$

where Q is the volumetric heat generation rate of the battery microelement ($W \cdot m^{-3}$), Q_{rev} is the reversible heat generation rate from electrochemical reactions per unit volume ($W \cdot m^{-3}$), Q_{act} is the heat generation rate from electrochemical polarization per unit volume ($W \cdot m^{-3}$), and Q_{ohm} is the heat generation rate due to Ohmic effects per unit volume ($W \cdot m^{-3}$).

The heat generated from reversible chemical reactions inside the battery is expressed by Eq. (3.6):

$$Q_{rev} = F\alpha J_{loc} T \frac{\partial U_{avg}}{\partial T} \tag{3.6}$$

where F is the Faraday constant (96,487 $C \cdot mol^{-1}$), α is the specific surface area of the electrode (m^{-1}), J_{loc} represents the local current density through the reaction interface per unit area ($mol \cdot m^{-2} \cdot s^{-1}$), and U_{avg} is the average open-circuit potential of the electrode against lithium (V, vs. Li/Li^+).

Polarization heat arises from charge imbalance at the solid–liquid interface of the reaction medium. It is expressed as the product of the local exchange current density J_{loc} and the overpotential η of the interface under non-equilibrium conditions, as shown in Eq. (3.7):

$$Q_{act} = F\alpha J_{loc}(\phi_s - \phi_1 - U_{avg}) \tag{3.7}$$

$$\eta = \phi_s - \phi_1 - U_{avg} \tag{3.8}$$

As described in Eq. (3.8), the overpotential represents the deviation of the cathode or anode from the equilibrium electrode potential. When the current density varies, two factors affect the overpotential. The variation of electrode potential with current is associated with the rate at which charge carriers cross the phase boundary between the electrode (an electronic conductor) and the electrolyte (an ionic conductor). This reaction rate is affected by the comparatively slower rate of mass transfer of reactants from the solution to the electrode or in the reverse direction.

Joule heat in the battery results from the migration of charged particles. It can be divided into Ohmic heat generated by electronic conduction and heat generated during the mass transfer of lithium ions in the liquid phase. This can be described by Eq. (3.9):

$$Q_{ohm} = \sigma_{eff}\nabla\phi_s \cdot \nabla\phi_s + k_{eff}\nabla\phi_l \cdot \nabla\phi_l + \frac{2k_{eff}RT}{F}(1 - t_+)\nabla(\ln c_l) \cdot \nabla\phi_l \tag{3.9}$$

where σ_{eff} is the effective solid-phase conductivity $(s \cdot m^{-1})$, ϕ_s is the solid-phase potential (V), k_{eff} is the effective ionic conductivity of the liquid phase $(s \cdot m^{-1})$, ϕ_l is the liquid-phase potential (V), R is the universal gas constant, t_+ is the lithium-ion transference number, and c_l is the ion concentration in the liquid phase $(mol \cdot m^{-3})$.

3.1.3 Analysis of Battery Heat Transfer and Dissipation Processes

According to the principles of heat transfer, the primary mechanisms of heat transfer include conduction, convection, and radiation. Since radiation involves heat transfer between two objects with different temperatures that are not in direct contact, through electromagnetic waves, its impact on heat transfer performance in battery cooling systems is relatively minor. Therefore, in studies of thermal models for lithium-ion batteries, the influence of thermal radiation is often neglected. This chapter focuses on discussing the heat transfer and heat dissipation conditions in the thermal modeling of lithium-ion batteries.

3.1.3.1 Determination of Heat Transfer Conditions

Thermal conduction refers to the transfer of heat energy within a material or between materials in contact, relying on the thermal motion of microscopic particles such as molecules, atoms, and free electrons, without macroscopic displacement. For instance, thermal conduction occurs between the electrodes, electrolyte, and current collectors inside a cell, as well as between current collector and battery casing, which may have a lower temperature. The heat conduction behavior in a battery can be described using Fourier's law. The heat flux per unit area per unit time is proportional to the temperature gradient in the direction perpendicular to the cross-section, expressed as:

$$q_f = -\lambda_k \frac{\partial T}{\partial n} \tag{3.10}$$

In this equation, the negative sign indicates that the direction of energy transfer points toward decreasing temperature, q_f represents the heat flux density, λ_k is the thermal conductivity, and $\partial T / \partial n$ is the temperature gradient in the direction normal to the isothermal surface of the electrode.

Numerically, the thermal conductivity is a material-specific parameter that quantifies its heat transfer capability, with units of $\mathrm{W \cdot m^{-1} \cdot K^{-1}}$. It equals the magnitude of the heat flux density vector under a unit temperature gradient. At the same temperature (e.g., room temperature of 20 °C), metals exhibit high thermal conductivity, gases have low thermal conductivity, and liquids fall between the two. For most materials, the thermal conductivity is considered to follow a linear approximation, expressed as $\lambda = \lambda_0(a + bT)$, where T is temperature, a, b and λ_0 are constants.

In the thermal characteristics of lithium-ion battery cells, the thermal conductivity of the cell is typically anisotropic. For example, in the thickness direction of a prismatic battery cell, the structure can be considered a series arrangement of multiple electrode units, each comprising a positive electrode, a separator, and a negative electrode. The thermal conductivity in this direction can be calculated using Eq. (3.11). In the plane direction, the system can be treated as a parallel arrangement from the positive electrode to the separator to the negative electrode, and its thermal conductivity can be calculated using (3.12) [4].

$$\lambda_x = \frac{L_x}{\sum_i \frac{L_x}{\lambda_i}} = \frac{L_x}{\frac{L_{xp}}{\lambda_p} + \frac{L_{xn}}{\lambda_n} + \frac{L_{xs}}{\lambda_s}} \tag{3.11}$$

$$\lambda_y = \lambda_z = \sum_i \frac{\lambda_i L_{x_i}}{L_x} = \frac{\lambda_p L_{x_p} + \lambda_n L_{x_n} + \lambda_s L_{x_s}}{L_x} \tag{3.12}$$

In these equations, λ_x, λ_y and λ_z are the thermal conductivity of the battery cell in the thickness direction (x) and the in-plane directions of the electrode (y and z), respectively, as is shown in Fig. 3.1. λ_p, λ_s and λ_n denote the thermal conductivity

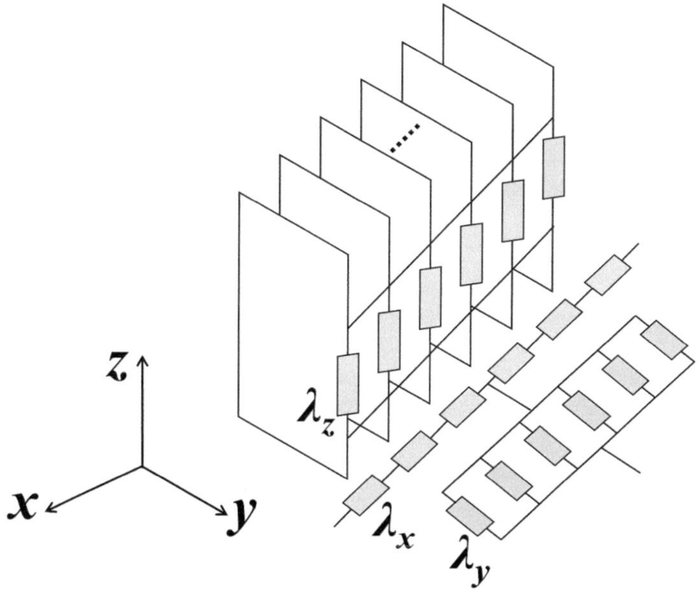

Fig. 3.1 Illustration of the thermal conductivity of the battery cell

of the positive electrode, separator, and negative electrode, respectively. L_{x_p}, L_{x_s} and L_{x_n} correspond to the thicknesses of the positive electrode, separator, and negative electrode, respectively. L_x refers to the overall thickness of the battery cell core.

The heat capacity of the cell core depends on the battery type, operating conditions, and ambient temperature. The heat capacity varies with temperature and state of charge (SOC) due to the properties of the electrode materials and the characteristics of the electrochemical reactions in the battery. The heat capacity of the battery can be measured using a calorimeter or estimated as a weighted average of the heat capacities of the constituent materials, as shown following:

$$C_p = \frac{1}{m}\sum_i C_i m_i \tag{3.13}$$

where C_p represents the equivalent specific heat capacity of the cell, m is the total mass of the cell, C_i is the mass of each constituent material of the cell, and m_i is the specific heat capacity of each constituent material.

3.1.3.2 Determination of Heat Dissipation Conditions

In the study of battery thermal characteristics, the effect of thermal radiation is often neglected. It is generally assumed that the heat dissipation between the battery surface and the environment is primarily determined by thermal convection. Thermal convection involves the transfer of heat through a moving medium, which occurs exclusively in fluids. Since the molecules in a fluid also exhibit random motion, thermal convection is invariably accompanied by heat conduction. Convection heat transfer can be classified into two categories: natural convection and forced convection. The calculation of convection heat transfer is typically based on Newton's law of cooling, expressed as

$$q_n = hS_{cell}\Delta T \tag{3.14}$$

Here, q_n represents the heat dissipation power, h represents the convection heat transfer coefficient, S_{cell} denotes the contact area between the battery surface and the surrounding environment, ΔT is the temperature difference between the battery surface and the fluid. The magnitude of the convection heat transfer coefficient depends not only on the physical properties of the fluid and the shape and arrangement of the heat exchange surface but also on the fluid flow velocity.

3.2 Battery AC Heat Generation Mechanism Based on Impedance Analysis

Electrochemical impedance spectroscopy (EIS) is extensively employed to investigate the correlation between the impedance properties of power batteries and diverse external factors, such as ambient temperature [5–7] and SOC [7, 8]. However, to meet the requirements of linearity and time-invariance in impedance measurements, small current or voltage perturbations are generally applied. As a result, the influence of excitation current amplitude on impedance characteristics is seldom considered. Gaberscek et al. [9, 10] studied the effects of different excitation currents on $LiFePO_4$-based battery systems. Excitation modes were divided into current excitation and voltage excitation. Their research showed high-frequency impedance is unaffected by excitation current amplitude, satisfying the conditions of linearity and time-invariance. Nevertheless, the mid-frequency impedance is highly sensitive to the excitation current amplitude.

To systematically investigate batteries' heat generation mechanisms, this section adopts two distinct approaches: full-cell and half-cell analyses. The full-cell experiments aim to explore the heat generation characteristics and overall impedance behavior under the complete battery configuration. This section investigates three different LIB chemistries, including $LiFePO_4/C$, $LiMn_2O_4/C$, and $LiNi_{1-x-y}Co_xAl_y/C$, to comprehensively elucidate their heat generation behavior. In contrast, the half-cell experiments utilize a reference electrode to decouple the interactions between the positive and negative electrodes, focusing on the individual contributions of electrode materials to heat generation. The combination of these two methodologies offers a multidimensional perspective for comprehensively understanding the heat generation mechanisms in batteries.

3.2.1 *Experimental Design for Impedance Test with Different Current Amplitudes*

In the AC heating method, excitation current amplitude and power battery impedance are among the important factors affecting the temperature rise. To further investigate the influence of current amplitude on the AC impedance characteristics of power batteries, this section designs an experimental matrix to deeply study the AC impedance characteristics of power batteries under high-current excitation. The following introduces the test samples, test platform, and test procedures.

3.2.1.1 Test Samples and Platforms

1) Full cells

For the full cells experiment, test samples are three types of batteries with different material systems: 8Ah $LiFePO_4$-based batteries (Sample B), 8Ah pouch-type lithium manganese oxide ($LiMn_2O_4/C$) batteries (Sample F), and NCA/graphite batteries (Sample G) with a nominal capacity of 1 Ah. The specific parameters of sample B are shown in Chap. 1. The detailed specifications for sample F and G are provided in Table 3.1.

For sample B and F, the single-cell power battery high-current AC impedance characteristic test platform consists of an electrochemical workstation (Solartron SI 1287, 1255B) and a power amplifier (PBI250-10). The frequency range for AC impedance measurements is 100 kHz–0.01 Hz, with excitation current amplitudes of 1.5, 2, 4, 6, 8, 12, and 16 A. The temperature of the power battery is adjusted using a thermostatic chamber, where the battery is placed and allowed to equilibrate to match the chamber's temperature. During the testing process, the temperature range is from −20 to 40 °C. Thermocouples are attached to the battery surface, and temperature data is collected using a temperature acquisition unit (LR8510) and a wireless temperature receiver unit (LR8410-30). The detailed test parameter matrix is shown in Table 3.2.

Table 3.1 Specification of sample F and G

	Sample F	Sample G
Material system	$LiMn_2O_4/C$	$LiNi_{1-x-y}Co_xAl_y/C$
Nominal capacity (Ah)	8	1
Nominal voltage	3.7	3.7
Charging cutoff voltage	4.2	4.2
Dimensions (mm, length × width × thickness)	140 × 115 × 8	40 × 60 × 3
Manufacturer	CITIC GUOAN MGL	LI-FAN TECHNOLOGY

Table 3.2 Parameter matrix for AC impedance characterization at high currents

Parameters	Values
SOC (%)	50
Environmental temperature (°C)	−20
Frequency sweep range	10 kHz–1 Hz; 10–1 Hz
Excitation current amplitude (A)	1.5, 2, 4, 6, 8, 12, 16

For sample G, the experimental setup comprises a host computer, a Maccor 4000 battery testing system, an EnergyLab XM electrochemical workstation, and a Parter PHV1402 temperature chamber. The applied AC current amplitudes include 0.013, 0.027, 0.135, 0.27, 0.54, and 1 A. To ensure consistency between tests, a 30-min resting period is maintained when switching between different amplitude conditions. After each test, the voltage stability is assessed by comparing it with the target OCV. If any deviation is observed, corrective adjustments are made to realign the voltage with the desired value.

2) Half cells

In order to delve deeper into the changes in the positive and negative impedance of the battery, this chapter also assembled half cells using the active materials from Sample G for further analysis. The testing encompasses half cells, symmetric coin cells, and full coin cells. The symmetric coin cells include graphite symmetric cells, NCA symmetric cells, and lithium symmetric cells. The experimental arrangement for half-cell testing is composed of a host computer, a NEWARE CT-4008Tn-5V10mA-164 charge/discharge system designed for coin cells, an EnergyLab XM electrochemical workstation, and a DONG&YE DY-T-20C temperature-controlled chamber.

The assembly of coin cells initiates with the activation of the sample battery via four charge–discharge cycles at 25 °C. Each cycle comprises three consecutive stages: a CCCV charging process, a 1-h stabilization period, and a CC discharging phase. In the CCCV charging stage, the battery is first charged at a rate of 1 C until it reaches 4.2 V, at which point it switches to the CV mode. The charging process is considered complete once the current drops below 0.02 C. During the CC discharging phase, the cell is discharged at a rate of 1 C until it reaches 3 V cut-off voltage. Additionally, a 1-h relaxation period is introduced between consecutive cycles. Following activation, a 0.5 SOC is set, and the battery is subsequently disassembled in an argon-filled MBraun glove box. Single-sided coated electrodes are prepared by removing the active material from one side of the electrode using N-methyl-2-pyrrolidone (NMP). The cleaned electrodes are then punched into circular shapes with a diameter of 14 mm using a precision disc cutter. These electrodes are subsequently rinsed multiple times with dimethyl carbonate (DMC) and dried in a vacuum chamber. Half cells are assembled using the punched electrodes along with thin lithium sheets (16 mm in diameter). Similarly, graphite symmetric cells, NCA symmetric cells, and full coin cells are constructed using the punched electrodes, while lithium symmetric cells are composed of two thin lithium sheets. The coin cells incorporate 18 mm-diameter separators sourced from Celgard. The electrolyte is formulated by dissolving 1 M lithium hexafluorophosphate ($LiPF_6$) in an equal-weight mixture of ethylene carbonate (EC) and diethyl carbonate (DEC).

Once assembled, the coin cells undergo a 12-h resting period before activation. The half cells are then cycled three times under CC charge–discharge conditions using a 0.4 mA current. The voltage cut-off ranges for the anode and cathode half cells are set to 0.01–2 V and 2.8–4.5 V, respectively. In addition to the half cells, lithium symmetric cells are also subjected to an activation process identical to that of the half cells.

EIS tests are conducted at two temperatures, 25 and − 10 °C. Prior to testing, the cells are conditioned in a temperature chamber at the designated temperature for at least three hours to ensure thermal equilibrium. The EIS measurements span an excitation frequency range from 0.01 Hz to 1000 kHz. For half-cell testing, the selection of AC amplitude is determined based on the cell's capacity. Given the proportional relationship between electrode areas in full and half cells, the capacity of a half cell is estimated to be approximately 3.69 mAh. To comprehensively analyze how different AC amplitudes affect the impedance behavior of the graphite anode and NCA cathode, excitation amplitudes of 0.05, 0.1, 0.5, 1, 2, and 3 mA are employed. It is important to note that these amplitude values represent RMS values as interpreted by the electrochemical workstation, rather than peak values. Furthermore, the EIS testing procedures for symmetric cells and full coin cells are identical to those applied to half cells.

3.2.1.2 The SOC-OCV Relationship Between the Full Cell and Half Cells

Before performing EIS tests on full and half cells at varying SOC, it is essential to first establish the SOC-OCV correlation between them. Based on the study by Ge et al. [11], the OCV of the full cell can be calculated as follows:

$$\begin{aligned} OCV(SOC) = {} & U_c((x_{c,100\%} - x_{c,0\%})SOC + x_{c,0\%}) \\ & - U_a((x_{a,100\%} - x_{a,0\%})SOC + x_{a,0\%}) \end{aligned} \tag{3.15}$$

where $OCV(SOC)$ defines the dependency of the full cell's OCV on its SOC. In this study, this relationship is modeled using a polynomial function:

$$OCV(SOC) = a_0 + \sum_{i=1}^{i=9} a_i SOC^i \tag{3.16}$$

where a_0 and a_i correspond to the various terms in the polynomial. U_c and U_a denote NCA cathode and graphite anode's OCV, respectively. These voltages are influenced by the normalized lithium content in the electrodes, represented as:

$$U_c(x_c) = b_0 + \sum_{i=1}^{i=9} b_i x_c^i \tag{3.17}$$

$$U_a(x_a) = c_0 + \sum_{i=1}^{i=8} (c_{2i-1}\cos(ix_a c_{17}) + c_{2i}\sin(ix_a c_{17})) \tag{3.18}$$

b_0, b_i, c_0, c_{2i-1}, c_{2i}, and c_{17} represent the different terms in the equation. In this context, x_c and x_a are assigned values of 0 when the cathode's OCV is 4.3 V and the anode's OCV is 2 V. Conversely, x_c and x_a are set to 1 when the OCV of the cathode is 2.8 V and that of the anode is 0.01 V. The lithium content of the cathode is denoted as $x_{c,100\%}$ and $x_{c,\,0\%}$ at 100% and 0% SOC of the full cell, respectively. Similarly, the lithium content of the anode at 100 and 0% SOC is represented as $x_{a,100\%}$ and $x_{a,0\%}$, respectively.

To determine $x_{c,100\%}$, $x_{c,\,0\%}$, $x_{a,100\%}$, and $x_{a,0\%}$, it is essential to first obtain the OCV-SOC curve of the full cell and the OCV-x curves for the half cells. When charge and discharge currents are small, the battery voltage closely approximates the OCV [12]. Thus, the OCV-SOC curve for the full cell and the OCV-x curves for the half cells can be derived using CC charge and discharge tests, with currents of 50 mA (0.05 C) for the full cell and 70 μA for the half cells. The temperature throughout the testing process is maintained at 25 °C. As shown in Fig. 3.2a–c, despite using small currents for both charging and discharging, differences between the charge and discharge OCV-SOC curves and OCV-x curves persist due to polarization effects. To precisely derive the OCV-SOC curve of the full cell and OCV-x curves for the half cells, the true OCV is calculated as the mean of the charge and discharge OCVs at the corresponding SOC. Consequently, the full cell's true OCV-SOC curve and the half cells' OCV-x curves are represented in Fig. 3.2c, a, and b by the orange lines, respectively. These curves are then fitted using Eqs. (3.16)–(3.18), with the fitting results showing good accuracy, as seen in Fig. 3.2a–c. Finally, a genetic algorithm is employed to identify $x_{c,100\%}$, $x_{c,\,0\%}$, $x_{a,100\%}$, and $x_{a,0\%}$, yielding the values of 0.0276, 0.836, 0.61022, and 0.13626, respectively. Fig. 3.2d compares the experimental OCV values with those obtained from the identified model, indicating that the determined OCV-SOC curve closely matches the full cell's true OCV-SOC curve.

This research examines the impedances of the graphite anode and NCA cathode at various AC amplitudes, corresponding to full cell SOCs of 0.2, 0.5, and 0.8. The OCV values of the graphite anode and NCA cathode, based on the SOC-OCV relationship between the full cell and half cells, are provided in Table 3.3.

3.2.1.3 DRT Calculation of Battery Impedance

Due to the close proximity of the time constants for the electrochemical processes within the cell, distinguishing each individual process directly from the Nyquist plot is challenging. However, by calculating the DRT from the impedance test, the various electrochemical processes within the cell can be efficiently and rapidly separated [14].

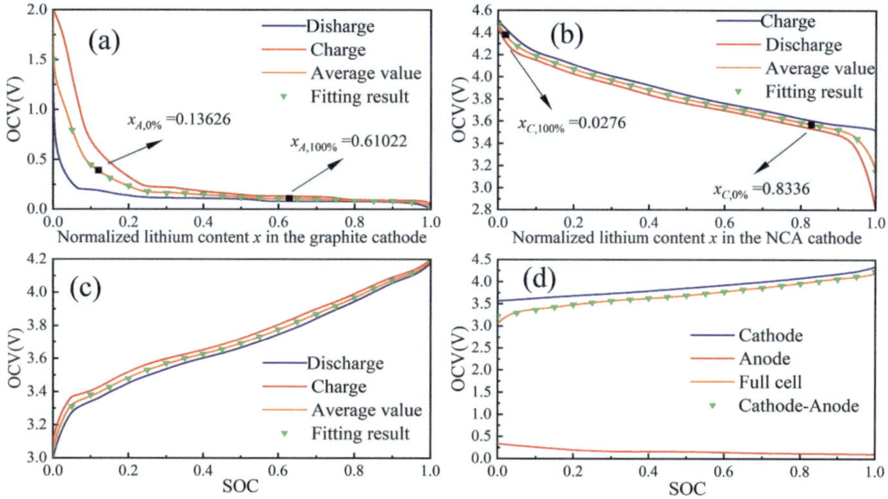

Fig. 3.2 Experimental data and fitting results of the OCV curves, along with the SOC-OCV relationship for the full and half cells. **a** The anode half cell's OCV-x curves, **b** The cathode half cell's OCV-x curves, **c** The full cell's OCV-SOC curves, **d** SOC-OCV relationship of the full and half cells [13]

Table 3.3 OCV of the full cell and half cells at SOC of 0.2, 0.5 and 0.8 [13]

	Full cell SOC		
	0.2	0.5	0.8
Full cell OCV	3.485 V	3.695 V	3.960 V
Anode OCV	0.198 V	0.160 V	0.123 V
Cathode OCV	3.683 V	3.856 V	4.082 V

The ideal polarization process is modeled by an *R*/*C* parallel circuit, which appears as a full semicircle on the Nyquist plot and as a Dirac impulse in the relaxation time function plot. The idealized charge separation process is depicted through a parallel *R*/*C* configuration. On the Nyquist diagram, this configuration appears as a perfect semicircle, while in the relaxation time spectrum, it is represented by a Dirac delta peak. Additionally, the height of this peak corresponds to the resistance *R* [15]. The characteristic time constant τ for the parallel *R/C* configuration t is given by

$$\tau = \frac{1}{2\pi f} = RC \tag{3.19}$$

The parallel *R*/*C* circuit's impedance is expressed as

$$Z_{RC} = \frac{R}{1 + j2\pi f \tau} = \frac{R}{1 + j2\pi fRC} \tag{3.20}$$

Owing to the intricate microstructure of the cell electrodes, the actual polarization is typically modeled by the parallel *R/Q* parallel circuit, where *Q* represents a constant phase element. Unlike the ideal scenario, the resistance *R* is determined by integrating the area under the peak. For real polarization process, the DRT function, $\gamma(\tau)$, can be viewed as a series of *R/C* circuits with varying relaxation times [16]. Consequently, the impedance of the actual polarization process can be described as [13]

$$Z_{DRT} = \sum_{n=1}^{N} \frac{R_n}{1 + j2\pi f \tau_n} = \int_{\tau_1}^{\tau_N} \frac{\gamma(\tau)}{1 + j2\pi f \tau} d\tau \tag{3.21}$$

The free software DRTtool [16] is used to calculate the DRT. In order to obtain consistent and reliable DRT results, the parameters in the software are kept uniform across all EIS tests.

3.2.2 Results and Analysis

3.2.2.1 Full Cells

1) Impedance Test Results

Figure 3.3 shows the AC impedance spectra obtained under different excitation current amplitudes when the ambient temperature is between 10 and 40 °C. A comparison of the four plots reveals that impedance is primarily influenced by temperature. As the temperature decreases, the impedance increases significantly. When the ambient temperature exceeds 10 °C, variations in current amplitude mainly affect the low-frequency diffusion process. From the perspective of electrode processes, low-frequency impedance is mainly influenced by the ion concentration gradient and is related to the diffusion distance (particle diameter) [17]. The reason for the reduction in the impedance arc under high-current excitation is that a larger ion concentration gradient forms on the surface of the electrode particles inside the cell, leading to a decrease in lithium-ion diffusion resistance, which manifests as a reduction in the low-frequency impedance arc. However, the mid- and high-frequency impedance of power batteries is essentially unaffected by changes in current amplitude. To further investigate and analyze, the AC impedance spectra under high-current excitation at ambient temperatures of 0, −10, and −20 °C are examined, with the results shown in Figs. 3.4, 3.5 and 3.6.

In Figs. 3.4, 3.5 and 3.6, the solid line represents the full frequency range from 100 kHz to 0.01 Hz, while the dashed line indicates a narrower frequency range from 10 to 1 Hz. Analysis of the figures reveals that when the power battery operates in a low-temperature environment (<10 °C), a critical impedance point exists. This point divides the mid-frequency impedance arc into two sections: when the excitation frequency is higher than this critical frequency, the impedance is unaffected by the excitation current amplitude; however, when the excitation frequency is lower than

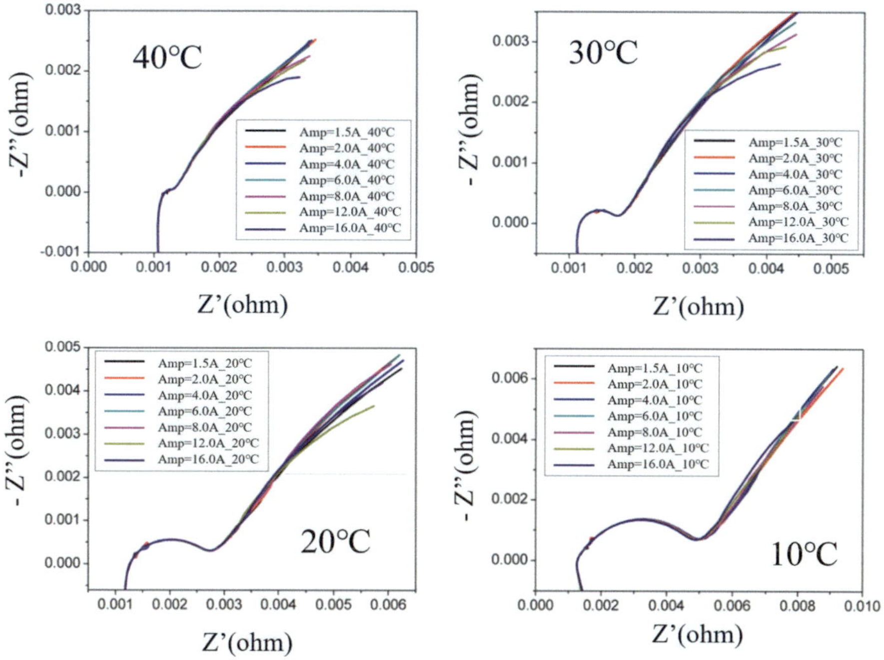

Fig. 3.3 Impedance map of sample B ($LiFePO_4$/C cells) at different excitation current amplitudes from 10 to 40 °C [18]

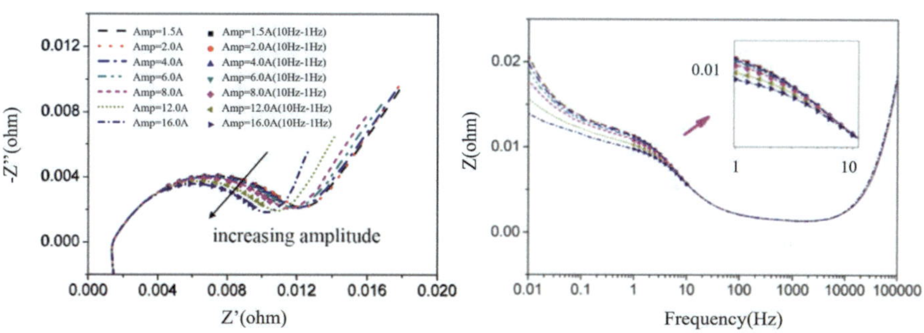

Fig. 3.4 Impedance map of sample B ($LiFePO_4$/C cells) cell with different excitation current amplitudes at ambient temperature of 0 °C [18]

this point, the impedance arc shows a systematic reduction with increasing excitation current amplitude.

From the Bode plots in Figs. 3.4, 3.5, 3.6, the variation in impedance values is more clearly observable. For sample B, the $LiFePO_4$/C power battery, the critical impedance point is approximately 10 Hz. Unlike the battery's characteristics at room temperature, the mid-frequency impedance of the power battery in a low-temperature

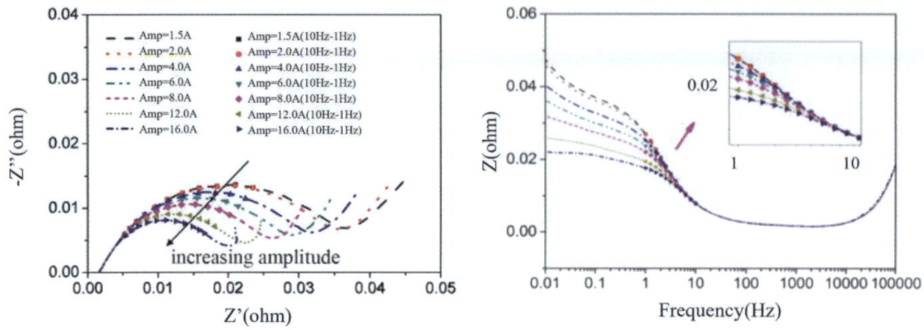

Fig. 3.5 Impedance map of sample B ($LiFePO_4/C$ cells) cell with different excitation current amplitudes at ambient temperature of −10 °C [18]

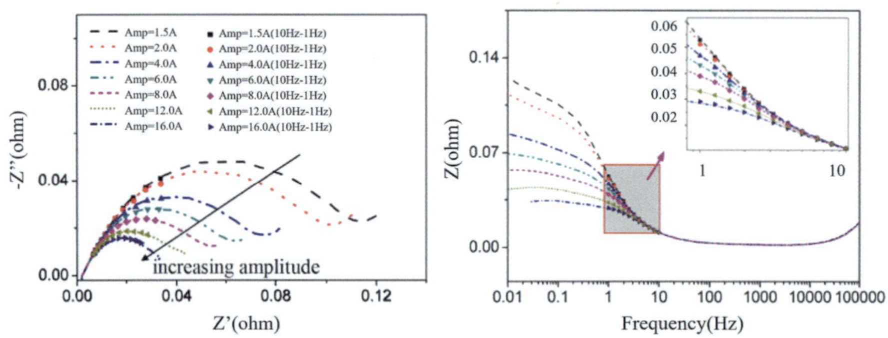

Fig. 3.6 Impedance map of sample B ($LiFePO_4/C$ cells) cell with different excitation current amplitudes at ambient temperature of −20 °C [18]

environment is jointly influenced by the ambient temperature and excitation current amplitude.

2) The Influence of Internal Pressure on Impedance Arc

To further investigate the observed variation in the mid-frequency impedance arc with excitation current, another power battery sample, sample F ($LiMn_2O_4/C$), was tested under high-current excitation, with the ambient temperature is −20 °C. The tested $LiMn_2O_4$ cell is packaged in a soft pouch with an aluminum-plastic film, which prevents internal pressure accumulation during battery operation, especially under AC excitation caused by charging and discharging processes [19]. The test results are shown in Fig. 3.7, where the solid line represents a sweep frequency range from 10 kHz to 1 Hz, and the dashed line represents a range from 10 to 1 Hz. The results for sample F ($LiMn_2O_4/C$) are consistent with those of the $LiFePO_4$-based power battery, showing that the mid-frequency impedance arc decreases systematically with increasing excitation current amplitude. Since the soft pouch aluminum-plastic packaging of the battery prevents internal pressure buildup during charging and

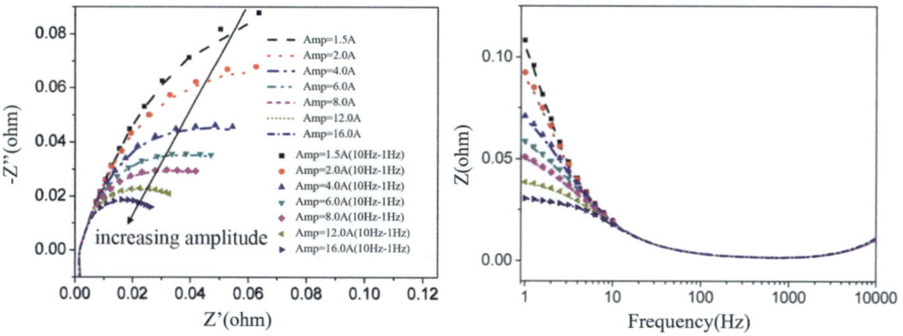

Fig. 3.7 Impedance map of sample F ($LiMn_2O_4$/C cells) cell with different excitation current amplitudes at ambient temperature of −20 °C [18]

discharging, it can be concluded that the reduction in the mid-frequency impedance arc with increasing excitation current amplitude is not caused by internal pressure changes within the battery.

3) The Influence of SOC Variation on Impedance Arc

Analyzing the impact of SOC variation on impedance spectra during the testing process, it was found that during AC impedance testing of the power battery, under the maximum test current amplitude and the lowest frequency excitation (16 A and 0.01 Hz), the SOC fluctuation range of the power battery was calculated to be ±1.96% using the ampere-hour integration method. Furthermore, based on previous studies [20, 21], for $LiFePO_4$-based power batteries, the low-frequency impedance in AC impedance testing is significantly affected by SOC, while the mid-frequency arc remains essentially unaffected by SOC changes. Therefore, it can be concluded that the reduction in the mid-frequency arc with increasing excitation current amplitude at low temperatures is not caused by SOC variation during the battery testing process.

4) The Influence of Temperature Variation on Impedance Arc

During charge/discharge processes, especially under high current conditions, the impact of temperature on power batteries is significant [22]. Temperature is a critical factor influencing the impedance variation of power batteries. As temperature increases, the impedance of power batteries decreases significantly. The following focuses on analyzing during the testing process, the influence of temperature variation on the impedance arc of power batteries.

From the Bode plots in Figs. 3.4, 3.5, 3.6 and 3.7, it is observed that the mid-frequency impedance critical point is around 10 Hz. When the frequency drops below 10 Hz, the mid-frequency impedance arc begins to split and shrink. It is reasonably hypothesized that if this phenomenon is caused by internal temperature changes in the battery, then the temperature rise within the battery accumulates when the frequency exceeds 10 Hz. To validate this hypothesis, a comparative test experiment was designed in this section. Keeping the test environmental temperature, battery

SOC, and other test conditions constant, the frequency sweep range was narrowed, and frequency sweep tests were conducted only between 10 and 1 Hz for sample B and F. The test results are shown in Figs. 3.4, 3.5, 3.6 and 3.7, which show the phenomenon of the impedance arc shrinking with increasing current persists, and the results within the narrow frequency range are consistent with those in the full frequency range.

Firstly, from a numerical analysis perspective, correlation analysis was conducted to calculate the correlation coefficients of the real and imaginary parts of the impedance under different temperatures and excitation currents for both narrow frequency excitation (10–1 Hz) and full frequency excitation (100 kHz–0.01 Hz). The calculated correlation coefficients for both the real and imaginary parts of the impedance exceeded 0.999, indicating an excellent positive correlation between the experimental results of the two frequency sweep ranges. To further quantify the differences in impedance measured within the narrow frequency range (10–1 Hz) and the full frequency range (100 kHz–0.01 Hz) as shown in Figs. 3.4, 3.5, 3.6 and 3.7, a parameter $\sigma(x, y)$ is defined to describe the deviation between the impedance modulus measured in the two frequency sweep ranges. The expression for $\sigma(x, y)$ is given as follow

$$\sigma(x, y) = \frac{\sum_{z=1}^{z=10} \left\{ \frac{\sqrt{\left| \begin{array}{l} (\mathrm{Re}^2[Z(x, y, z)] + \mathrm{Im}^2[Z(x, y, z)]) - \\ \mathrm{Re}^2[Z_{10\mathrm{Hz}-1\mathrm{Hz}}(x, y, z)] + \mathrm{Im}^2[Z_{10\mathrm{Hz}-1\mathrm{Hz}}(x, y, z)] \end{array} \right|}}{\sqrt{\mathrm{Re}^2[Z(x,y,z)] + \mathrm{Im}^2[Z(x,y,z)]}} \right\}}{n} \tag{3.22}$$

Here, x is the ambient temperature, y and z represent the current amplitude and frequency of the excitation, and n represents the number of excitation frequency points. $Z(x, y, z)$ denotes the impedance measurements within the full frequency range, while $Z_{10\mathrm{Hz}-1\mathrm{Hz}}(x, y, z)$ represents the impedance measurements within the short frequency range (10 Hz–1 Hz). The parameter $\sigma(x, y)$ is defined as the deviation in impedance modulus measurements between the two frequency sweep ranges.

After computational analysis, Fig. 3.8 illustrates the impedance modulus deviations under different temperatures and excitation current amplitudes for the two frequency sweep ranges. In the figure, the lithium iron phosphate (LFP) battery system is labeled "LFP," and the lithium manganese oxide (LMO) battery system is labeled "LM". From Fig. 3.8, it can be clearly observed that the impedance modulus deviations under different temperatures and excitation currents are minimal for both frequency sweep ranges. The deviations for the LFP battery system are around 0.1, while those for the LMO battery system are around 0.2. Overall, the deviations decrease as the excitation current increases. Furthermore, the deviations for the LMO battery system are larger than those for the LFP battery system. The difference in these deviations is mainly determined by the impedance characteristics of the batteries; in general, smaller impedance values correspond to smaller deviations between the two frequency sweep ranges.

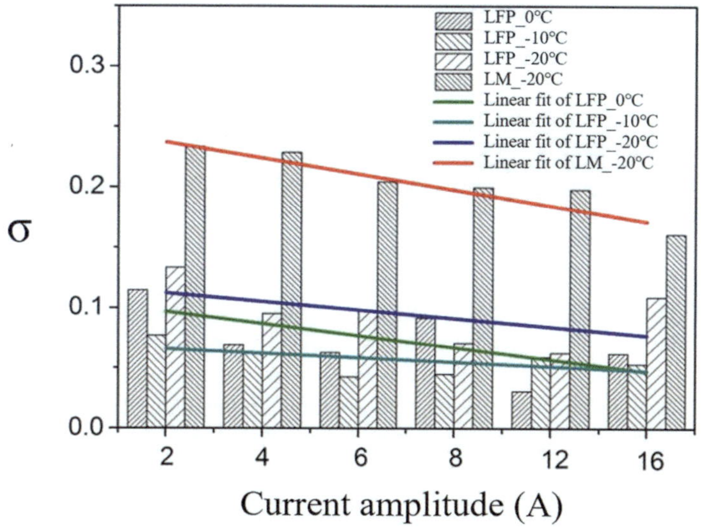

Fig. 3.8 Impedance mode deviation values for different temperatures and excitation current amplitudes in two sweep ranges [18]

To comprehensively analyze the influence of temperature in test results, surface temperature monitoring was conducted on sample B during full-frequency range testing. The temperature rise results are listed in Table 3.4, which shows that the temperature rise during testing is minimal. The maximum temperature rise occurs under the condition of a −20 °C ambient temperature and a 16 A excitation current, with a value of 1.4 °C. This chapter assumes that the battery operates in an adiabatic environment, where all heat generated is converted into the internal temperature rise. By transforming Eqs. (3.1) and (3.2), Eq. (3.23) is derived.

$$\Delta T' = qt/m/C_p = I_{RMS}^2 R_{ave} t/m/C_p \tag{3.23}$$

where in Eq. (3.23), $\Delta T'$ represents the mean temperature rise in an adiabatic environment, q is the heat generation rate, m is the mass, C_p is the equivalent specific

Table 3.4 Summary of measured temperature change $\Delta T(K)$ and theoretical calculated temperature rise $\Delta T'(K)$ [18]

Ambient temperature	6 A		8 A		12 A		16 A	
	ΔT	$\Delta T'$	ΔT	$\Delta T'$	ΔT	$\Delta T'$	ΔT	$\Delta T'$
10 °C	/	0.1	/	0.2	/	0.4	0.2	0.6
0 °C	/	0.2	/	0.3	0.4	0.6	0.7	0.9
−10 °C	/	0.4	0.2	0.5	0.4	1.0	0.7	1.4
−20 °C	0.25	0.7	0.5	0.9	0.8	1.8	1.4	2.2

heat capacity, $I_{RMS} = I/\sqrt{2}$ is the RMS current, and t is the duration of the current (9 min, which is the total time for frequency sweep testing across the full frequency range). Considering that the battery impedance varies with excitation frequency, the calculation uses the maximum impedance value R_{ave} within the frequency range, specifically at the turning point between mid and low frequencies. Table 3.4 provides the theoretical temperature rise values calculated from this approach, alongside the experimentally measured temperature rise values ΔT.

Even if all the heat generated is converted into temperature rise, the difference between ΔT and $\Delta T'$ is minimal. This small difference can be attributed to two factors: (1) The lithium iron phosphate battery system exhibits excellent power characteristics and has a relatively low internal resistance, leading to minimal heat generation during testing; (2) The battery's high thermal capacity allows the generated heat to be absorbed by itself. Based on the above experimental results and theoretical temperature rise calculations, it can be concluded that the phenomenon of the mid-frequency impedance arc shrinking as the temperature decreases is not caused by internal temperature changes within the battery.

Through the testing and theoretical analysis in the preceding sections, it has been demonstrated that the shrinking mid-frequency impedance phenomenon observed during high-current excitation testing is not due to variations in the battery's SOC state, internal pressure, or internal temperature changes. The next section will investigate the influence of amplitude on the impedance of the positive and negative electrodes individually, and focus more on the theoretical factors contributing to variations in the mid-frequency impedance arc.

3.2.2.2 Half Cells

1) Impedance Test Results

Figure 3.9 presents the impedance characteristics of NCA full cell across various AC amplitudes. The ultra-high-frequency impedance, which lies beneath the real axis, indicates inductive behavior. However, this aspect is not the primary focus of this study. Consequently, the analysis here is limited to the impedance features above the imaginary axis. In Fig. 3.9a–i, the low-SOC impedance shows only a slight reduction in the mid- and low-frequency regions at 25 °C. As the AC amplitude increases, the impedance arcs for medium and high SOC show no substantial variation. Although the mid- and low-frequency impedances of the low-SOC arc exhibit some variation with different AC amplitudes, these variations are relatively small. Moreover, no significant differences are observed in the Bode plots at 25 °C in Fig. 3.9. At −10 °C, however, the mid- and low-frequency impedances are more significantly influenced by varying AC amplitudes, regardless of the SOC level. In Fig. 3.9j, m, and p, the impedance arcs contract noticeably with the increase in AC amplitude when the frequency is below approximately 10 Hz.

Moreover, the Bode plots at −10 °C in Fig. 3.9 indicate that the impedance decreases as the arcs contract, with the phase value remaining unchanged. Results

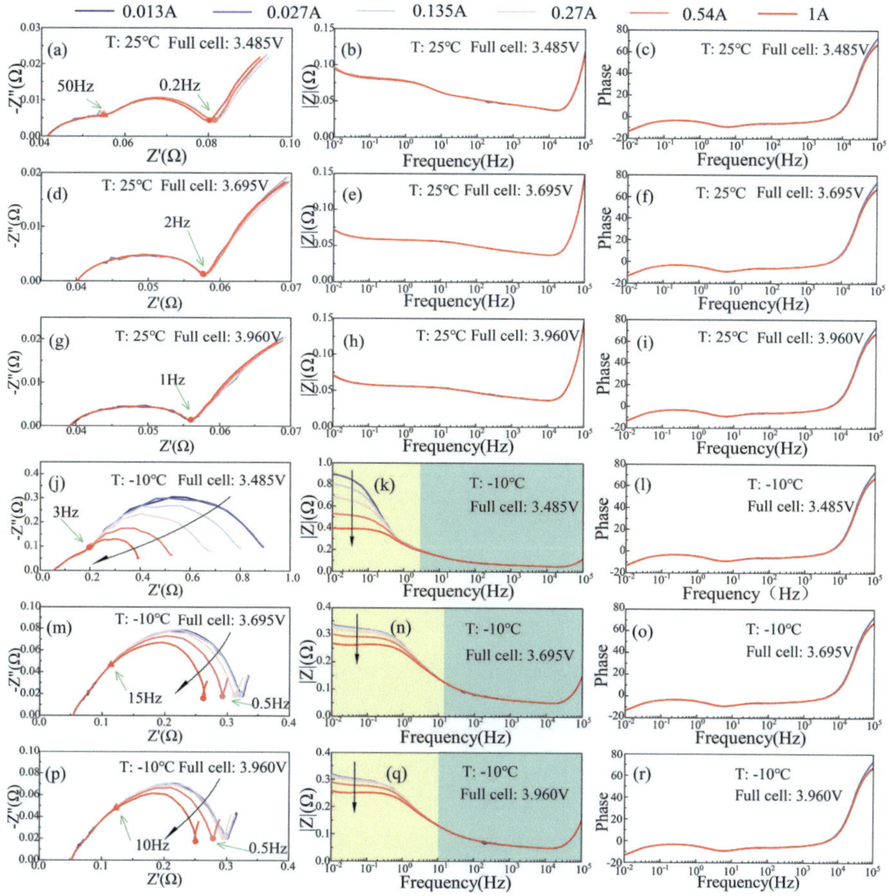

Fig. 3.9 Full cell impedance response under varying AC amplitudes. **a–i** At 25 °C, **j–r** At −10 °C [13]

of full cell align with the observations described above. Figure 3.9 shows the experimental results of the graphite half cell. In Fig. 3.10a, d, and g, the high-frequency impedance arc remains relatively unchanged at 25 °C. For medium-frequency impedances below 400 Hz, the impedance arc exhibits a slight shrinkage, though this is not significant. Notably, in the low-frequency region—below 2 Hz at 0.198 V and 0.160 V, and 5 Hz at 0.123 V—the impedance arc tends to shift toward the lower-left.

In the frequency range where the impedance arc contracts, the impedance value decreases with increasing AC amplitude, as shown in Fig. 3.10b, e, and h. In the high-frequency range, the AC amplitude has little effect on the impedance. In Fig. 3.10c, f, and i, a noticeable increase in phase value is observed when the frequency is between 1 and 10 Hz, as the impedance arc shrinks. While this change is subtle,

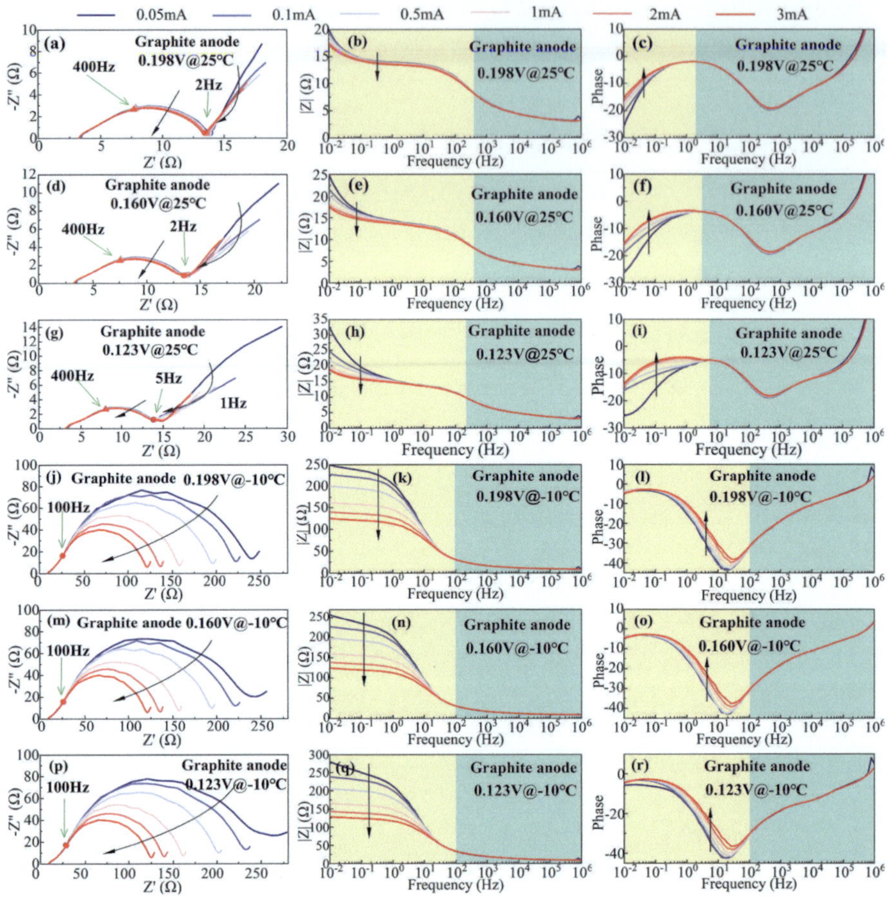

Fig. 3.10 Graphite anode impedance response under varying AC amplitudes. **a**–**i** At 25 °C, **j**–**r** At −10 °C [13]

the impedance at 0.123 V is higher. This is consistent with findings in the literature [23], and may be attributed to the graphite anode lithium content. As the SOC of the full cell rises, the graphite anode potential decreases. During the phase transition of lithium-intercalated graphite, the resistance increases even more.

Figure 3.10j–r presents the experimental results of graphite anode at −10 °C under varying AC amplitudes. A specific frequency, denoted as *b*, separates the impedance response into two distinct regions. Above this frequency, the impedance remains largely unchanged by variations in the AC amplitude. However, below frequency *b*, the impedance progressively reduces as the AC amplitude increases, with the impedance value decreasing and the phase value rising gradually. Notably, the frequency *b* is approximately 100 Hz.

Figure 3.11 presents NCA half cell's test results, showcasing a notable distinction from Fig. 3.10. Upon comparing the Nyquist plots, it is evident that the NCA cathode impedance exhibits two distinct arcs, resembling the findings of Steinhauer et al. [24] and Stiaszny et al. [25], which could be attributed to the lithium intercalation mechanism in the NCA cathode. Fig. 3.11a, d, and g show how the impedance arc of the NCA cathode changes with different AC amplitudes at 25 °C. Like the graphite anode, when the NCA cathode has a relatively low amount of embedded lithium (at voltages of 3.856 and 4.082 V), the AC amplitude primarily affects the low-frequency impedance. Additionally, Fig. 3.11 indicate that within the frequency range where the impedance arc shrinks, the impedance contracts as the AC amplitude increases, while the phase broadens. At −10 °C, the NCA impedance variation follows a similar trend to that of the graphite impedance in Fig. 3.11j–r. Moreover, there is also a frequency *b* for the NCA cathode, with a value approximately equal to 100 Hz, similar to that of the graphite anode. However, a key difference with the NCA cathode is that the phase variation with amplitude is separated into two distinct intervals: one spanning approximately 0.01 Hz to 3 Hz, and the other between 10 and 100 Hz.

2) The Impact of AC Amplitudes on Impedance Test

From the results, it is evident that the variations in full-cell impedance across different AC amplitudes are affected by both anode and cathode. AC amplitude primarily affects the low-frequency region of the impedance at 25 °C. Specifically, when the SOC of the full cell is 0.2, the NCA cathode impedance cathode rises significantly and becomes more responsive to the AC amplitude. Therefore, at low SOC levels, the NCA cathode is likely the dominant factor influencing impedance variation with AC amplitude at 25 °C. At −10 °C, both the mid- and low-frequency impedances of the graphite anode and NCA cathode are strongly affected by the AC amplitude, irrespective of the lithium content in the electrodes. The trend of impedance change at −10 °C mirrors that seen at 25 °C. However, the high-frequency impedance remains unchanged regardless of the AC amplitude. Thus, the variation in impedance with AC amplitude at −10 °C in full cell is mainly governed by NCA cathode and graphite anode.

As demonstrated in previous studies [13, 26–28], the impedance of batteries is generally influenced by SOC, temperature, etc. However, with analysis of lithium iron phosphate and lithium manganate full cells, it was found that changes in SOC and temperature do not significantly affect the dominant impedance arc. For NCA batteries, the SOC variation in the full cell and half cells is 1.39% and 1.13%, respectively, when the AC amplitude is at its maximum and the frequency is 0.01 Hz. Given that the SOC variation is negligible, the impedance is not influenced by SOC changes. In terms of temperature, the half-cell temperature is tracked during impedance testing and remains virtually constant at both 25 and −10 °C. The temperature of full cell remains stable at 25 °C. At −10 °C, there is a maximum temperature rise of the full cell approximately 1.5 °C with an SOC of 0.2, an AC amplitude of 1A and a frequency of 0.01 Hz. Due to minimal temperature increase, the impedance change due to this slight rise in temperature is insignificant. Based on this assessment, the conclusion is that the variations in impedance are primarily caused by the different AC amplitudes.

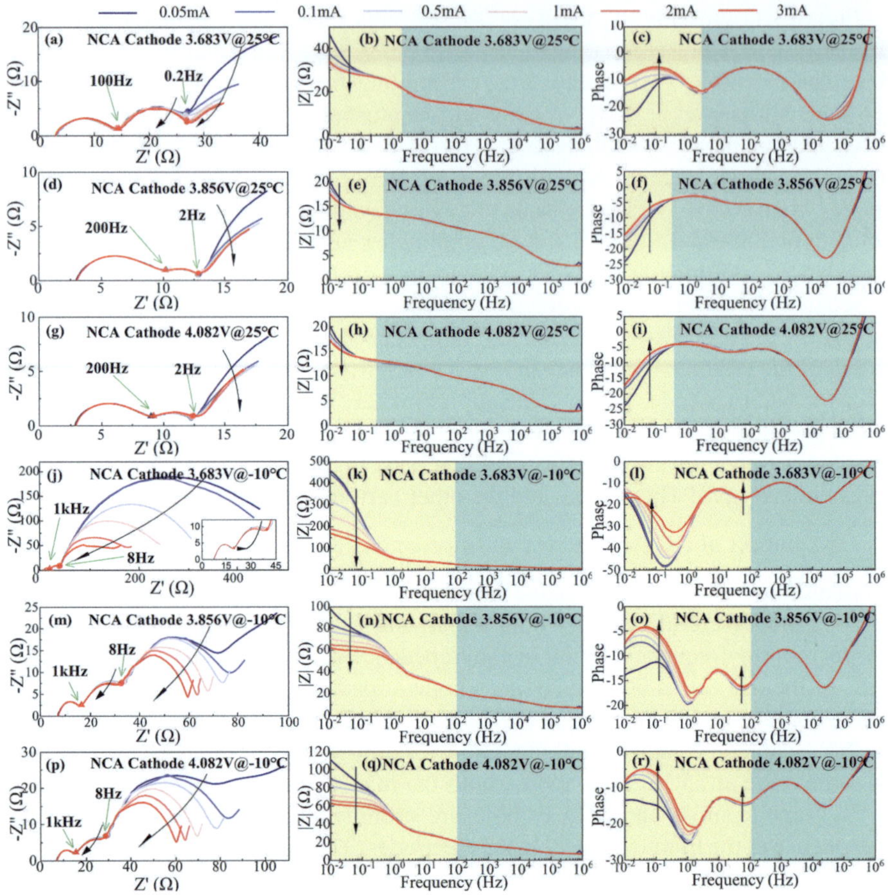

Fig. 3.11 NCA cathode impedance response under varying AC amplitudes. **a–i** At 25 °C, **j–r** At −10 °C [13]

3) The Impact of Voltage Harmonics on Impedance Test

Moreover, with AC amplitude increasing, the battery's nonlinear nature become more evident. This is primarily manifested as the presence of harmonic components in the voltage response [29–31]. Impedance measurements can be conducted in either galvanostatic or potentiostatic mode, utilizing the fundamental frequency components of voltage and current [32]. Consequently, analyzing how voltage harmonics influence the impedance test is essential in this study. In Figs. 3.10 and 3.11, the influence of AC amplitude on mid- and low-frequency impedance becomes especially significant at −10 °C. To analyze this, the voltage response is analyzed for harmonic components at different AC amplitudes under −10 °C conditions. To ensure extensive coverage of the mid- and low-frequency range, the harmonic behavior l is evaluated at four distinct frequencies. For full cells, the selected frequencies for analysis are 10,

0.5, 0.1, and 0.01 Hz. The graphite half-cell measurements are conducted at 50, 2, 0.1, and 0.01 Hz, while the NCA half-cell is examined at 10, 0.5, 0.05, and 0.01 Hz. The voltage and current sampling rate is configured to be 1000 times the frequency used in harmonic analysis. Furthermore, the Fast Fourier Transform (FFT) technique is utilized to break down the voltage harmonic components [33]. This decomposition process converts the time-domain voltage signal into the frequency domain, allowing the extraction of the amplitude Y1 corresponding to the fundamental frequency as well as the amplitudes Y_n (n $\geq$ 2) of the higher harmonics. Harmonics of higher order can be disregarded. For this reason, the proportion K is defined as the fundamental frequency relative to the total voltage response, which is $K = Y \Big/ \sum_{n=1}^{n=5} Y_n$.

The proportion K for both minimum and maximum AC amplitudes of full and half cells at -10 °C is shown in Table 3.5. In both full and half cells, the fundamental frequency response is predominant. As the frequency decreases, the value of K at the maximum AC amplitude gradually decreases. The smallest K is 83.5% for full cell, 87.9% for graphite half cell, and 78.7% for NCA half cell. To delve deeper into the correlation between impedance attenuation and harmonic distortion effects, two quantitative metrics (M and N) are established. M represents the ratio of the impedance magnitude |Z| between maximum and minimum AC amplitude conditions, while N evaluates the contrast of K values at 0.01 Hz under identical amplitude extremes. As tabulated in Table 3.6, N substantially exceeds M providing quantitative evidence against higher-order harmonics being the dominant driver of impedance reduction. Complementary data in Table 3.5 further demonstrate sustained K-values above 85% across test conditions, reaffirming the fundamental frequency component's predominance in the composite voltage response.

4) The Impact of AC Amplitudes on Kinetic Processes

The DRT technique is employed for impedance analysis to explore the influence of varying AC amplitudes on the battery's kinetic processes. Previous studies have established that different electrochemical processes occur across distinct frequency domains [24, 34]. Specifically, the ultra-high frequency region is associated with inductive effects from metallic components in the cell. Interfacial phenomena, such as lithium-ion diffusion in the electrolyte, ion transport across the SEI/CEI layer, and charge transfer reactions, predominantly take place in the high- and mid-frequency ranges. In contrast, lithium-ion diffusion within the solid phase is a much slower process, occurring at low frequencies. Notably, the DRT method is incapable of capturing inductive effects or solid-phase diffusion behavior [15].

The DRT evaluation strategically focuses on intermediate and high-frequency regimes, as ultra-high/low frequency components—readily distinguishable in Nyquist representations but beyond this investigation's scope—are deliberately excluded. Figure 3.12 depicts this methodological filtering: ultra-high frequencies are eliminated, with low-frequency DRT outcomes visually de-emphasized through gray shading. Three characteristic relaxation modes (designated F_1–F_3) emerge in complete cell configurations. Notably, the F_3 signature becomes undetectable under specific operational parameters, particularly when correlated with thermal conditions

Table 3.5 A *K* of fundamental frequency response in the voltage response at −10 °C [13]

Full cell								
Voltage	10 Hz		0.5 Hz		0.1 Hz		0.01 Hz	
	13 mA	1 A	13 mA	1 A	13 mA	1 A	13 mA	1 A
3.960 V	97.9%	99.6%	99.0%	96.5%	97.6%	93.7%	97.9%	94.6%
3.695 V	94.9%	99.0%	97.7%	96.0%	98.3%	93.4%	96.6%	93.9%
3.484 V	97.2%	99.3%	98.3%	97.0%	99.1%	89.7%	98.7%	83.5%
Graphite half cell								
Voltage	50 Hz		2 Hz		0.1 Hz		0.01 Hz	
	0.05 mA	3 mA	0.05 mA	3 mA	0.05 mA	3 mA	0.05 mA	3 mA
0.198 V	99.3%	99.3%	99.5%	93.0%	99.6%	88.4%	99.5%	89.1%
0.160 V	99.5%	99.6%	99.3%	93.5%	99.4%	89.0%	98.5%	88.3%
0.123 V	99.3%	99.4%	99.2%	92.7%	99.3%	88.4%	97.0%	87.9%
NCA half cell								
Voltage	10 Hz		0.5 Hz		0.05 Hz		0.01 Hz	
	0.05 mA	3 mA	0.05 mA	3 mA	0.05 mA	3 mA	0.05 mA	3 mA
4.028 V	93.7%	99.5%	98.0%	96.9%	97.1%	93.0%	94.3%	92.3%
3.856 V	96.4%	99.4%	97.4%	97.1%	97.3%	93.4%	96.7%	93.0%
3.683 V	96.3%	98.6%	97.4%	98.1%	98.3%	86.0%	96.4%	78.7%

Table 3.6 Results of *M* and *N* [13]

	Full cell	Graphite half cell	NCA half cell
M	0.44	0.46	0.37
N	0.85	0.91	0.82

and SOC variations at 50 and 80% charge states. Half-cell tests identified four distinct relaxation peaks for both graphite (A_1, A_1', A_2, A_3) and NCA (C_1, C_1', C_2, and C_3) electrodes. Figure 3.13 contrasts symmetric and full coin-cell impedance/DRT data to elucidate kinetic disparities between industrial-scale and coin-cell architectures, including lithium electrode interference. At 25 °C, symmetric and full cells mirror full/half-cell amplitude stability, except lithium symmetric configurations. These exhibition marked low-frequency amplitude sensitivity, contrasting with minimal mid-frequency shifts.

Experimental observations demonstrate notable AC amplitude dependence at − 10 °C, with impedance spectra undergoing progressive contraction under elevated excitation magnitudes. DRT analysis across cell configurations shows distinct polarization characteristics: complete coin cell assemblies and symmetric graphite / NCA cells consistently present four relaxation signatures, in contrast to Li-based symmetric counterparts which manifest only three identifiable response modes. According to the literature [35], as the time constant increases, the primary electrochemical processes occurring within a lithium-ion battery include conduction, SEI/

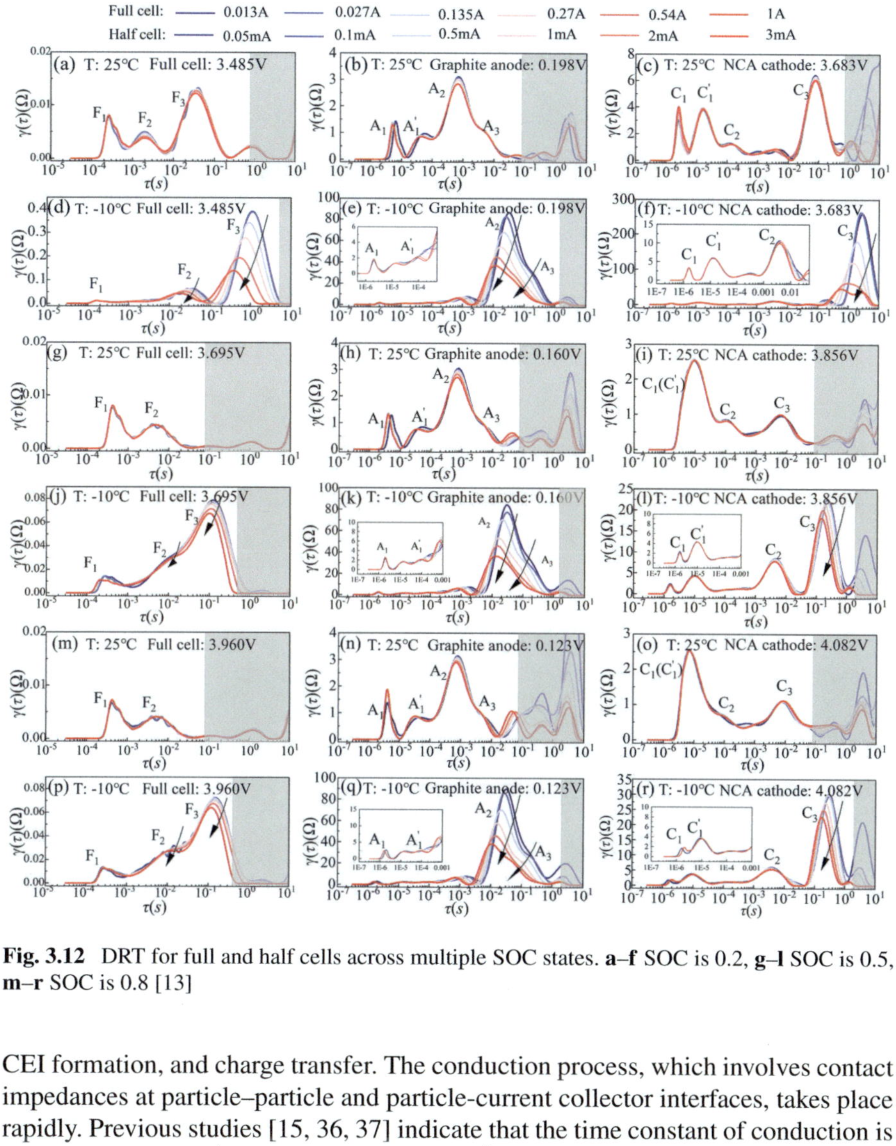

Fig. 3.12 DRT for full and half cells across multiple SOC states. **a–f** SOC is 0.2, **g–l** SOC is 0.5, **m–r** SOC is 0.8 [13]

CEI formation, and charge transfer. The conduction process, which involves contact impedances at particle–particle and particle-current collector interfaces, takes place rapidly. Previous studies [15, 36, 37] indicate that the time constant of conduction is typically below 10^{-4} s for coin cells and 10^{-3} s for full cells. Notably, this process remains largely unaffected by variations in temperature and SOC. Consequently, the conduction process can be attributed to peak F_1 in the full cell, peaks A_1 and A_1' in the graphite anode, and peaks C_1 and C_1' in the NCA cathode.

Existing research [35] identifies three principal lithium-ion battery mechanisms correlated with increasing time constants: interfacial conduction, SEI/CEI development, and charge transfer dynamics. The conduction mechanism-involving rapid interfacial resistance at particle/collector junctions—demonstrates characteristic time constants below 10^{-3} s (industrial cells) and 10^{-4} s (coin cells) [15, 36, 37],

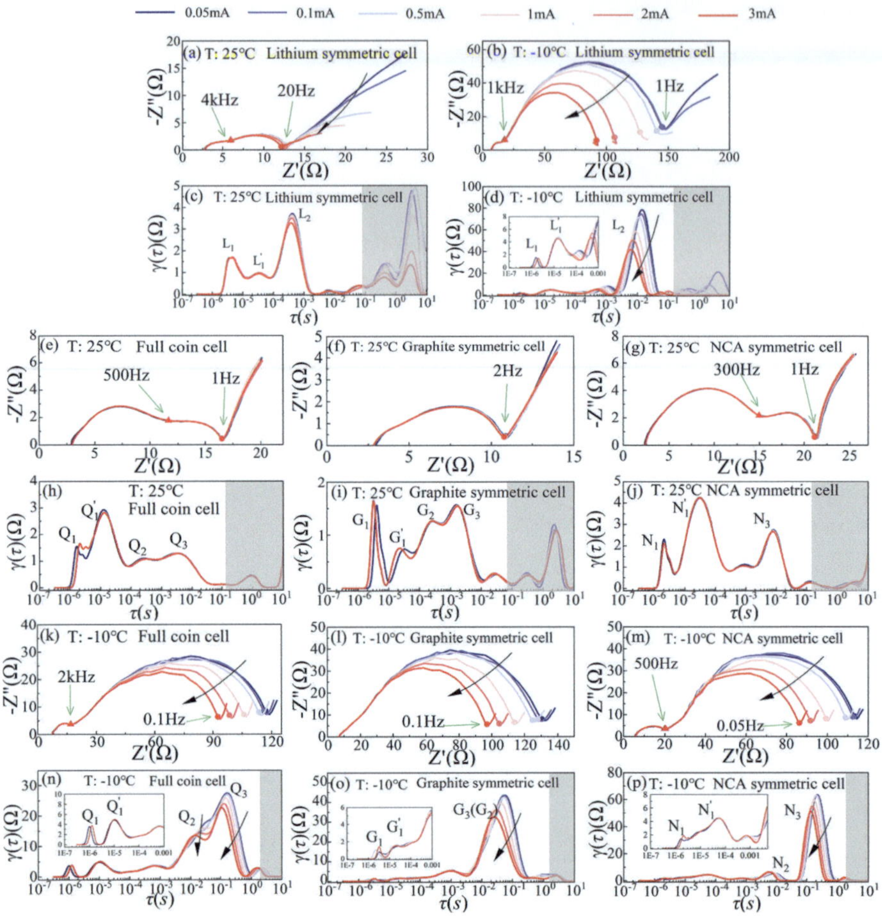

Fig. 3.13 The impedance characteristics and corresponding DRT analysis for various cell configurations under different AC amplitudes: **a–d** lithium symmetric cell, **e**, **h**, **k**, **n** full coin cell, **f**, **i**, **l**, **o** graphite symmetric cell, and **g**, **j**, **m**, **p** NCA symmetric cell [13]

exhibiting minimal sensitivity to thermal or SOC variations. Comparative analysis links these stable conduction phenomena to specific spectral features: F_1 peaks in full-cell configurations, dual A_1/A_1' signatures in graphite anodes, and paired C_1/C_1' responses in NCA cathodes. Furthermore, the presence of two distinct peaks in the conduction process of half-cells is likely originate from interfacial contact heterogeneities between constituent materials in coin-type assemblies.

As illustrated in Fig. 3.12, anode and cathode also exhibit peaks within the 10^{-4} s to10^{-3} s range at 25 °C, labeled A_2 and C_2, respectively. These peaks do not correspond to the conduction process, creating a discrepancy when compared to the full cell. Similarly, Fig. 3.13 reveals the presence of Q_2 in full cell within the same time constant range. A comparative analysis of Fig. 3.13h, i, and j indicates that Q_2 stems from G_2 in the graphite symmetric cell. According to previous studies [24, 38], the SEI process time constant at room temperature falls approximately between 1.6×10^{-4} s and 10^{-2} s. Consequently, G_2 in the graphite symmetric cell can be attributed to the SEI process. Likewise, CEI process in NCA exhibits a time constant around 1.6×10^{-4} s [23]. Since CEI layers are generally thinner than SEI layers [39], their corresponding time constant is shorter, making the CEI process less prominent in the DRT results of Fig. 3.13j. C_2's time constant aligns with CEI process in Fig. 3.12i, yet it is not directly classified as such due to the presence of the half-cell lithium electrode.

Detailed examination of Fig. 3.13 reveals lithium symmetric cells exhibit three relaxation signatures: conduction-associated L_1/L_1' and a supplementary peak L_2 (10^{-4}~10^{-3} s). Prior studies [40, 41] attribute L_2 to Li^+ migration across lithium metal's solid electrolyte interphase (SEI). This correlation suggests NCA cathode peak C_2 at 25 °C originates from lithium electrode interactions, while graphite anode peak A_2 represents a composite response involving both graphite and lithium SEI dynamics. Consequently, full-cell peak F_1 inherently incorporates graphite SEI contributions. Comparative analysis identifies charge transfer as the dominant kinetic process in medium/high-frequency regimes, demonstrating superior temporal persistence among interfacial phenomena.

Research [14, 23, 24, 38] indicates that under ambient conditions, graphite anodes exhibit charge transfer time constants of 3.2×10^{-3}~10^{-1} s, contrasting with NCA cathodes' 5.3×10^{-2}~3.98×10^{-1} s range. Within these temporal boundaries, relaxation signatures A_3 (graphite) and C_3 (NCA) correspond conclusively to electrode-specific charge transfer dynamics. This correlation finds experimental validation in symmetric cell configurations (Fig. 3.13i and j), where analogous peaks G_3 (graphite) and N_3 (NCA) emerge, with direct A_3–G_3 and C_3–N_3 parallels confirming the association. At 25 °C, full-cell signature F_2 demonstrates minimal SOC dependence within 10^{-3}~10^{-2} s, showing temporal alignment with graphite's A_3 signatures. Conversely, NCA's C_3 displays SOC-modulated behavior: confined to 10^{-3} s~10^{-2} s at 50–80% SOC but expanding significantly (10^{-2} s~3×10^{-1} s) at 20% SOC. This SOC-dependent temporal expansion mirrors observations from Shafiei Sabet et al. [23], who established lithium occupancy effects on NCA charge transfer kinetics. Crucially, the full-cell F_3 signature's emergence within NCA's characteristic timeframe confirms F_3's representation of NCA charge transfer mechanisms.

At −10 °C, the electrochemical reactions within the cell slow down significantly, leading to an increase in reaction time constants. The impedance variations in both full and symmetric cells, as illustrated in Fig. 3.13, highlight these changes at −10 °C. Specifically, the lithium symmetric cell exhibits peak L_2 within the time range of 3×10^{-3} to 3×10^{-2} s, which aligns closely with the time constants of peaks A_2 and C_2 at −10 °C in Fig. 3.12. Similarly, the graphite symmetric cell in Fig. 3.13 displays a peak spanning 3×10^{-3} s to 3×10^{-1} s. Given that the lithium metal SEI process closely resembles that of graphite [41], it is reasonable to expect that the SEI process in graphite should also fall within 3×10^{-3} s to 3×10^{-2} s range at −10 °C. The temporal overlap between graphite's SEI formation and charge transfer mechanisms [23] suggests their combined manifestation in a singular relaxation signature, corresponding to peaks A_2/A_3 in Fig. 3.12. At 25 °C, graphite's A_2 peak reflects interfacial contributions from both graphite and lithium metal SEI layers. Low-temperature analysis (−10 °C) reveals an NCA-specific CEI-related peak N_2 (2×10^{-3}~2×10^{-2} s) alongside charge transfer signature N_3 in symmetric cells (Fig. 3.13), directly mirroring full-cell peaks C_2/C_3 (Fig. 3.12l). These correlations establish C_2 as a composite indicator of NCA CEI and lithium SEI processes, while C_3 exclusively represents NCA charge transfer kinetics. Comparative evaluation of full-cell configurations demonstrates near-identical −10 °C behavior except for conduction process variations. At 20% SOC, F_2 spans 4×10^{-3}~10^{-1} s, maintaining comparable temporal distribution at higher SOCs (50–80%) despite partial obscuration by F_3 overlap (Fig. 3.12j and p). This regime exhibits temporal alignment between graphite's A_2 and F_2, with partial congruence from NCA's C_2. At −10 °C, full-cell peak F_2 arises from the synergistic interaction between graphite anode SEI development and NCA cathode CEI formation. Under 20% SOC conditions, F_3 demonstrates extended temporal persistence compared to higher SOC states, paralleling the SOC-dependent behavior of peak C_3. Figure 3.12d delineates F_3's temporal domain as spanning 10^{-1}~5 s, while Fig. 3.12e and f reveal concurrent temporal convergence of graphite's A_3 and NCA's C_3 signatures. This pattern persists across 50–80% SOC conditions. Critical examination of Fig. 3.12j, k, l, p, q, and r confirms F_3's composite nature, representing the combined electrochemical responses of both A_3 and C_3 mechanisms.

Table 3.7 outlines the physical interpretations and peak assignments across cell configurations, with data for the full coin cell and graphite/NCA symmetric cells corresponding to a 50% SOC. This clearly established peak categorization forms the basis for probing AC amplitude-dependent variations in electrochemical reaction dynamics.

Table 3.7 Summary of different DRT peaks [13]

	Conduction process	SEI process	CEI process	Anode charge transfer	Cathode charge transfer
Graphite anode	A_1/A_1'	A_2	–	A_3	–
NCA Cathode	C_1/C_1'	–	C_2	–	C_3
Graphite symmetric cell	G_1/G_1'	G_2	–	G_3	–
NCA symmetric cell	N_1/N_1'	–	N_2(−10 °C)	–	N_3
Lithium symmetric cell	L_1/L_1'	L_2	L_2	–	–
Full coin cell	Q_1/Q_1'	Q_2	Q_2(−10 °C)	Q_3	Q_3
Full cell	F_1	F_1(25 °C) F_2(−10 °C)	F_2(−10 °C)	F_2(25 °C) F_3(−10 °C)	F_2(25 °C, SOC:0.5 and 0.8)/F_3

3.2.3 Coupling Law of Impedance with Current and Temperature

It is generally recognized that the high-frequency component of electrochemical impedance in batteries is primarily caused by the ionic conductivity of the solid–liquid phases and the interfacial impedance of SEI. This component is significantly influenced by the battery's cycle life [42] and low-temperature conditions [43]. For the tested samples, it can be assumed that when the frequency exceeds the critical point, the ionic conductivity and interfacial impedance are only affected by other state variables and are independent of the excitation current magnitude. Gaberscek [10] using a 200 mV AC voltage excitation for impedance testing, concluded that this impedance is caused by interfacial contact resistance and remains unaffected by the excitation conditions.

The mid-frequency impedance arc of a battery represents the electrochemical reaction process, with the associated impedance typically denoted as R_{ct}, which reflects the charge transfer resistance. To ensure the linearity and time-invariance requirements of impedance spectrum measurements, small currents or voltages are generally used for perturbation. However, in AC heating methods aimed at achieving rapid and efficient temperature rise, larger currents are often employed to excite the battery. At low temperatures, the internal resistance is dominated by R_{ct}. Understanding the variation of R_{ct} during AC heating is crucial for exploring the mechanisms of heat generation in this context. In previous experiments, a phenomenon was observed for the first time in which the mid-frequency impedance arc gradually shrinks under high-current excitation.

To investigate the coupling relationships between impedance, current, and temperature from a reaction mechanism perspective, differences in excitation current amplitude were analyzed. Unlike the charge transfer resistance obtained from impedance spectrum measurements using small currents or voltages, a new parameter, R_{ct_spec}, is introduced in this study to represent the charge transfer resistance derived from the measured impedance spectrum.

Figures 3.14 and 3.15 illustrate the effects of temperature and excitation current amplitude on R_{ct_spec}. From Fig. 3.14, it can be observed that under low-current excitation, the charge transfer resistance decreases exponentially with increasing temperature. However, under high-current excitation, R_{ct_spec} in Fig. 3.14b no longer follows an exponential relationship with temperature. Figure 3.15 shows that R_{ct_spec} decreases progressively with increasing excitation current, and this effect is particularly pronounced under low-temperature conditions. The following sections systematically discuss and analyze these experimental phenomena based on the Arrhenius equation and the Butler–Volmer equation.

The relationship between charge transfer resistance and temperature has been extensively studied by various researchers [5, 44]. In summary, it can be described using the Arrhenius empirical equation, which is defined in this section as:

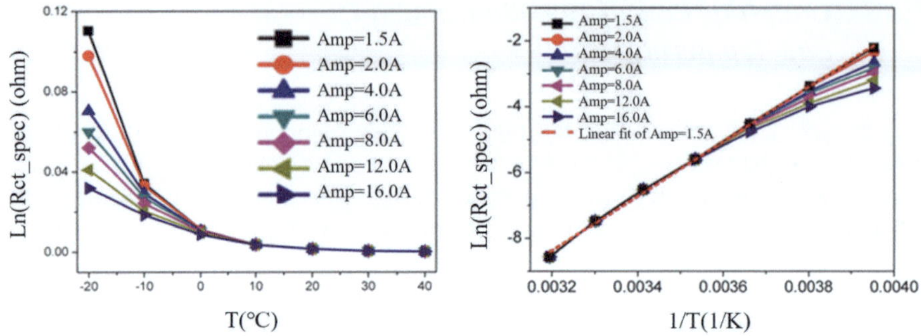

Fig. 3.14 The effect of excitation current on charge migration resistance [18]

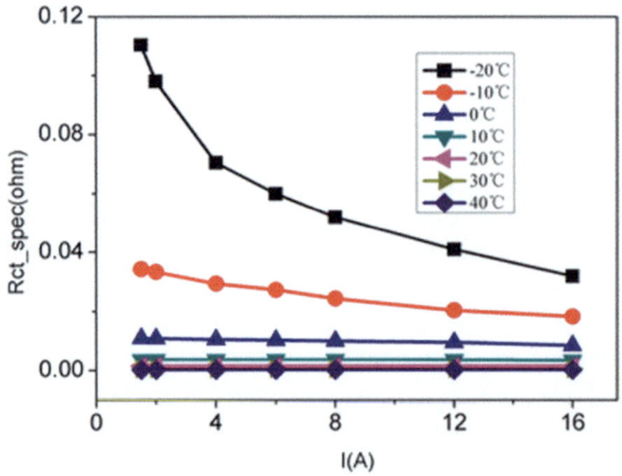

Fig. 3.15 The effect of temperature on charge migration resistance [18]

$$R_{ct\_spec}(t) = A_4 \exp(\frac{Ea}{RT}) \tag{3.24}$$

where Ea represents the activation energy of the electrochemical reaction, A_4 is the pre-exponential factor for the charge transfer resistance, and T is the temperature. From Fig. 3.14b, it is observed that when the excitation current is 1.5 A or 2 A, the relationship between charge transfer resistance and temperature satisfies the exponential form described by the equation. However, when the excitation current exceeds 2 A, especially under low-temperature conditions, the Eq. (3.24) fails to accurately describe the effect of temperature on impedance. Hence, it is necessary to further explore the relationship between excitation current amplitude and impedance and modify the equation accordingly.

The current at the electrode reaction interface can be calculated using the Butler–Volmer equation. One form of its expression is:

$$I_F = Si_0\left[\exp(\frac{\alpha_a F\eta}{RT}) - \exp(-\frac{\alpha_c F\eta}{RT})\right] \tag{3.25}$$

where I_F is the Faradaic charge transfer current, S is the active surface area of the electrode, α_a and α_c are the anodic and cathodic transfer coefficients, F is the Faraday constant, R is the gas constant, and η is the overpotential for the electrode reaction.

The overpotential η is expressed in Eq. (3.8). For the convenience of further derivation and analysis, the overpotential for electrode reactions is redefined as:

$$\eta = \phi_s - \phi_l - E_{eq}(SOC) \tag{3.26}$$

where ϕ_s is the solid-phase potential of the electrode, ϕ_l is the liquid-phase potential of the electrode, and $E_{eq}(SOC)$ is the equilibrium potential of the electrode, which is related to SOC.

The Butler–Volmer equation effectively describes the electrode kinetics at the electrode–electrolyte interface. Waag et al. [45, 46] and other researchers first proposed an approximation method for the Butler–Volmer equation and used it for the online estimation of dynamic battery state parameters.

According to this method, $R_{ct}(I)$ can be described as:

$$R_{ct}(I_F) = \frac{\eta}{I_F} \tag{3.27}$$

Using the mathematical transformation of $\sinh(x) = (e^x - e^{-x})/2$ the Butler–Volmer equation can be reformulated as:

$$I_F = 2Si_0 \sin(\frac{\alpha F}{RT}\eta) \tag{3.28}$$

where $\alpha = \alpha_a = \alpha_c = 0.5$.

Through further transformation of $\sinh^{-1}(x) = \ln(x + \sqrt{x^2 + 1})$, the above equation can be expressed as

$$\frac{\alpha F}{RT}\eta = \ln[\frac{1}{2Si_0}I_F + \sqrt{(\frac{1}{2Si_0}I_F)^2 + 1}] \tag{3.29}$$

From Eq. (3.27), the charge transfer resistance R_{ct} can be obtained as:

$$R_{ct} = R_{ct,0}\frac{\ln[\frac{1}{2Si_0}I_F + \sqrt{(\frac{1}{2Si_0}I_F)^2 + 1}]}{\frac{1}{2Si_0}I_F} \tag{3.30}$$

where $R_{ct,0}$ is defined as a parameter dependent on temperature but independent of current [47]. To simplify the equation, $R_{ct,0}$ is replaced by Eq. (3.24), leading to the following relationship

$$R_{ct} = A \cdot \exp(\frac{Ea}{RT}) \cdot \frac{\ln[\frac{1}{2Si_0} I_F + \sqrt{(\frac{1}{2Si_0} I_F)^2 + 1}]}{\frac{1}{2Si_0} I_F} \tag{3.31}$$

According to porous electrode theory [47–49], the exchange current density i_0 is independent of the applied current and can be described as:

$$i_0 = BFk \tag{3.32}$$

where k is the chemical reaction rate constant, and B is a function related to lithium-ion concentration.

The reaction rate constant k is described using the Arrhenius law, which for the convenience of derivation in this section, is defined as:

$$k = A_3 \cdot \exp\left(\frac{-Ea}{RT}\right) \tag{3.33}$$

Based on the above analysis, Eq. (3.31) can be further expressed as:

$$R_{ct} = f(T, I_F) = A \cdot \left\{ \frac{\ln\left[\frac{1}{2SA/BF} \cdot \exp\left(\frac{Ea}{RT}\right) \cdot I_F + \sqrt{\left(\frac{1}{2SA/BF} \cdot \exp\left(\frac{Ea}{RT}\right) \cdot I_F\right)^2 + 1}\right]}{\frac{1}{2SA/BF} I_F} \right\} \tag{3.34}$$

$$R_{ct\_spec} = f(T, I) = A \cdot \left\{ \frac{\ln\left[\frac{1}{2SA/BF} \cdot \exp\left(\frac{Ea}{RT}\right) \cdot I + \sqrt{\left(\frac{1}{2SA/BF} \cdot \exp\left(\frac{Ea}{RT}\right) \cdot I\right)^2 + 1}\right]}{\frac{1}{2SA/BF} I} \right\} \tag{3.35}$$

From the impedance spectra, it is observed that at mid-frequency arc and the low-frequency inflection point, the imaginary part of the impedance approaches zero. The non-Faradaic current through the electric double layer can thus be neglected. Therefore, it can be approximated that I_F is approximately equal to the excitation current, leading to Eq. (3.35).

As illustrated in Figs. 3.14 and 3.15, Eq. (3.35) provides an accurate description of the coupling relationship between charge transfer resistance, temperature, and excitation current. Fig. 3.16 describes a comparison between the experimental data and the simulated results calculated using Eq. (3.35). In Fig. 3.16b, ΔR_{ct_spec} represents the error between the experiment and the equation. Table 3.8 summarizes the evaluation of the fit between the experimental data and the derived Eq. (3.35). From Fig. 3.16 and Table 3.8, it is concluded that Eq. (3.35) effectively describes the phenomenon of the mid-frequency arc contraction observed in Figs. 3.4 and 3.7.

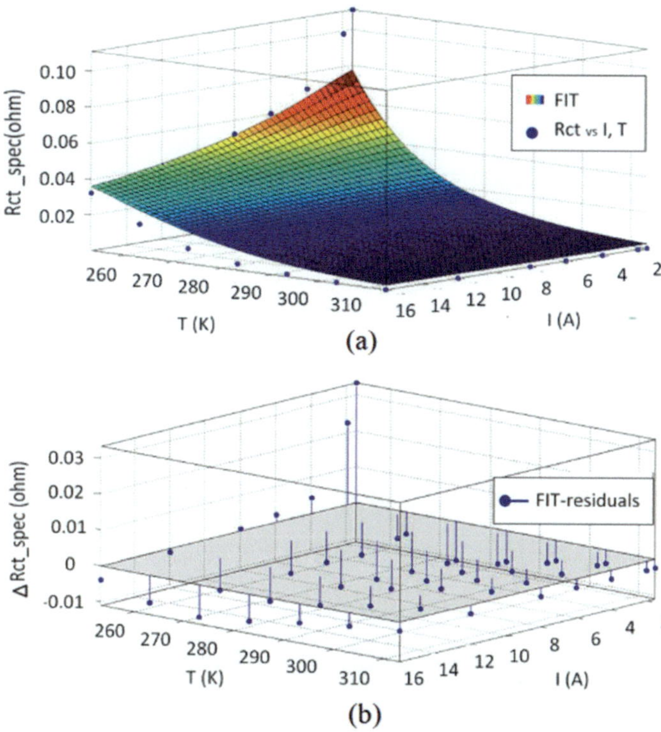

Fig. 3.16 Comparison of test experiments and simulation of derivation formulae [18]

Table 3.8 Evaluation for fitting the simulation relationship between test experiments and derivation equations [18]

Metrics	Values
Sum of squared errors (SSE)	0.003752
Root mean square error (RMSE)	0.009031
R-square	0.8748
Adjusted-R-square	0.8694

It provides an accurate representation of the coupling relationship between charge transfer resistance, excitation current, and temperature.

To validate the accuracy of the aforementioned formula, the relationship between the impedance and current amplitude for NCA full cell and its positive and negative half cells was fitted using Eq. (3.35), as shown in Fig. 3.17. The results confirm the correctness of the deduction.

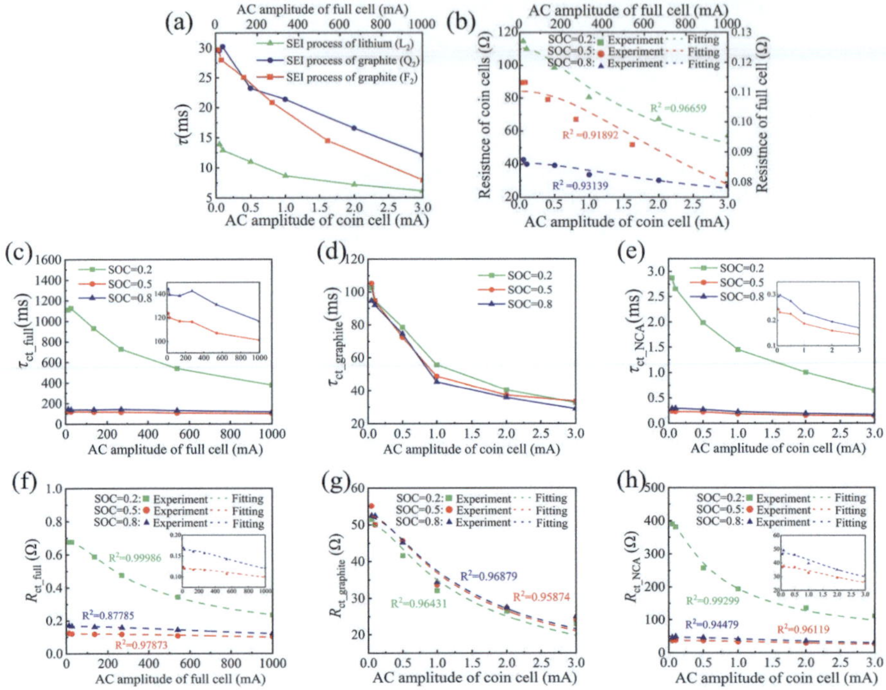

Fig. 3.17 Variations in τ and R with different AC amplitudes for interfacial processes. **a, b** the SEI processes for lithium and graphite under −10 °C; **c, f** the charge transfer process of the full cell under −10 °C; **d, g** the charge transfer process for graphite under −10 °C; **e, h** the charge transfer process for NCA under −10 °C [13]

3.3 Effect of AC Heating Parameters on Battery Temperature Rise

The principle of heat production in batteries and the nonlinear characteristics of electrode processes at low temperatures are analyzed above to provide a theoretical basis for AC heating. This section will discuss in detail the effect of different AC heating parameters on the battery temperature rise. Currently, there are many AC heating waveforms. Relevant studies have shown that compared with other multiple heating waveforms, bidirectional pulse current (BPC) has the highest heating efficiency. Therefore, BPC are directly utilized in this section for analysis. The traditional BPC heating waveform is symmetrical, with the same charging and discharging times and amplitudes in one cycle, which makes it easier to implement in heating devices. However, the heating effect of asymmetric BPC has not been studied much. Unlike the traditional symmetric BPC, the asymmetric BPC has different charging and discharging times and amplitudes in one cycle. Therefore, the study of heating parameters on battery temperature rise in this chapter is divided into two parts: symmetric BPC heating and asymmetric BPC heating, according to the different heating waveforms. Since asymmetric BPC heating is difficult to realize using current heating equipment, only a model is used to analyze the effect of its parameters on the battery heating power.

3.3.1 Experimental Design for Different AC Heating Parameters

3.3.1.1 Experimental Design for Symmetric BPC Heating

Below are the test samples and experimental procedures for symmetric BPC heating. The power battery samples used in the study are sample E, a 30 Ah lithium iron phosphate ($LiFePO_4$) battery, and sample H, an 8Ah $LiNi_{1/3}Co_{1/3}Mn_{1/3}O_2$ battery. The specific parameters of sample E are shown in Chap. 2. Detailed specifications of sample H are shown in Table 3.9. For Sample E, internal temperature is measured by embedding a thermocouple at the geometric center, while five thermocouples are affixed to the surface to monitor surface temperatures.

First, impedance tests were conducted on both power battery samples under high-current excitation at low temperatures to obtain their impedance characteristics. Based on these characteristics, the excitation parameters for alternating current (AC) heating—frequency, amplitude, and voltage protection thresholds—were selected.

For the 30 Ah $LiFePO_4$ battery (Sample E), the excitation frequencies were set to 300, 30, 10, 2, 1, and 0.1 Hz, with excitation current amplitudes of 20, 40, and 60 A. Considering the differences between AC excitation and constant-current charge/discharge in electrode processes and other aspects, tests were conducted with different voltage protection thresholds to analyze the effect of these limits on temperature rise. The upper and lower voltage protection thresholds were set to 3.7 V/2.5 V, 3.8 V/2.4 V, and 3.9 V/2.3 V, with 3.7 V/2.5 V being the typical threshold for constant-current charge/discharge tests in $LiFePO_4$ systems. The SOC was set to 50% during the tests.

For the 8Ah $LiNi_{1/3}Co_{1/3}Mn_{1/3}O_2$ battery (Sample H), the AC heating excitation frequencies were 300, 100, 30, 10, 5, and 1 Hz, with current amplitudes of 24, 40,

Table 3.9 Specification of sample H

	Sample H
Material system	$LiNi_{1/3}Co_{1/3}Mn_{1/3}O_2/C$
Nominal capacity (Ah)	8
Nominal voltage	3.7
Charging cutoff voltage	4.2
Dimensions (length × width × thickness, mm)	140 × 115 × 8
Manufacturer	CITIC GUOAN MGL

Table 3.10 Values of different heating excitation parameters for symmetric BPC heating methods

Test subject	Adjustable factor		Fixed factor
Sample E	Excitation frequency (Hz)	300, 30, 10, 2, 1, 0.1	Waveform: Square wave; SOC: 50%; $T_{Am} = -25$ °C
	Excitation current amplitude (A)	20, 40, 60	
	Voltage limit threshold (V)	3.7/2.5, 3.8/2.4, 3.9/2.3	
Sample H	Excitation frequency (Hz)	300, 100, 30, 10, 5, 1	
	Excitation current amplitude (A)	24, 40, 64	
	Voltage limit threshold (V)	4.2/2.75, 4.3/2.65, 4.2/2.25	

and 64 A. The upper and lower voltage protection thresholds were set to 4.2 V/2.75 V, 4.3 V/2.65 V, and 4.4 V/2.55 V, where 4.2 V/2.75 V represents the typical threshold for constant-current charge/discharge tests in this system. Table 3.10 gives the additional experimental parameter matrices.

3.3.1.2 Experimental Design for Asymmetric BPC Heating

1) Experiment Platform

Figure 3.18 illustrates the experimental setup and tests conducted in this study. The battery testing platform comprises Hongrun OHR-XH710 (temperature acquisition recorder), Binder MK56 (temperature chamber), a host computer, EnergyLab XM (electrochemical workstation), and Solartron Analytical (power booster). The test samples are 1 Ah nickel–cobalt-manganese (NCM)/graphite lithium-ion battery (Sample I), operating within a voltage range of 3–4.2 V. The temperature acquisition recorder tracks the battery's surface temperature, whereas the EnergyLab XM conducts both EIS and asymmetric BPC tests. Given the restricted current range of the EnergyLab XM, a power booster is employed to augment current capability.

2) EIS Test across Diverse Temperatures and AC Amplitudes

Impedance is influenced by temperature and current amplitude. To capture the parameters of the fractional-order ECM, EIS tests are conducted over various temperatures and AC amplitudes. Before starting the EIS tests, 20% SOC is set. As noted in reference [51], lithium plating is more likely under high-rate charging and low-temperature conditions. Moreover, a higher SOC can increase the risk of lithium plating, as highlighted in reference [52]. Since battery impedance is affected by

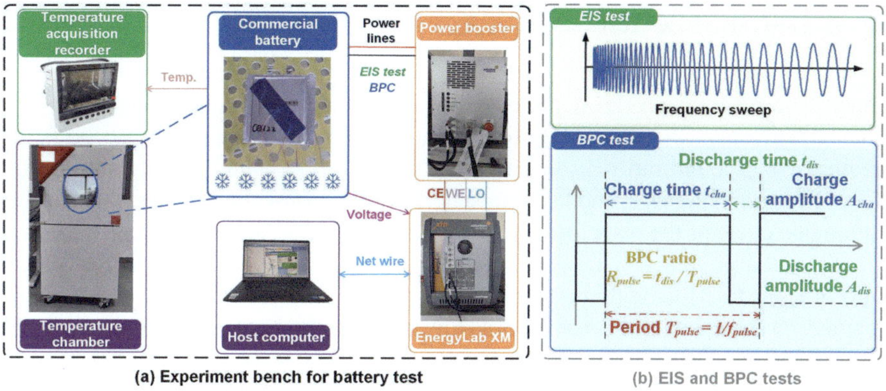

Fig. 3.18 Experiment platform and tests [50]

both temperature and current amplitude, EIS tests are carried out across a range of temperatures and AC amplitudes to extract the parameters of the fractional-order equivalent circuit model. Prior to conducting the EIS tests, the SOC of the battery is set at 20%. According to reference [51], lithium plating is more prone to occur under high-rate charging and low-temperature conditions. Additionally, a high SOC further exacerbates the risk of lithium plating [52]. Considering that the EIS testing procedure includes a charging phase, lithium plating could potentially occur during high-current EIS tests. So, a lower SOC is chosen. Nevertheless, an overly low SOC might lead to over-discharge while testing. To strike a balance, 20% SOC of is selected. The preparation process begins by charging the battery to its full capacity using a CCCV charging method, with a charge current set at 1 A and a termination current of 0.02 A, while holding the voltage constant at 4.2 V. After a 30-min resting period, the battery is discharged at a constant current of 1 A until the discharged capacity reaches 0.8Ah, corresponding to an SOC of 20%. Given that the actual capacity of the battery is 1.05 Ah, the true SOC is approximately 23.8%. The impact of SOC variations is not a focus of this study, and since the discrepancy is only 3.8%, it is considered negligible. Impedance measurements are conducted in galvanostatic mode within a frequency range of 0.1–10^6 Hz. Many studies [53, 54] have explored AC heating strategies for batteries under low-temperature conditions, where the maximum heating temperature is generally around 10 °C. The minimum testing temperature is usually selected based on the expected operational range of the battery. Furthermore, prior research has shown that battery parameters are commonly calibrated at 5 °C intervals. Given that the battery under investigation has a minimum discharge temperature of −15 °C, this study selects test temperatures of 10, 5, 0, −5, −10, and −15 °C. Before each impedance test, the battery is stabilized at the target temperature for 2 h. Impedance tests are performed at various AC amplitudes, specifically 0.25, 0.65, 1, 1.5, 2, and 2.5 A. To maintain consistency in testing, a 1-h break is observed between consecutive impedance tests conducted at the same temperature but with varying amplitudes.

3). Asymmetric BPC Test

Prior to initiating the asymmetric BPC test, the critical parameters must be predetermined. These parameters are depicted in Fig. 3.18, which encompass charge time t_{cha}, discharge time t_{dis}, charge amplitude A_{cha}, and discharge amplitude A_{dis}, period T_{pulse}, frequency f_{pulse}, and ratio R_{pulse}. The definitions of T_{pulse} and f_{pulse} are straightforward. The parameters t_{cha} and t_{dis} represent the durations of charging and discharging within one cycle of the waveform, respectively, while A_{cha} and A_{dis} denote the corresponding charge and discharge magnitudes. Additionally, R_{pulse} is defined as the ratio of f_{pulse} to T_{pulse} within a single waveform cycle. An accurate electro-thermal coupled model is essential for reliably quantifying the heat generation inside the battery. The precision of the equivalent circuit model can be gauged by examining the deviation between the model-predicted terminal voltage and the experimentally measured terminal voltage. Additionally, the validity of the heat generation calculation is verified by comparing the temperature rise predicted by the electro-thermal coupled model with the experimentally measured temperature rise. Correspondingly, the asymmetric BPC test is divided into two categories: voltage testing and temperature testing. For voltage testing, the test temperatures are set at 10, −5, and −15 °C, with the specific parameters for different asymmetric BPCs at each temperature detailed in Table 3.11. The testing procedure encompasses three distinct asymmetric BPC frequencies: 1, 50, and 100 Hz. For each frequency, two unique asymmetric BPC profiles—designated as BPC I and BPC II—are utilized. BPC I is characterized by a discharge current of 3.5 A and a charge current of 1.5 A, with a pulse ratio of 0.5. Conversely, BPC II is defined by a discharge and charge current of 2.5 A each, along with a pulse ratio of 0.7. During the voltage tests, the EnergyLab XM records both current and voltage data at a sampling rate of 100 times the respective asymmetric BPC frequency. Additionally, a 2-h stabilization period follows each temperature adjustment, and a 30-min interval is maintained between consecutive asymmetric BPC tests at the same temperature. Using the 10 °C test scenario as a reference, the battery is first acclimated in a 10 °C environment for 2 h to reach thermal equilibrium before the test begins. Next, the BPC I @ 1 Hz test is carried out using the EnergyLab XM and the power booster. Thanks to the high sampling frequency of the EnergyLab XM, ample current and voltage data can be collected within just a few seconds.

Table 3.11 The asymmetric BPCs for voltage testing [50]

Frequency f_{pulse} (Hz)	Ratio R_{pulse}	Discharge amplitude A_{dis} (A)	Charge amplitude A_{cha} (A)	Abbreviation
1	0.5	3.5	1.5	I @ 1 Hz
	0.7	2.5	2.5	II @ 1 Hz
50	0.5	3.5	1.5	I @ 50 Hz
	0.7	2.5	2.5	II @ 50 Hz
100	0.5	3.5	1.5	I @ 100 Hz
	0.7	2.5	2.5	II @ 100 Hz

Table 3.12 The asymmetric BPCs for temperature testing [50]

Frequency f_{pulse} (Hz)	Ratio R_{pulse}	Discharge amplitude A_{dis} (A)	Charge amplitude A_{cha} (A)	Abbreviation
1	0.5	2.5	2.5	III @ 1 Hz
	0.7	1.5	3.5	IV @ 1 Hz
10	0.5	2.5	2.5	III @ 10 Hz
	0.7	1.5	3.5	IV @ 10 Hz
100	0.5	2.5	2.5	III @ 100 Hz
	0.7	1.5	3.5	IV @ 100 Hz

After completing the BPC I @ 1 Hz test, the battery is given a 30-min rest period to ensure stabilization. The BPC II @ 1 Hz test is subsequently conducted using the same approach. Table 3.11 lists the same procedure is applied to the remaining BPCs. After completing the BPC II @ 100 Hz test, voltage testing at 10 °C is finished. The temperature is then adjusted to the other specified values for further testing. Once measurements at −5 and −15 °C are completed, the voltage testing is considered concluded. In contrast to voltage testing, temperature testing only requires obtaining the temperature rise under the various asymmetric BPCs listed in Table 3.12. High-frequency sampling is unnecessary. The asymmetric BPCs utilized in the temperature tests feature three distinct frequencies. For each frequency, two variants of asymmetric BPCs are deployed: BPC III and BPC IV. The ratio for BPC III is set at 0.5, while that for BPC IV is 0.7. For BPC III, both the discharge and charge amplitudes are 2.5 A, whereas for BPC IV, the discharge and charge amplitudes are 1.5 A and 3.5 A, respectively. The temperature testing follows a structured procedure. First, the cell is placed under −13 °C for 2 h to ensure thermal equilibrium. Subsequently, the BPC III @ 1 Hz is applied using the EnergyLab XM and the power booster, while the battery surface temperature is simultaneously recorded. After 30 min, the heating process is terminated, and the cell rests for 1 h before undergoing the BPC IV @ 1 Hz test. The same procedure is followed for the remaining asymmetric BPCs. Upon completion of the BPC IV @ 100 Hz test, the temperature testing is considered concluded.

3.3.2 *Results and Analysis*

3.3.2.1 Symmetric BPC Heating

1) Analysis of Temperature Rise Variation under Different Heating Parameters

Figure 3.19 illustrates the temperature rise curves for the two lithium-ion power battery samples under 30 and 1 Hz excitation. The horizontal axis represents heating time, while the vertical axis shows the average temperature obtained from surface

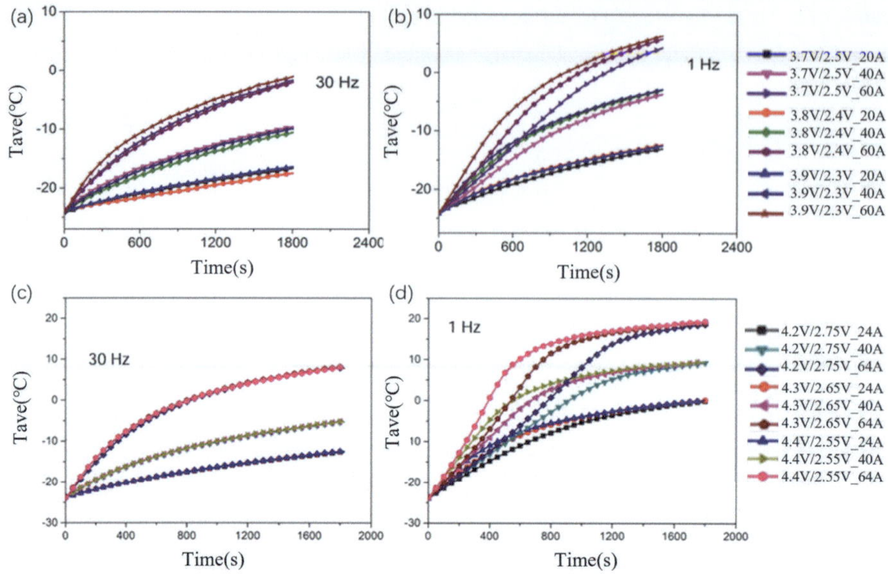

Fig. 3.19 Temperature rise of AC heated batteries under different parameter excitations. **a** Temperature rise profile of 30 Ah $LiFePO_4$/C battery under 30 Hz excitation, **b** Temperature rise profile of 30 Ah $LiFePO_4$/C battery under 1 Hz excitation, **c** Temperature rise profile of 8Ah $LiNi_{1/3}Co_{1/3}Mn_{1/3}O_2$ battery under 30 Hz excitation, **d** Temperature rise profile of 8Ah $LiNi_{1/3}Co_{1/3}Mn_{1/3}O_2$/C battery under 1 Hz excitation

thermocouples. In the initial stage of AC heating, the battery temperature rises rapidly. As heating continues, the rate of temperature increase gradually decreases. The temperature rise significantly increases with higher excitation current. For sample E, the 30 Ah $LiFePO_4$ battery, at 1 Hz excitation frequency and 60 A current, the temperature rises from −25 to 5 °C within 1800s. For sample H, the 8Ah $LiNi_{1/3}Co_{1/3}Mn_{1/3}O_2$ battery, under the same frequency and a current of 64 A, the temperature increases from −25 °C to nearly 20 °C. Since the heat generation rate is proportional to the root-mean-square (RMS) value of the excitation current, increasing the excitation current amplitude effectively enhances the heating rate. Under low-frequency range (1 Hz), the temperature rise curves under different voltage protection thresholds show significant differences for both battery samples. Narrow voltage protection thresholds result in poorer heating performance. This is mainly because of the high impedance under low-frequency excitation. When large current excitation is applied, the voltage protection threshold is quickly triggered, causing the current amplitude to be adjusted and reducing the heating effect.

Figure 3.20 shows the corrected current waveforms in the initial stage for sample E at 1 Hz and 60 A under different voltage protection thresholds. From 3.7 V/2.5 V to 3.9 V/2.3 V, the correction of the excitation current decreases as the voltage protection thresholds widen. When the thresholds are set to 3.9 V/2.3 V, the upper limit is no longer triggered, as indicated by the waveform in the black frame of Fig. 3.20.

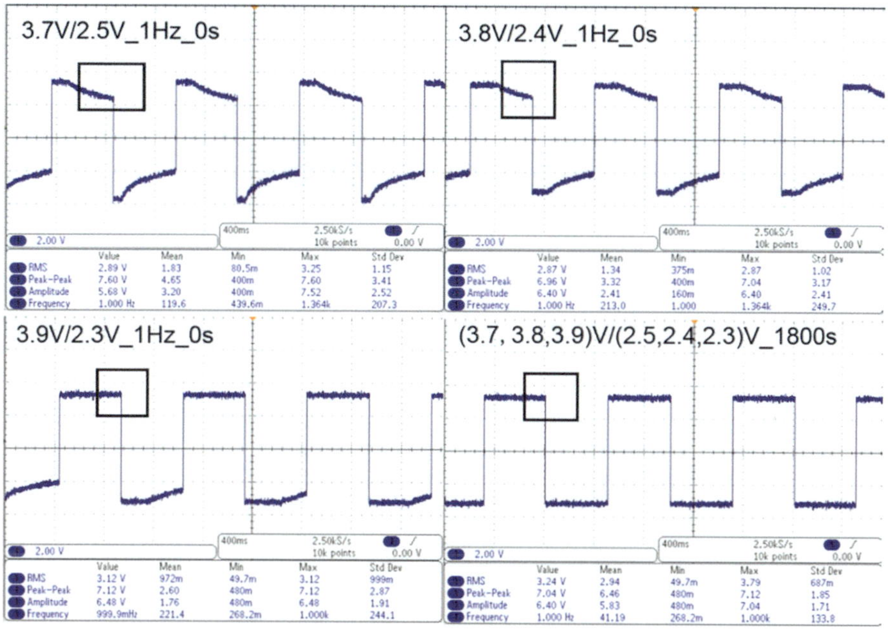

Fig. 3.20 Corrected waveform of the excitation current waveform of sample E at different voltage limiting thresholds

Widening the voltage protection thresholds helps to increase the effective RMS value of the excitation current, thereby improving the heating performance. Sample H exhibits the same pattern, which is not repeated here. As the temperature rises, its impedance decreases, reducing the effect of voltage protection corrections until they are no longer active. This aligns with the consistent temperature rise behavior observed at the end of heating (1800 s) in Fig. 3.19.

To further analyze the comprehensive effects of excitation current, frequency, and voltage protection thresholds on the temperature rise of lithium-ion power batteries, 3D relationship diagrams of the temperature rise results for samples C and D are shown in Figs. 3.21 and 3.22. These figures provide a clear depiction of the following conclusions. The excitation current amplitude has the most pronounced effect on the AC heating temperature rise, with high-current excitation promoting rapid heat accumulation and temperature increase. Due to the relatively high impedance of batteries at low frequencies, low-frequency excitation achieves better heating effects compared to high-frequency excitation. However, low-frequency high-current excitation often triggers voltage protection thresholds, significantly reducing the heating effect.

As shown in Fig. 3.21a and Fig. 3.22a, during the initial heating stage, when the battery's internal resistance is high at low temperatures, the current is adjusted by the voltage protection thresholds. This results in identical effective current inputs and thus similar temperature rise outcomes under different excitation amplitudes. With

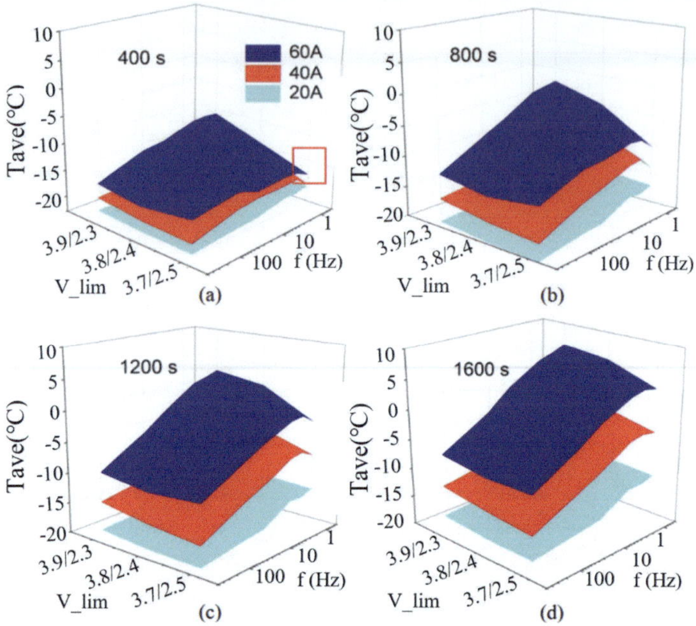

Fig. 3.21 Three-dimensional plot of the effect of three elements of sample E cell excitation on temperature rise

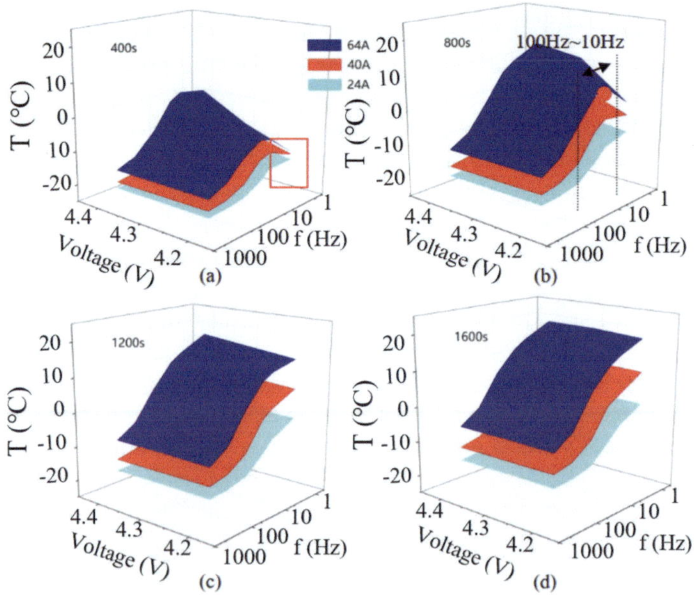

Fig. 3.22 Three-dimensional plot of the effect of three elements of sample H cell excitation on temperature rise

increasing battery temperature and decreasing internal resistance, the differences in temperature rise caused by varying voltage protection thresholds gradually diminish.

In general, the AC heating temperature rise is jointly influenced by excitation current amplitude, excitation frequency, and voltage protection thresholds. Under fixed-frequency excitation, at the 800-s mark (Fig. 3.22b), the optimal heating effect is observed in the frequency range of 100–10 Hz. Comparing Figs. 3.21 and 3.22 shows that sample H (8Ah $LiNi_{1/3}Co_{1/3}Mn_{1/3}O_2$) achieves better temperature rise performance than sample E (30 Ah $LiFePO_4$), which is primarily attributed to differences in their impedance characteristics, volume, and thermal capacity.

2) Experimental Analysis of Temperature Gradient between Interior and Exterior

The AC heating method relies on the heat generation of cells, theoretically providing better temperature uniformity. In this study, the internal and external temperature rise during AC heating was investigated through experimental testing. Using an in-situ thermocouple measurement method, the internal temperature of the battery was measured in real time during the AC heating process. Fig. 3.23 shows the internal and external temperature rise curves of sample E, a $LiFePO_4$-based battery, under excitation conditions of 30 and 1 Hz at a current of 40 A.

From Fig. 3.23, it can be concluded that even under high current excitation, the surface temperature of the battery exhibits good uniformity, with a surface temperature difference of less than 2 °C. The internal thermocouple reflects the true internal temperature state of the battery, showing that the internal–external temperature difference is slightly larger than the surface temperature difference, with a temperature range of approximately 5 °C. The internal structure of this battery is a laminated pouch battery. The surface of the battery is enclosed by an aluminum-plastic film, which has a good heat transfer coefficient with the environment. In the thickness direction of the battery, it can be regarded as a series connection of multiple battery electrode units, where each unit is composed of a series connection of a positive electrode, separator, and negative electrode, with the thermal conductivity calculated by (Eq. 3.11) [4]. In the planar direction, the battery can be regarded as a parallel connection of positive

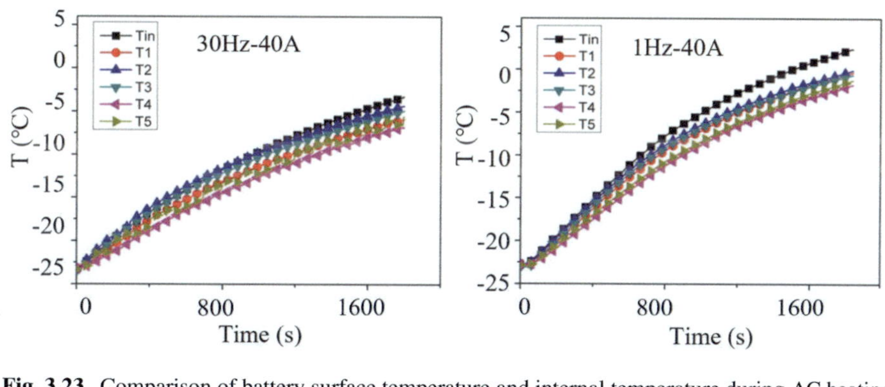

Fig. 3.23 Comparison of battery surface temperature and internal temperature during AC heating process

electrode, separator, and negative electrode, with the thermal conductivity calculated by (Eq. 3.12). The thermal conductivity in the thickness direction is determined by the series connection of the laminated electrode materials, while the current collectors made of copper and aluminum exhibit higher thermal conductivity in the planar direction. The thermal interaction between the battery and its surroundings, as well as the thermal conductivity differences in different directions within the battery, are the main causes of the temperature difference.

It can be concluded that low-temperature AC heating method provides good temperature uniformity in X–Y planar direction but results in a certain internal–external temperature difference in the thickness direction. In practical applications for high-capacity batteries, monitoring the temperature solely through the battery surface is limited. During the AC heating process, there is a potential risk of excessively high internal temperatures within the battery. Thus, in the development of AC heating systems, approaches for real-time estimation of the battery's internal temperature must be taken into account.

3.3.2.2 Asymmetric BPC Heating

1) Determination of Asymmetric BPC Parameters

The complexity of the asymmetric BPC definition in Fig. 3.18, which includes numerous parameters, complicates the analysis of heating power variations under diverse settings. In practical AC heating scenarios, two main approaches are typically utilized. The first approach involves AC heating with energy injection, such as heating battery packs via charging devices [55–58]. The second category comprises AC heating without energy injection, where circuit elements are utilized to configure the heating system, frequently employing controlled switching components to facilitate AC heating within batteries [59–62]. For AC heating schemes with energy injection, maintaining a balance between discharge and charge ampere-hours within a single cycle is not a strict theoretical requirement. For AC heating methods that do not involve energy injection, maintaining a balance between discharge and charge ampere-hours per cycle is ideal to optimize energy efficiency. In the realm of applications of EVs, such energy-injection-free AC heating approaches tend to be more feasible and practical for implementation. Moreover, AC heating schemes that involve energy injection can also adopt a charge–discharge balancing strategy. Therefore, this study investigates the impact of asymmetric BPC parameter variations on heating power, under the condition of equal charge/discharge ampere-hours. This condition can be mathematically represented as:

$$A_{cha}t_{cha} = A_{dis}t_{dis} \tag{3.36}$$

Given the constraints specified in Eq. (3.36), asymmetric BPC can be defined by three parameters uniquely: frequency f_{pulse}, ratio R_{pulse}, and discharge amplitude A_{dis}. Additionally, the failure boundaries of the battery must be considered when designing heating strategies. The heating power corresponding to different asymmetric BPC parameters must adhere to the constraints set by these failure boundaries. While several methods have been suggested for developing AC heating strategies, the establishment of constraint boundaries is still a subject of ongoing discussion. At present, the acknowledged failure mechanisms associated with low-temperature AC heating encompass lithium plating, over-charging and discharging. [54, 63, 64]. One approach to mitigating these failure modes is to regulate the battery's polarization voltage, particularly by constraining the charge and discharge cut-off voltages. However, the relationships between lithium plating and terminal voltage, overcharge and terminal voltage, and over-discharge and terminal voltage under high-frequency AC conditions remain unclear.

This research encompasses a broad spectrum of terminal voltages by examining charge cut-off voltages between 4.2 and 5 V, as well as discharge cut-off voltages ranging from 2 to 3 V. While the fundamental parameters of the asymmetric BPC encompass f_{pulse}, R_{pulse} and A_{dis}. Given a fixed frequency and ratio, there exists a maximum allowable A_{dis} such that ensures the terminal voltage hits the constraint boundary, which corresponds to the maximum attainable heating power.

In addition to A_{dis}, two additional key parameters that warrant attention are f_{pulse} and R_{pulse}. The R_{pulse} ranges from 0 to 1, with 0 representing full charging and 1 indicating full discharging. Consequently, these extreme values are excluded from consideration. In this study, the ratio analysis spans from 0.05 to 0.95, with heating power calculations performed at intervals of 0.05. Moreover, the choice of frequency range is informed by the battery's EIS results. To capture the full impedance spectrum, four characteristic frequencies—0.5, 5, 100, and 10,000 Hz—are selected. Figure 3.24 depicts how these frequencies are distributed within the EIS spectra at various temperatures. For each impedance measurement, the current amplitude is fixed at 0.25 A.

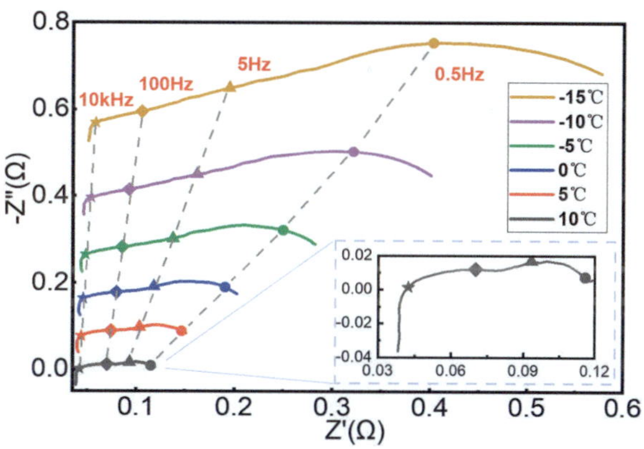

Fig. 3.24 Distribution of the studied frequencies over the impedances [50]

2) Heating Power under Various Asymmetric BPC Parameters

The thermal output of the battery, measured by its heating power, can be determined using the parameters of the asymmetric BPC outlined in Fig. 3.18. The heating power q of the battery under asymmetric BPC conditions is given by the expression [50]

$$q = q_{dis} + q_{cha} = \frac{A_{dis}^2 Z'_{dis} t_{dis}}{T_{pulse}} + \frac{A_{cha}^2 Z'_{cha} t_{cha}}{T_{pulse}} \tag{3.37}$$

where q_{dis} and q_{cha} represent the discharge and charge heating power, respectively.

Figure 3.25 depicts the variation of heating power with R_{pulse} under different frequencies and temperatures, across various discharge cut-off voltages. The corresponding legends for the 0.5, 5, and 100 Hz cases range from 0 to 150 W, whereas those for the 10 kHz scenario extend from 0 to 300 W. The black areas in the plots signify instances where the heating power exceeds the maximum value indicated in the legend. For a more detailed analysis of the heating power distribution, Fig. 3.26 offers a finer breakdown for each case. Values within the black regions can be further examined in Fig. 3.26. Across all subplots in Figs. 3.25 and 3.26, it is evident that lowering the discharge cut-off voltage, while keeping R_{pulse} constant, leads to higher heating power. This is because a lower cut-off voltage necessitates a higher discharge current to reach the threshold.

Furthermore, as indicated by Eq. (3.36), the charge current increases correspondingly, leading to an increase in heating power. Under a consistent discharge cut-off voltage condition, the battery's heating power generally follows an upward trend before declining as R_{pulse} increases. This fluctuation in heating power is primarily driven by the interplay of currents, along with the respective durations. To clarify this behavior, consider the results for 100 Hz at −5 °C with a discharge cut-off voltage of 2.7 V, as presented in Table 3.13. When combined with Eq. (3.37), the magnitudes of the currents and the discharge/charge times are substantial. While variations in these parameters exist, their effect on heating power is relatively insignificant. Therefore, it is primarily the discharge/charge currents and their corresponding durations that influence the battery's heating power. As R_{pulse} increases, the current decreases, causing the heating power in the discharge phase to rise, followed by a subsequent decrease. It is noteworthy that the discharge heating power peaks when R_{pulse} is 0.8. For values of R_{pulse} below 0.8, the discharge current does decrease, but the reduction is relatively small. Additionally, the discharge time per cycle increases, leading to an increase in heating power in the discharge phase.

However, after R_{pulse} exceeds 0.8, while the discharge time continues to increase, the current decreases substantially, resulting in a decrease in discharge heating power., which can be explained by the fact that the increase in discharge heating power is primarily due to the longer discharge time, whereas the reduction in discharge heating power is due to the significant decrease in discharge current. Furthermore, as R_{pulse} increases, Table 3.13 shows that the charge current increases, causing the charge heating power to rise initially, before eventually declining. The peak charge heating power is observed at R_{pulse} of 0.85. Analogous to the analysis of discharge heating

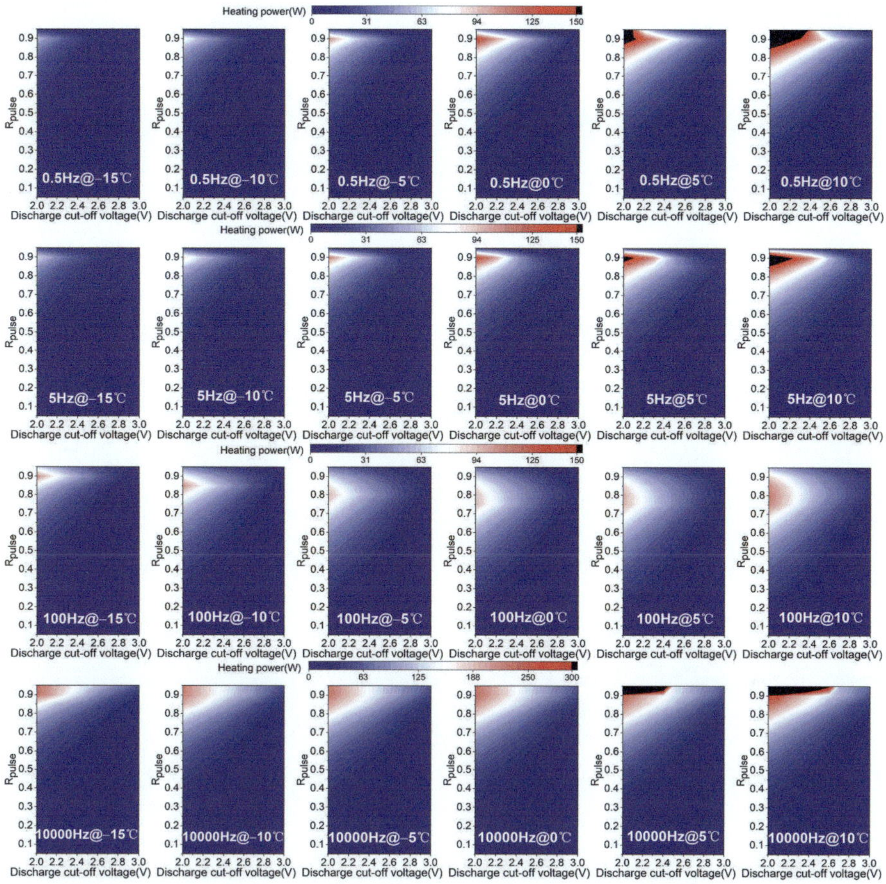

Fig. 3.25 Overview of the relationship between R_{pulse} and heating power under varying frequency and temperature at different discharge cut-off voltages

power, the increase in charge heating power stems from the higher charge current, whereas the decline is driven by the reduced charge duration. Ultimately, the net battery heating power is determined by the collective variations in both discharge and charge heating power.

The thermal output characteristics exhibit intricate dependencies on frequency and temperature, which are further influenced by alterations in R_{pulse} and discharge termination voltage thresholds. As depicted in Fig. 3.25, the areas marked in white and red expand as the temperature increases at a constant frequency, indicating an increase in regions with higher heating power. This suggests that, for a given discharge cut-off voltage and R_{pulse}, the battery's heating power tends to increase with rising temperature. This behavior can be attributed to the variation in battery resistance with temperature. Figure 3.24 demonstrates an inverse correlation between temperature and resistance, indicating that maintaining identical discharge termination voltages

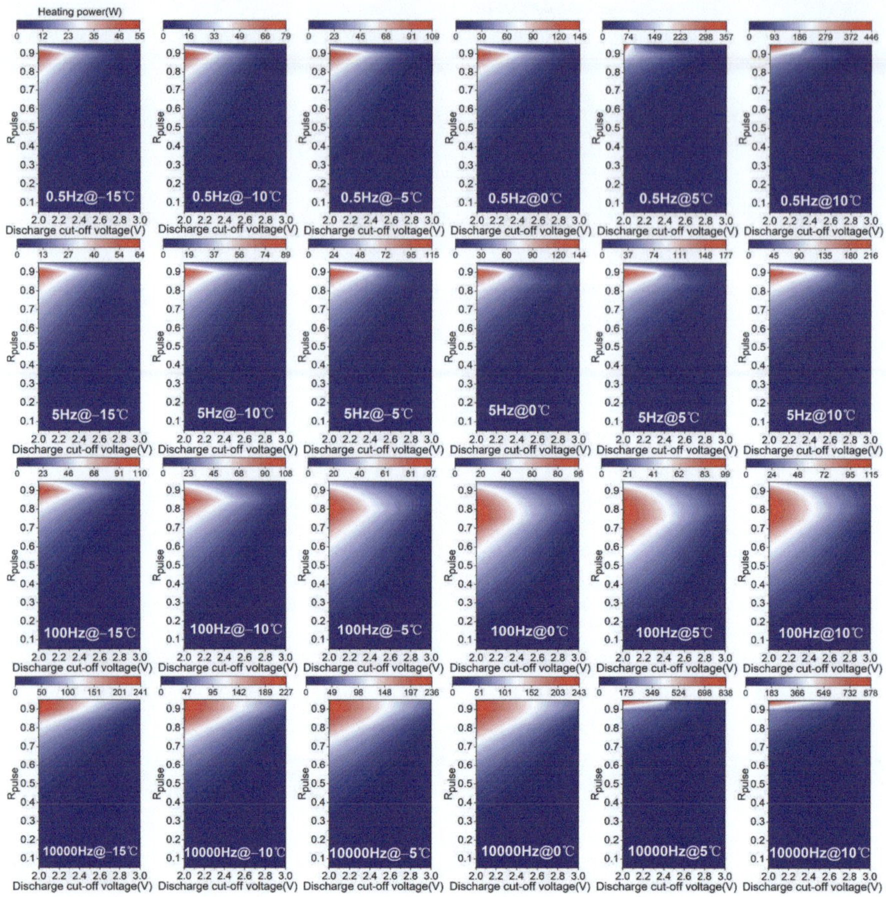

Fig. 3.26 Detailed view of the relationship between R_{pulse} and heating power under varying frequency and temperature at different discharge cut-off voltages

demands greater current at elevated temperatures. This observation confirms the dominant role of current in determining thermal output levels, with heating capacity showing direct proportionality to current magnitude. Additional insights emerge from Fig. 3.26, where the peak heating power-temperature relationship displays non-uniform behavior. Specifically, for 0.5 Hz operation, the optimal R_{pulse} maintains a stable value of 0.9 throughout subzero temperatures (−15 to 0 °C), but rises to 0.95 when temperatures reach 5 °C and above.

Along with this change, max heating power gradually rises as temperature rises. On the other hand, at a frequency of 100 Hz, the R_{pulse} value associated with maximum heating power tends to decrease as the temperature increases, although the variation in maximum heating power is relatively small. Additionally, the variations in R_{pulse} values for maximum heating power at different frequencies, while maintaining the same discharge cut-off voltage, differ at temperatures of −15 and 5 °C. Notably,

Table 3.13 Changes in the parameters of Eq. (3.37) at −5 °C with varying Rpulse values under a 2.7 V discharge cut-off voltage, when 100 Hz asymmetric BPC frequency is set [50]

R_{pulse}	0.05	0.10	0.15	0.20	0.25	0.30	0.35	0.40	0.45	0.50
A_{dis} (A)	14.49	13.80	13.89	13.07	12.81	12.58	12.38	12.20	12.04	11.88
Z'_{dis} (Ω)	0.0619	0.0624	0.0628	0.0631	0.0633	0.0635	0.0637	0.0638	0.0640	0.0641
q_{dis} (W)	0.65	1.19	1.69	2.15	2.60	3.02	3.42	3.80	4.17	4.53
A_{cha} (A)	0.76	1.53	2.36	3.27	4.27	5.39	6.67	8.13	9.84	11.88
Z'_{cha} (Ω)	0.0841	0.0827	0.0823	0.0796	0.0766	0.0738	0.0711	0.0687	0.0664	0.0641
q_{cha} (W)	0.047	0.18	0.39	0.68	1.05	1.50	2.06	2.73	3.54	4.53
q (W)	0.70	1.36	2.08	2.83	3.64	4.52	5.47	6.53	7.71	9.06
R_{pulse}	0.55	0.60	0.65	0.70	0.75	0.80	0.85	0.90	0.95	
A_{dis} (A)	11.74	11.61	11.48	11.36	11.25	11.14	9.21	6.49	3.44	
Z'_{dis} (Ω)	0.0643	0.0644	0.0645	0.0647	0.0648	0.0649	0.0672	0.0715	0.079	
q_{dis} (W)	4.87	5.21	5.53	5.84	6.15	6.45	4.84	2.71	0.89	
A_{cha} (A)	14.35	17.41	21.32	26.51	33.75	44.57	52.18	58.43	65.33	
Z'_{cha} (Ω)	0.0620	0.0600	0.0580	0.0560	0.0541	0.0522	0.0513	0.0506	0.0501	
q_{cha} (W)	5.75	7.27	9.22	11.82	15.41	20.74	20.93	17.29	10.68	
q (W)	10.62	12.48	14.75	17.66	21.56	27.19	25.77	20.00	11.57	

when the temperature is below 0 °C, Fig. 3.25 indicates that the area covered by high heating power, shown in white and red, expands as the frequency increases at a given temperature, suggesting an increase in regions with higher heating power. Conversely, under 100 Hz operation, the optimal R_{pulse} for peak thermal output demonstrates an inverse temperature dependence, decreasing with rising temperatures despite minimal variation in maximum power output. Furthermore, comparative analysis reveals frequency-dependent discrepancies in optimal R_{pulse} values under identical discharge voltage constraints, particularly evident between −15 and 5 °C conditions. Subzero temperature observations (Fig. 3.25) show a distinct pattern: the high-output zones (white/red regions) exhibit progressive expansion with frequency elevation at fixed temperatures, indicating enhanced coverage of areas capable of superior heat generation. However, this trend becomes less noticeable once the temperature rises above 0 °C. At temperatures above this threshold, the areas shaded in white and red initially decrease before expanding as the frequency increases. When frequency is 100 Hz, this zone exhibits the most limited coverage, with significantly reduced thermal output capacity. This discrepancy can be explained by changes in the battery's resistance. For instance, Table 3.14 illustrates the variations in parameters, as described in Eq. (3.37), across varying asymmetric BPC frequency conditions under both −15 °C and 10 °C, with a discharge cut-off voltage of 2.2 V and the R_{pulse} value of 0.85. Both Z'_{dis} and Z'_{cha} show a gradual decrease as frequency increases at both −15 and 10 °C. However, the reduction in Z'_{dis} and Z'_{cha} is more pronounced at −15 °C compared to 10 °C, which results in a more substantial increase in currents at the lower temperature. Consequently, the battery's heating power at − 15 °C increases more noticeably as the frequency rises. At a temperature of 10 °C, the change in currents between 0.5 and 100 Hz is not significant, and the reduction in Z'_{dis} leads to a decrease in heating power. However, when the frequency reaches 10,000 Hz, both Z'_{dis} and Z'_{cha} are substantially reduced, which results in a significant increase in both discharge and charge currents. Consequently, the battery's thermal output experiences a secondary rise. Detailed examination demonstrates that under subzero conditions, the frequency-induced resistance decline becomes markedly more significant, causing current parameters to dominate heat generation characteristics. As the frequency increases, the battery heating power gradually increases. However, when the temperature is 0 °C or higher, the reduction in battery resistance with frequency becomes less significant. In this case, the heating power is primarily influenced by currents, along with the parameters Z'_{dis} and Z'_{cha}. Furthermore, the pattern of maximum heating power variation across different frequencies mirrors the changes in maximum heating power with varying temperatures, showing no distinct trend, as illustrated in Fig. 3.26.

The frequency- and temperature-dependent correlation between R_{pulse} and heating power under varying charge termination voltages is depicted in Fig. 3.27. Legend scales (0–130 W) accompany each plot, with shaded regions marking where heating power surpasses the maximum indicated value. Complementary data for individual cases appear in Fig. 3.28. Analysis of both figures reveals a consistent pattern: for any given R_{pulse} value, elevating the charge termination voltage results in systematically greater heating power generation.

Table 3.14 Variations in the parameters from Eq. (3.37) at temperatures of −15 and 10 °C, considering different asymmetric BPC frequencies, with a discharge cut-off voltage of 2.2 V and R_{pulse} set to 0.85 [50]

Temperature (°C)	−15				10			
f_{pulse} (Hz)	0.5	5	100	10,000	0.5	5	100	10,000
A_{dis} (A)	6.13	7.12	10.91	18.89	17.74	18.14	18.26	28.37
Z'_{dis} (Ω)	0.2085	0.1578	0.0956	0.0637	0.0765	0.0652	0.0522	0.0435
q_{dis} (W)	6.66	6.79	9.67	19.31	20.47	18.24	14.79	29.74
A_{cha} (A)	34.73	40.34	61.81	107.04	100.54	102.81	103.49	160.79
Z'_{cha} (Ω)	0.1062	0.0960	0.0742	0.0589	0.0587	0.0536	0.0433	0.0414
q_{cha} (W)	19.21	23.44	42.52	101.30	89.03	85.05	69.54	160.71
q (W)	25.87	30.24	52.19	120.61	109.51	103.29	84.33	190.45

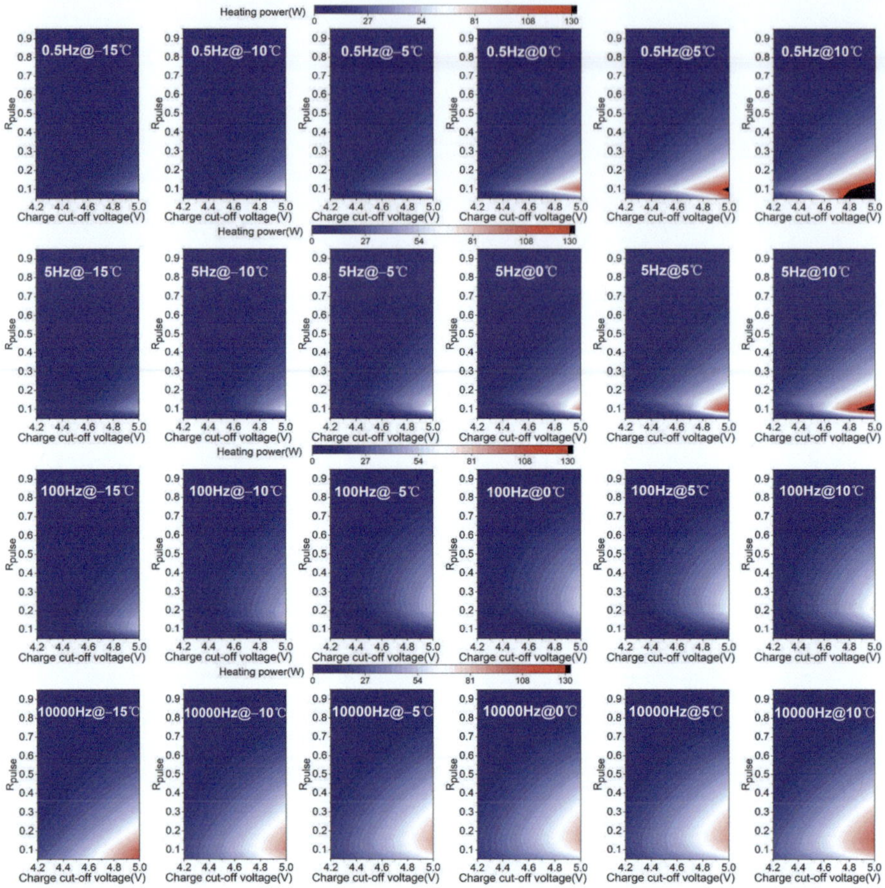

Fig. 3.27 Overview of the relationship between R_{pulse} and heating power under varying frequency and temperature at different charge cut-off voltages

This phenomenon parallels the discharge voltage relationship—increased charge voltage demands higher current to satisfy operational limits, consequently amplifying thermal power output. In most cases, the heating power initially increases with R_{pulse} and then decreases as R_{pulse} continues to rise, though the heating power is concentrated at lower R_{pulse} values in contrast to the discharge cut-off voltage. For further clarification, Table 3.15 presents the results for 100 Hz @ −5 °C with a charge cut-off voltage of 4.8 V, where the heating power is seen to increase and then decrease as R_{pulse} increases, consistent with the observed trends for both discharge and charge heating power. Furthermore, the peak values of the battery heating power, discharge heating power, and charge heating power occur at R_{pulse} values of 0.25, 0.2, and 0.45, respectively. Consistent with the analysis of the discharge cut-off voltage, Z'_{dis} and Z'_{cha}'s influence on heating power is relatively minimal, so the discussion will focus primarily on the charge and discharge currents and their respective durations.

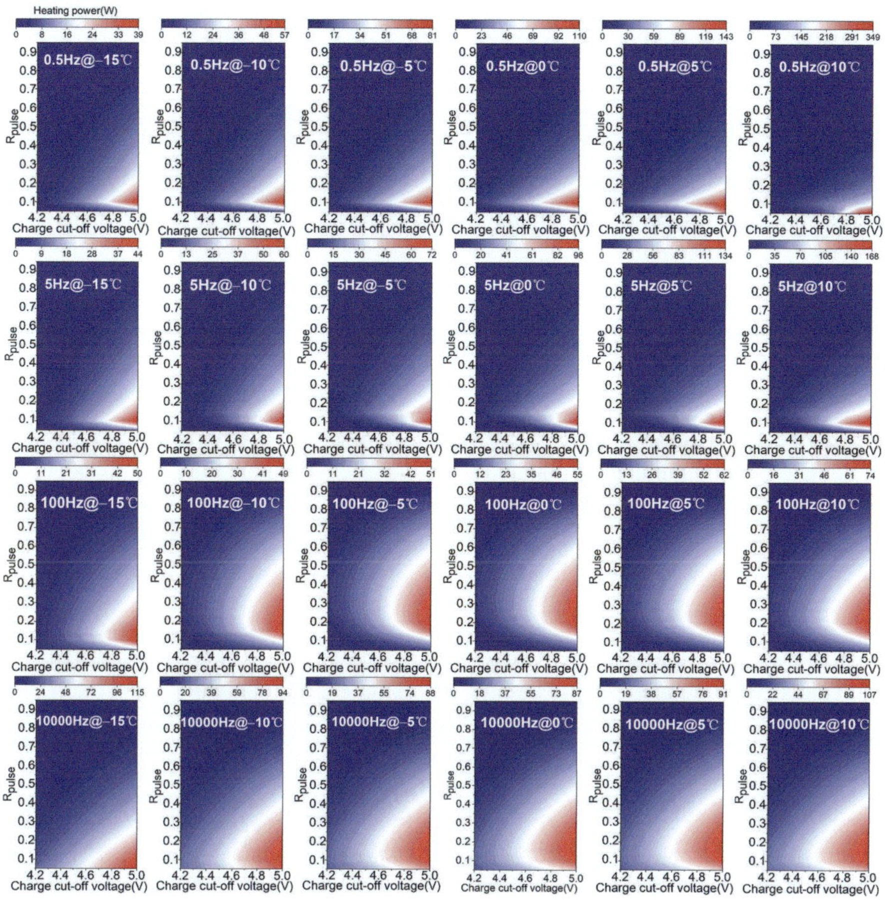

Fig. 3.28 Detailed view of the relationship between R_{pulse} and heating power under varying frequency and temperature at different charge cut-off voltages

Regarding discharge heating power, as R_{pulse} increases, the heating power mirrors the change in discharge current. As a result, the heating power during discharge primarily depends on the current magnitude. On the charging side, higher R_{pulse} values produce a dual effect: the charging current progressively grows while the charging duration shortens. Initially, the boost in charging heating power stems from current enhancement. Yet beyond a certain point, the substantial reduction in charging time becomes the prevailing factor, causing the charging heating power to decline with further R_{pulse} elevation.

Like the discharge cut-off voltage, the impact of temperature and frequency on heating power varies significantly. As shown in Fig. 3.27, the region of high heating power grows as the temperature increases at a constant frequency. Consequently, for a fixed cut-off voltage and R_{pulse}, the heating power tends to increase as the temperature rises. This behavior is consistent with the effect seen at the cut-off voltage

Table 3.15 Parameter variations according to Eq. (3.37) at −5 °C, displaying different R_{pulse} values under a charge cut-off voltage of 4.8 V when the asymmetric BPC frequency is set to 100 Hz [50]

R_{pulse}	0.05	0.10	0.15	0.20	0.25	0.30	0.35	0.40	0.45	0.50
A_{dis} (A)	43.79	48.18	51.54	48.93	42.83	37.02	31.83	27.30	23.35	19.88
Z'_{dis} (Ω)	0.0523	0.0517	0.0513	0.0516	0.0525	0.0534	0.0546	0.0558	0.0571	0.0586
q_{dis} (A)	5.02	12.01	20.45	24.72	24.06	21.97	19.35	16.63	14.02	11.59
A_{cha} (A)	2.31	5.35	9.10	12.23	14.28	15.86	17.14	18.20	19.10	19.88
Z'_{cha} (Ω)	0.0823	0.0739	0.0673	0.0638	0.0621	0.0609	0.0601	0.0595	0.0590	0.0586
q_{cha} (W)	0.42	1.91	4.73	7.64	9.49	10.74	11.48	11.83	11.85	11.59
q (W)	5.43	13.91	25.19	32.36	33.55	32.71	30.83	28.46	25.87	23.19
R_{pulse}	0.55	0.60	0.65	0.70	0.75	0.80	0.85	0.90	0.95	
A_{dis} (A)	16.83	14.12	11.69	9.51	7.53	5.73	4.08	2.58	1.21	
Z'_{dis} (Ω)	0.0603	0.0622	0.0643	0.0668	0.0696	0.0730	0.0771	0.0820	0.0832	
q_{dis} (A)	9.40	7.44	5.72	4.23	2.96	1.92	1.09	0.49	0.12	
A_{cha} (A)	20.57	21.18	21.72	22.19	22.59	22.91	23.14	23.22	22.93	
Z'_{cha} (Ω)	0.0583	0.0580	0.0578	0.0576	0.0574	0.0573	0.0572	0.0572	0.0573	
q_{cha} (W)	11.10	10.41	9.54	8.51	7.33	6.02	4.60	3.08	1.51	
q (W)	20.50	17.85	15.26	12.73	10.29	7.94	5.69	3.58	1.62	

of discharging, where the reduction in battery resistance with rising temperature leads to an increase in current for the same cut-off voltage of charging, thereby raising the heating power. Additionally, Fig. 3.28 does not show a clear pattern between maximum heating power and temperature. For example, at 0.5 Hz, when the temperature is below 5 °C, the R_{pulse} value for maximum heating power at a given charge cut-off voltage is 0.1. However, at 10 °C, R_{pulse} decreases to 0.05, and the maximum heating power rises with temperature. At a frequency of 10 kHz, the R_{pulse} associated with maximum heating power increases with temperature as well. At a constant temperature, the frequency distribution pattern of high heating power resembles that seen at the discharge cut-off voltage. Below 0 °C, the area with high heating power broadens as the frequency increases. However, when the temperature is 0 °C or higher, the high heating power distribution first declines and then rises with frequency. This behavior can be explained by the variations in battery parameters shown in Table 3.16, where the effects of different asymmetric BPC frequencies at −15 and 10 °C are illustrated with a charge cut-off voltage of 4.8 V and R_{pulse} of 0.2. Whether at −15 °C or 10 °C, resistance generally drops as frequency increases.

However, at −15 °C, the reduction in resistance is more pronounced, leading to a more noticeable increase in current. As a result, battery heating power steadily increases with frequency at −15 °C. In contrast, at 10 °C, the increase in current from 0.5 to 100 Hz is minimal, while the resistance decreases, resulting in a decrease in battery heating power. When the frequency reaches 10 kHz, the current slightly increases, causing the battery heating power to rise once again. To summarize, under −15 °C conditions, the heating power exhibits a frequency-dependent enhancement, which is predominantly governed by the applied current magnitude. At 10 °C, both current and resistance play significant roles in determining the heating power. Furthermore, the change in maximum heating power with frequency is shown as well, as shown in Fig. 3.28, follows a similar trend to that observed for the discharge cut-off voltage, with no clear pattern emerging in either case.

Figures 3.25 and 3.27 analyze the influence of frequency and temperature on the relationship between R_{pulse} and heating power under different discharge and charge cut-off voltages, respectively. However, in practical applications of the AC heating strategy, it is crucial to consider cut-off voltages simultaneously of both charging and discharging. As shown in Fig. 3.29, the dependence of Rpulse on heating power is analyzed under the combined influence of charge and discharge cut-off voltages at −5 °C and 100 Hz. The plotted curves indicate the upper limit of attainable heating power for each cut-off voltage condition. Since the actual heating power is constrained by the lower of the two cut-off voltages, the point where the charge and discharge curves intersect (e.g., the green markers in Fig. 3.29) defines the peak heating power permissible within the given terminal voltage limits. When R_{pulse} falls below its value at the intersection point, the discharge cut-off voltage becomes the limiting factor for peak heating power. Conversely, when R_{pulse} rises above the intersection value, the charge cut-off voltage imposes the upper boundary on achievable heating power. The two graphs on the right side of Fig. 3.29 show

Table 3.16 The variations of parameters in Eq. (3.37) at temperatures of −15 and 10 °C for different asymmetric BPC frequencies, with a charge cut-off voltage of 4.8 V and an R_{pulse} value of 0.2 [50]

Temperature (°C)	−15				10			
f_{pulse} (Hz)	0.5	5	100	10,000	0.5	5	100	10,000
A_{dis} (A)	19.59	23.44	35.59	59.85	61.52	63.33	64.69	84.00
Z'_{dis} (Ω)	0.1294	0.1110	0.0795	0.0608	0.0626	0.0560	0.0449	0.0426
q_{dis} (W)	9.93	12.20	20.13	43.52	47.37	44.93	37.55	60.15
A_{cha} (A)	4.90	5.86	8.90	14.96	15.38	15.83	16.17	21.00
Z'_{cha} (Ω)	0.2292	0.1669	0.0991	0.064	0.0784	0.0665	0.0530	0.0435
q_{cha} (W)	4.40	4.59	6.27	11.47	14.84	13.34	11.10	15.35
q (W)	14.32	16.78	26.41	54.99	62.21	58.27	48.65	75.50

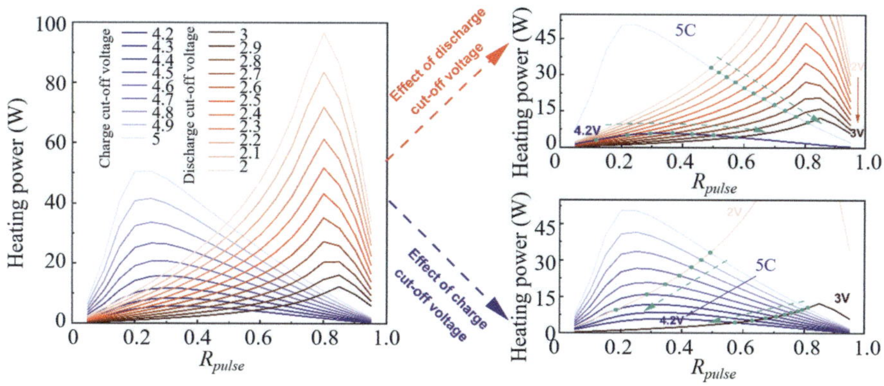

Fig. 3.29 Relationship between R_{pulse} and heating power changes with varying charge and discharge cut-off voltages at a temperature of −5 °C and a frequency of 100 Hz [50]

the effect of the discharge and charge cut-off voltages on the intersection point's location. When the discharge cut-off voltage increases, the intersection gradually shifts to a higher R_{pulse}. On the other hand, reducing the charge cut-off voltage leads to the intersection moving toward a lower R_{pulse}. Furthermore, the maximum heating power does not correspond to an R_{pulse} value of 0.5. For instance, when cut-off voltage of charging is 5 V and cut-off voltage of discharging is 3 V, the maximum heating power occurs at an R_{pulse} value of approximately 0.8.

The point where the heating power curves for the charge and discharge cut-off voltages intersect indicates the maximum heating power. Consequently, the intersection provides insight into how varying frequencies and temperatures influence heating power, while adhering to the constraints of both charge and discharge cut-off voltages. In this study, charge and discharge cut-off voltages are fixed at 4.2 and 3 V. Consequently, the maximum heating power under these voltage constraints is used as a benchmark to assess the influence of frequency and temperature, as depicted in Fig. 3.30. Figure 3.30a and b show the maximum heating power and its associated R_{pulse} values across different frequencies and temperatures. It is clear that for a given frequency, maximum heating power increases with rising temperature. Likewise, at a fixed temperature, higher frequencies lead to a steady increase in maximum heating power. Fig. 3.30b indicates that the R_{pulse} corresponding to max heating power is generally around 0.55. However, at a frequency of 0.5 Hz and temperatures of either 5 °C or 10 °C, the R_{pulse} for the maximum heating power is 0.5.

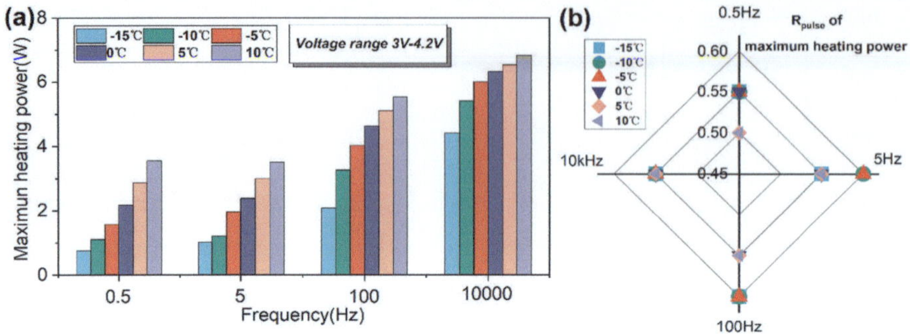

Fig. 3.30 The maximum heating power and its associated R_{pulse} values across different frequencies and temperatures within the 3–4.2 V voltage range are presented as follows. **a** Maximum heating power, **b** R_{pulse} corresponding to the peak heating power [50]

3.4 Conclusions

This chapter investigates the AC heating technology for LIBs under low-temperature conditions. The focus includes the battery heat generation mechanism, the impedance-based analysis of the battery heat generation mechanism, and the effect of different parameters of the AC heating method on battery heat generation.

Based on the basic theory of battery thermal characteristics and the derivation of the Bernardi equation, a method for calculating the heat generation rate of a micro-unit in a power battery is derived. The reversible heat generation rate due to electrochemical reactions, the heat generation rate due to electrochemical polarization, and the heat generation rate due to Ohmic losses in a micro-unit of the power battery are presented. Additionally, the boundary conditions for the heat generation and heat dissipation of the battery thermal characteristics are determined based on basic heat transfer theory.

To study the AC heating heat generation mechanism of the battery, AC impedance testing experiments under high-current excitation were designed for batteries with various material systems. The experiments revealed that the mid-frequency impedance of the power battery gradually decreased as the excitation current amplitude increased. This phenomenon was found to be unrelated to the internal pressure, SOC, or internal temperature changes of the battery. To further study the impedance variation of the battery's positive and negative electrodes, half-cell experiments using NCA cathodes and graphite anodes were also designed. DRT analysis revealed that the shrinkage of the impedance arc stemmed from decreases in both the time constants and resistances associated with the SEI/CEI processes and charge transfer mechanisms. Additionally, the contributions of the anode and cathode to the overall cell impedance varied with SOC. At low SOC, charge transfer impedance was primarily influenced by the cathode, whereas at medium to high SOC, it was jointly affected by both the anode and cathode. As for SEI/CEI impedance, since the cathode exhibited a negligible CEI process, the anode's SEI impedance played a dominant role.

This chapter also investigates the effects of excitation parameters of symmetric and asymmetric BPCs on the temperature rise characteristics of batteries. For symmetric BPC heating, a comprehensive analysis was conducted on the effects of three parameters—excitation frequency, current amplitude, and voltage protection threshold—on the heating results of LIBs of various material systems and specifications. Among these, the current amplitude has the most significant impact on the AC heating results. High-amplitude excitation currents facilitate rapid heat accumulation and temperature rise. Due to the relatively high impedance of batteries at low frequencies, low-frequency excitation generally results in better heating performance than high-frequency excitation. However, large currents at low frequencies can trigger the battery's voltage protection threshold, significantly reducing the heating efficiency. As the battery temperature rises and its internal resistance decreases, the differences in temperature rise caused by varying voltage protection thresholds gradually diminish. For asymmetric BPC heating, an analysis of three parameters—percentage of discharge time in a cycle (R_{pulse}), frequency, and voltage thresholds—was conducted on the maximum heating power of LIBs at different temperatures. Under the discharge cut-off voltage constraint, higher heating powers are observed in the range of larger R_{pulse} values. Under the constraint of charge cut-off voltage, higher heating power is primarily observed when R_{pulse} is relatively small. The symmetrical BPC method does not yield the highest heating power. The correlation between maximum heating power and R_{pulse} is jointly affected by both charge and discharge cut-off voltages. Specifically, an increase in the discharge cut-off voltage results in a higher R_{pulse} for peak heating power, whereas a lower charge cut-off voltage reduces it. When the voltage range is maintained between 3 and 4.2 V, the optimal R_{pulse} for maximum heating power is approximately 0.55. At a fixed frequency, heating power grows with rising temperature, while an increase in frequency further enhances heating power.

References

1. Bernardi D, Pawlikowski E, Newman J. A General Energy Balance for Battery Systems [J]. Journal of The Electrochemical Society, 1985, 132(1): 5–12.
2. Kumaresan K, Sikha G, White R E. Thermal Model for a Li-Ion Cell [J]. Journal of The Electrochemical Society, 2008, 155(2): A164.
3. Gu W B, Wang C Y. Thermal-Electrochemical Modeling of Battery Systems [J]. Journal of The Electrochemical Society, 2000, 147(8): 2910.
4. Chen S C, Wan C C, Wang Y Y. Thermal analysis of lithium-ion batteries [J]. Journal of Power Sources, 2005, 140(1): 111–124.
5. Cho H M, Choi W S, Go J Y, et al. A study on time-dependent low temperature power performance of a lithium-ion battery [J]. Journal of Power Sources, 2012, 198: 273–280.
6. Zhang S S, Xu K, Jow T R. A new approach toward improved low temperature performance of Li-ion battery [J]. Electrochemistry Communications, 2002, 4(11): 928–932.

7. Gomez J, Nelson R, Kalu E E, et al. Equivalent circuit model parameters of a high-power Li-ion battery: Thermal and state of charge effects [J]. Journal of Power Sources, 2011, 196(10): 4826–4831.
8. Jiménez Gordon I, Grugeon S, Débart A, et al. Electrode contributions to the impedance of a high-enesrgy density Li-ion cell designed for EV applications [J]. Solid State Ionics, 2013, 237: 50–55.
9. Gaberscek M, Dominko R, Jamnik J. The meaning of impedance measurements of LiFePO4 cathodes: A linearity study [J]. Journal of Power Sources, 2007, 174(2): 944–948.
10. Gaberscek M, Moskon J, Erjavec B, et al. The Importance of Interphase Contacts in Li Ion Electrodes: The Meaning of the High-Frequency Impedance Arc [J]. Electrochemical and Solid-State Letters, 2008, 11(10): A170.
11. Ge H, Aoki T, Ikeda N, et al. Investigating Lithium Plating in Lithium-Ion Batteries at Low Temperatures Using Electrochemical Model with NMR Assisted Parameterization [J]. Journal of The Electrochemical Society, 2017, 164(6): A1050-A1060.
12. Xu L, Lin X, Xie Y, et al. Enabling high-fidelity electrochemical P2D modeling of lithium-ion batteries via fast and non-destructive parameter identification [J]. Energy Storage Materials, 2022, 45: 952–968.
13. Huang R, Wang X, Jiang B, et al. Revealing the electrochemical impedance characteristics of lithium-ion battery (nickel-cobalt-aluminum vs. graphite) under various alternating current amplitudes [J]. Journal of Power Sources, 2023, 566: 232929.
14. Lu Y, Zhao C Z, Huang J Q, et al. The timescale identification decoupling complicated kinetic processes in lithium batteries [J]. Joule, 2022, 6(6): 1172–1198.
15. Zhu J, Knapp M, Liu X, et al. Low-Temperature Separating Lithium-Ion Battery Interfacial Polarization Based on Distribution of Relaxation Times (DRT) of Impedance [J]. IEEE Transactions on Transportation Electrification, 2021, 7(2): 410–421.
16. Wan T H, Saccoccio M, Chen C, et al. Influence of the Discretization Methods on the Distribution of Relaxation Times Deconvolution: Implementing Radial Basis Functions with DRTtools [J]. Electrochimica Acta, 2015, 184: 483–499.
17. Bae S, Song H D, Nam I, et al. Quantitative performance analysis of graphite-LiFePO4 battery working at low temperature [J]. Chemical Engineering Science, 2014, 118: 74–82.
18. Zhu J, Sun Z, Wei X, et al. Studies on the medium-frequency impedance arc for Lithium-ion batteries considering various alternating current amplitudes [J]. Journal of Applied Electrochemistry, 2016, 46(2): 157–167.
19. Peabody C, Arnold C B. The role of mechanically induced separator creep in lithium-ion battery capacity fade [J]. Journal of Power Sources, 2011, 196(19): 8147–8153.
20. Love C T, Virji M B V, Rocheleau R E, et al. State-of-health monitoring of 18650 4S packs with a single-point impedance diagnostic [J]. Journal of Power Sources, 2014, 266: 512–519.
21. Zhu J G, Sun Z C, Wei X Z, et al. A new lithium-ion battery internal temperature on-line estimate method based on electrochemical impedance spectroscopy measurement [J]. Journal of Power Sources, 2015, 274: 990–1004.
22. Fan J, Tan S. Studies on Charging Lithium-Ion Cells at Low Temperatures [J]. Journal of The Electrochemical Society, 2006, 153(6): A1081.
23. Shafiei Sabet P, Stahl G, Sauer D U. Non-invasive investigation of predominant processes in the impedance spectra of high energy lithium-ion batteries with nickel–cobalt–aluminum cathodes [J]. Journal of Power Sources, 2020, 472: 228189.
24. Steinhauer M, Risse S, Wagner N, et al. Investigation of the Solid Electrolyte Interphase Formation at Graphite Anodes in Lithium-Ion Batteries with Electrochemical Impedance Spectroscopy [J]. Electrochimica Acta, 2017, 228: 652–658.

25. Stiaszny B, Ziegler J C, Krauß E E, et al. Electrochemical characterization and post-mortem analysis of aged LiMn2O4–Li(Ni0.5Mn0.3Co0.2)O2/graphite lithium ion batteries. Part I: Cycle aging [J]. Journal of Power Sources, 2014, 251: 439–450.
26. Manikandan B, Ramar V, Yap C, et al. Investigation of physico-chemical processes in lithium-ion batteries by deconvolution of electrochemical impedance spectra [J]. Journal of Power Sources, 2017, 361: 300–309.
27. Tippmann S, Walper D, Balboa L, et al. Low-temperature charging of lithium-ion cells part I: Electrochemical modeling and experimental investigation of degradation behavior [J]. Journal of Power Sources, 2014, 252: 305–316.
28. Wang X, Wei X, Dai H. Estimation of state of health of lithium-ion batteries based on charge transfer resistance considering different temperature and state of charge [J]. Journal of Energy Storage, 2019, 21: 618–631.
29. Zabara M A, Katırcı G, Ülgüt B. Non-linear harmonics in EIS of batteries with lithium anodes: Proper controls and analysis [J]. Electrochimica Acta, 2022, 429: 140969.
30. Harting N, Wolff N, Krewer U. Identification of Lithium Plating in Lithium-Ion Batteries using Nonlinear Frequency Response Analysis (NFRA) [J]. Electrochimica Acta, 2018, 281: 378–385.
31. Vidaković-Koch T, Miličić T, Živković L A, et al. Nonlinear frequency response analysis: a recent review and perspectives [J]. Current Opinion in Electrochemistry, 2021, 30: 100851.
32. Wang X, Wei X, Zhu J, et al. A review of modeling, acquisition, and application of lithium-ion battery impedance for onboard battery management [J]. eTransportation, 2021, 7: 100093.
33. Murbach M D, Hu V W, Schwartz D T. Nonlinear Electrochemical Impedance Spectroscopy of Lithium-Ion Batteries: Experimental Approach, Analysis, and Initial Findings [J]. Journal of The Electrochemical Society, 2018, 165(11): A2758–A2765.
34. Feng F, Yang R, Meng J, et al. Electrochemical impedance characteristics at various conditions for commercial solid–liquid electrolyte lithium-ion batteries: Part 1. experiment investigation and regression analysis [J]. Energy, 2022, 242: 122880.
35. Zhou X, Huang J, Pan Z, et al. Impedance characterization of lithium-ion batteries aging under high-temperature cycling: Importance of electrolyte-phase diffusion [J]. Journal of Power Sources, 2019, 426: 216–222.
36. Illig J, Ender M, Weber A, et al. Modeling graphite anodes with serial and transmission line models [J]. Journal of Power Sources, 2015, 282: 335–347.
37. Zhou X, Pan Z, Han X, et al. An easy-to-implement multi-point impedance technique for monitoring aging of lithium ion batteries [J]. Journal of Power Sources, 2019, 417: 188–192.
38. Katzer F, Danzer M A. Analysis and detection of lithium deposition after fast charging of lithium-ion batteries by investigating the impedance relaxation [J]. Journal of Power Sources, 2021, 503: 230009.
39. Nurpeissova A, Park D I, Kim S S, et al. Epicyanohydrin as an Interface Stabilizer Agent for Cathodes of Li-Ion Batteries [J]. Journal of The Electrochemical Society, 2016, 163(2): A171–A177.
40. Aurbach D. Review of selected electrode–solution interactions which determine the performance of Li and Li ion batteries [J]. Journal of Power Sources, 2000, 89(2): 206–218.
41. Chen X, Li L, Liu M, et al. Detection of lithium plating in lithium-ion batteries by distribution of relaxation times [J]. Journal of Power Sources, 2021, 496: 229867.
42. Li S E, Wang B, Peng H, et al. An electrochemistry-based impedance model for lithium-ion batteries [J]. Journal of Power Sources, 2014, 258: 9–18.
43. Momma T, Matsunaga M, Mukoyama D, et al. Ac impedance analysis of lithium ion battery under temperature control [J]. Journal of Power Sources, 2012, 216: 304–307.

44. Zhang S S, Xu K, Jow T R. Electrochemical impedance study on the low temperature of Li-ion batteries [J]. Electrochimica Acta, 2004, 49(7): 1057–1061.
45. Waag W, Fleischer C, Sauer D U. On-line estimation of lithium-ion battery impedance parameters using a novel varied-parameters approach [J]. Journal of Power Sources, 2013, 237: 260–269.
46. Fleischer C, Waag W, Heyn H M, et al. On-line adaptive battery impedance parameter and state estimation considering physical principles in reduced order equivalent circuit battery models [J]. Journal of Power Sources, 2014, 260: 276–291.
47. Meyers J P, Doyle M, Darling R M, et al. The Impedance Response of a Porous Electrode Composed of Intercalation Particles [J]. Journal of The Electrochemical Society, 2000, 147(8): 2930.
48. Doyle M, Fuller T F, Newman J. Modeling of Galvanostatic Charge and Discharge of the Lithium/Polymer/Insertion Cell [J]. Journal of The Electrochemical Society, 1993, 140(6): 1526–1533.
49. Doyle M, Newman J, Gozdz A S, et al. Comparison of Modeling Predictions with Experimental Data from Plastic Lithium Ion Cells [J]. Journal of The Electrochemical Society, 1996, 143(6): 1890–1903.
50. Huang R, Wei G, Jiang B, et al. Investigating the Effect of Different Bidirectional Pulsed Current Parameters on the Heat Generation of Lithium-Ion Battery at Low Temperatures [J]. Batteries, 2023, 9(9): 457.
51. Tomaszewska A, Chu Z, Feng X, et al. Lithium-ion battery fast charging: A review [J]. eTransportation, 2019, 1: 100011.
52. Konz Z M, McShane E J, McCloskey B D. Detecting the Onset of Lithium Plating and Monitoring Fast Charging Performance with Voltage Relaxation [J]. ACS Energy Letters, 2020, 5(6): 1750–1757.
53. Guo S, Xiong R, Shen W, et al. Aging investigation of an echelon internal heating method on a three-electrode lithium ion cell at low temperatures [J]. Journal of Energy Storage, 2019, 25: 100878.
54. Ge H, Huang J, Zhang J, et al. Temperature-Adaptive Alternating Current Preheating of Lithium-Ion Batteries with Lithium Deposition Prevention [J]. Journal of The Electrochemical Society, 2016, 163(2): A290–A299.
55. Montenegro L E G, Villegas Pico H N. AC-Heating and Fast-Charging Power Requirements of EV Battery Packs in Subzero Temperature [J]. IEEE Transactions on Transportation Electrification, 2023, 9(3): 3936–3946.
56. Cao W, Xu X, Wei Z, et al. Synergized Heating and Optimal Charging of Lithium-Ion Batteries at Low Temperature [J]. IEEE Transactions on Transportation Electrification, 2023, 9(4): 5002–5011.
57. Suh I S, Lee M, Kim J, et al. Design and experimental analysis of an efficient HVAC (heating, ventilation, air-conditioning) system on an electric bus with dynamic on-road wireless charging [J]. Energy, 2015, 81: 262–273.
58. Soares R, Djekanovic N, Wallmark O, et al. Integration of Magnified Alternating Current in Battery Fast Chargers Based on DC–DC Converters Using Transformerless Resonant Filter Design [J]. IEEE Transactions on Transportation Electrification, 2019, 5(4): 925–933.
59. Zhu C, Han J, Zhang H, et al. Modeling and Control of an Integrated Self-Heater for Automotive Batteries Based on Traction Motor Drive Reconfiguration [J]. IEEE Journal of Emerging and Selected Topics in Power Electronics, 2023, 11(1): 384–395.
60. Shang Y, Liu K, Cui N, et al. A Compact Resonant Switched-Capacitor Heater for Lithium-Ion Battery Self-Heating at Low Temperatures [J]. IEEE Transactions on Power Electronics, 2020, 35(7): 7134–7144.
61. Hu Z, Li Y, Liu F, et al. Thermal Characteristics Investigation of Lithium-Ion Battery Under High-Frequency AC Excitation in Low-Temperature Environment [J]. IEEE Transactions on Transportation Electrification, 2022, 8(1): 407–419.

62. Li Y, Gao X, Qin Y, et al. Drive circuitry of an electric vehicle enabling rapid heating of the battery pack at low temperatures [J]. iScience, 2021, 24(1): 101921.
63. Qin Y, Du J, Lu L, et al. A rapid lithium-ion battery heating method based on bidirectional pulsed current: Heating effect and impact on battery life [J]. Applied Energy, 2020, 280: 115957.
64. Ghassemi A, Hollenkamp A F, Chakraborty Banerjee P, et al. Impact of high-amplitude alternating current on LiFePO4 battery life performance: Investigation of AC-preheating and microcycling effects [J]. Applied Energy, 2022, 314: 118940.

Chapter 4
Modeling for AC Heating

Abstract Experimental heating tests are both time-consuming and costly, which makes model-based approaches essential for designing efficient alternating current (AC) heating. The development of rapid internal-heating strategies for lithium-ion batteries (LIBs) in low-temperature environments requires accurate simulation models. During the battery heating process, both electrical parameters and temperature can affect the heating performance. Therefore, integrating the electrical and thermal models is essential to establish an electro-thermal coupling framework. However, low-temperature operation significantly degrades battery performance due to the strong nonlinearity of electrode processes. These processes are highly sensitive to temperature, state of charge (SOC), and current amplitude. As a result, LIBs exhibit complex, time-varying, and nonlinear behaviors, making it challenging to accurately model heat generation and aging mechanisms under low-temperature conditions. Current research primarily focuses on two types of models: the electrochemical-thermal coupling model and the equivalent circuit-thermal coupling model. The electrochemical-thermal model, based on detailed electrochemical mechanisms (e.g., pseudo-two-dimensional models), can accurately describe internal battery states such as lithium-ion concentration gradients and lithium plating risks. However, its high computational complexity limits its real-time application. Conversely, the equivalent circuit-thermal model, which simplifies battery's electrical behavior using components like resistors and capacitors, offers computational efficiency and is suitable for real-time control. However, it lacks the ability to reveal internal electrochemical processes. Then, this chapter introduces the application of these two models in the field of AC heating. An electrochemical-thermal coupling model of LIBs was established to investigate the effects of low temperatures on battery behavior. Simulation results revealed that at low temperatures, the rate of electrochemical reactions, liquid-phase ionic conductivity, and diffusion coefficients all influence the impedance characteristics of the battery to varying degrees. It was also identified that lithium plating is the primary failure mode during low-temperature AC heating. However, the high computational requirements of the electrochemical model make it impractical for real-time application. To address this issue, this chapter proposes an equivalent circuit model (ECM) based on three-electrode battery. The reference electrode enables this model to effectively monitor lithium plating, but the structural

J. Zhu et al., *Alternating Current (AC) Heating for Lithium-Ion Batteries in Electric Vehicles*, https://doi.org/10.1007/978-981-96-9071-8_4

modifications required make it unsuitable for widespread execution. Therefore, an equivalent circuit-thermal coupling model based on full battery was also proposed, offering a critical tool for low-temperature AC heating optimization.

Keywords Lithium-ion battery · AC heating · Electrochemical thermal coupling model · Equivalent circuit model · Three-electrode battery

4.1 Lithium-Ion Battery Model

LIBs rely on precise modeling methods for in-depth research on performance optimization and safety management. LIB modeling is mainly divided into two categories: electrical models and thermal models, which are used to describe the electrical and thermal behaviors of batteries, respectively. Electrical models, based on their complexity and application scenarios, can be categorized into ECM and electrochemical models. ECM simulate the electrical characteristics of batteries, such as voltage, current, and capacity, by utilizing components like resistors and capacitors. This approach makes them well-suited for rapid simulation and real-time control. Electrochemical models are based on the internal electrochemical reaction mechanisms of batteries, describing processes such as diffusion, migration, and reaction kinetics of lithium ions (Li^+). They are particularly appropriate for in-depth analysis of internal battery behavior. And thermal models analyze the temperature distribution and thermal runaway risks of batteries through lumped parameter or distributed parameter methods, providing a basis for the design of thermal management systems. Electrochemical-thermal coupling models and equivalent circuit-thermal coupling models combine electrochemical mechanisms or equivalent circuits with thermal effects, respectively, to achieve more comprehensive performance predictions. Additionally, these models play an important role in the study of AC heating technology, offering theoretical support for understanding aging mechanisms, optimizing heating strategies, and improving heating efficiency and safety.

4.1.1 Electrochemical Thermal Coupling Model

The electrochemical-thermal coupling model utilizes electrochemical mechanisms to calculate the heat generation processes of the battery, and then combines the heat dissipation processes of the battery to calculate its temperature. The relationship between the model parameters and temperature is calibrated in an offline manner. During the calculation process, the relevant parameters of the electrochemical model are corrected by temperature, thereby achieving the coupling between the electrical and thermal characteristics of the batteries, as shown in Fig. 4.1.

The commonly used electrochemical mechanism model is the pseudo-two-dimensional (P2D) model established based on the porous electrode theory [1, 2].

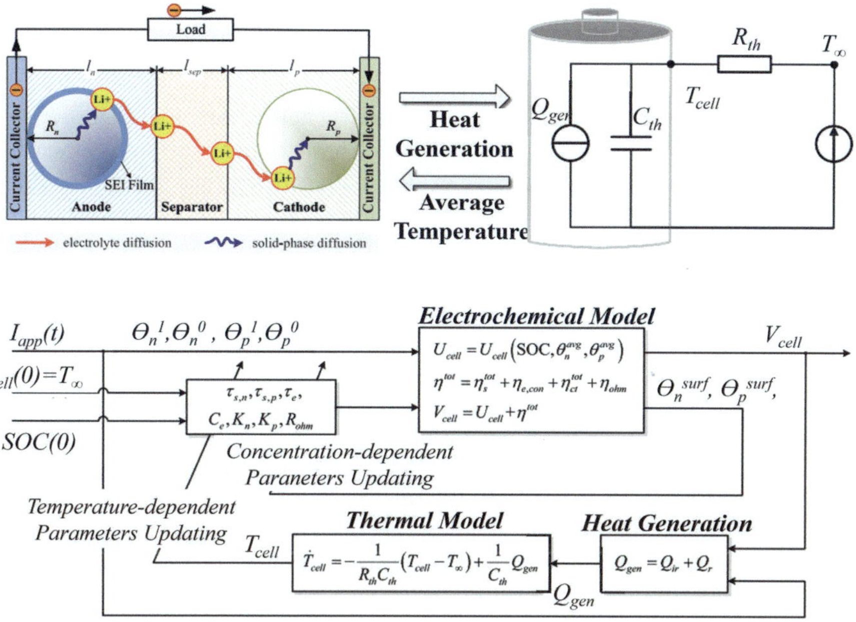

Fig. 4.1 The Principle of Electrochemical-Thermal Coupling Model of Lithium-Ion Batteries [4]

The P2D model, proposed by Newman et al. [3] is commonly used electrochemical models for LIBs. This model integrates porous electrode theory and concentrated solution theory. It accurately describes the diffusion and migration of lithium ions within the battery. Additionally, it captures the electrochemical reactions occurring at the surface of active particles. These processes are mathematically represented through a series of partial differential equations (PDEs) as shown below. Equations (4.1) to (4.5) describe the conservation of charge in the liquid phase, conservation of charge in the solid phase, conservation of mass in the liquid phase, conservation of mass in the solid phase, and the reaction at the solid–liquid interface, respectively.

$$\kappa^{eff}\frac{\partial^2\varphi_e}{\partial x^2}+\frac{2RT\kappa^{eff}(t^+-1)}{F}(1+\frac{dlnf_+}{dlnc_e})\frac{\partial^2 lnc_e}{\partial x^2}+j_f=0 \tag{4.1}$$

$$\sigma^{eff}\frac{\partial^2\varphi_e}{\partial x^2}=j_f \tag{4.2}$$

$$\varepsilon_e\frac{dc_e}{dt}=D^{eff}\frac{\partial^2 C_e}{\partial x^2}+(1-t_+)\frac{jf}{F} \tag{4.3}$$

$$\frac{\partial c_s}{\partial t}=\frac{D_s}{r^2}\frac{\partial}{\partial r}\left(r\frac{\partial c_s}{\partial r}\right) \tag{4.4}$$

$$j_f = a_s i_0[\exp(\frac{\alpha_a F}{RT}\eta) - exp(-\frac{\alpha_c F}{RT}\eta)] \tag{4.5}$$

The P2D model can describe the internal states within the LIBs, thus the electrochemical-thermal coupling model is often used to construct multi-dimensional numerical models of batteries, thereby guiding battery design and analysis of internal mechanisms. Hou et al. [5] constructed an electrochemical-thermal coupling model for a 155 Ah prismatic battery and analyzed the temperature distribution inside the battery. Wang et al. [6] used Comsol to build an electrochemical-thermal coupling model for a coin cell and investigated the impact of inconsistent electrode temperatures on lithium plating in the battery. Similarly, Mei et al. [7] also used the electrochemical-thermal coupling model to study the distribution of lithium-ion concentration inside the battery and analyzed the risk locations of lithium plating within the battery. In addition, Zhang et al. [8] designed a digital battery using the electrochemical-thermal coupling model to guide the design and manufacturing of high-power batteries. Gao et al. [9] improved the charging scheme based on the electrochemical-thermal model, proposing a health-aware fast charging strategy that balances charging time and battery aging. Cai et al. [2] formulated a three-dimensional electrochemical model for a 58Ah pouch-type LIBs and applied a particle swarm optimization algorithm to analyze the impact of tab positions on lithium plating, reaching similar conclusions to those in the literature [10].

The precision of electrochemical models often comes with high computational complexity and long computation times, which limit their use in real-time applications and large-scale battery management systems. Therefore, simplifying electrochemical models to reduce computational complexity is of great significance for enhancing the practicality and real-time performance of the models. There are three main approaches to simplifying the P2D model: simplification based on the model itself [11, 12], simplification based on parameters [13, 14], and simplification based on PDEs [15, 16].

(1) Simplification based on the Model Itself

Due to the large number of parameters and high computational complexity of electrochemical mechanism models, it is difficult to apply them online. To address this challenge, researchers have simplified the complex P2D model and constructed various simplified electrochemical models for the design of electro-thermal coupling models, such as the single particle model (SPM) and the lumped single particle model (LPM), etc. Li et al. [4] designed an electrochemical-thermal coupling model for lithium-ion batteries based on an improved SPM, and achieved the prediction of available capacity and power based on this model. Hwang et al. [12] also used the SPM to design an electrochemical-thermal coupling model that incorporates SEI growth, and based on this model, they designed a non-destructive charging strategy.

(2) Simplification based on Parameters

Many parameters of the P2D model are time-varying and difficult to measure accurately, so selecting some values that significantly affect the accuracy of model can more effectively parameterize the model of battery. This optimizes the model's predictive performance, ensuring the model better represents the battery's actual performance across different operating scenarios with lower computational demand. These parameters are often referred to as sensitive or dynamic parameters. Zhao et al. [13] utilized the Morris method to assess the impact of model parameters on simulated voltage and temperature. Their findings indicated that 12 parameters, such as the thickness of the two electrodes, porosity, and particle radius, can be considered constant in new batteries or during a single cycle. In contrast, 4 parameters, including the diffusion coefficients of the positive and negative electrodes and the reaction rate constants, exhibit complex variations under dynamic conditions. Li et al. [14] used a similar method to conclude that the solid-phase diffusion coefficient and reaction rate constant are sensitive parameters of the SPM and applied the simplified model to SOH estimation and remaining life prediction.

(3) Simplification based on Partial Differential Equations

PDEs are the main source of computational load in the P2D model, and using polynomial approximations is another common method for model simplification. Deng et al. [15] used a cubic polynomial in Eq. (4.6) to approximate the conservation of charge in the solid phase in Eq. (4.2). Han [16] used a quadratic polynomial in Eq. (4.7) to approximate the conservation of mass in the liquid phase in Eq. (4.3). These methods effectively reduce the computational load of the electrochemical model.

$$j_f(x) = ax^3 + bx^2 + cx + d, 0 \leq x \leq L \tag{4.6}$$

$$c_e(x,t) = \begin{cases} f_n(x,t) = a_1x^2 + a_2, 0 \leq x \leq L_n \\ f_{sep}(x,t) = a_3x^2 + a_2x + a_1, L_n \leq x \leq L_n + L_{sep} \\ f_p(x,t) = a_6(L-x)^2 + a_7, L_n + L_{sep} \leq x \leq L \end{cases} \tag{4.7}$$

4.1.2 Equivalent Circuit Thermal Coupling Model

The ECM simulates the macroscopic electrical characteristics of a battery by combining electronic components such as capacitors, resistors, and inductors. This method has the advantages of moderate design difficulty, ease of calculation, good simulation of both dynamic and static effects of the battery, and applicability to various types of batteries. Moreover, it can adjust the circuit according to actual battery test results to reduce system errors, making it highly practical. Standard equivalent circuit models in practice involve: the Rint model [17–19], Thevenin model [20–22], second-order and higher-order RC models [17, 23], and the PNGV model [24, 25], etc. Their comparisons are shown in the Table 4.1.

Compared with the electrochemical-thermal coupling model, the equivalent circuit thermal coupling model is directly based on the battery's current and voltage

Table 4.1 Common equivalent circuit models

Type	Circuit Diagram	Advantage	Disadventage
Rint model	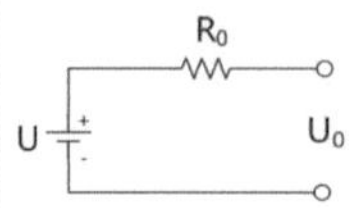	Simple structure, ease of calculation	Inability to simulate dynamic characteristics, low accuracy
Thevenin model	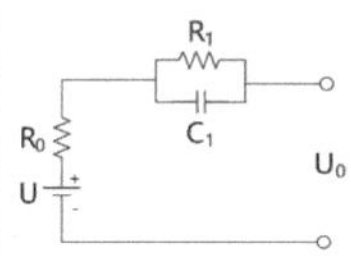	Reflection of polarization phenomena	Ignorance of overcharging and over-discharging
Second-order RC model	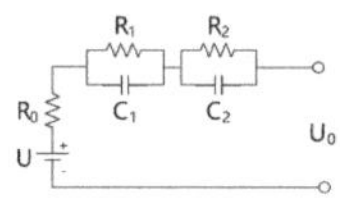	Accurate simulation of dynamic characteristics	Difficulty in parameter identification, high computational load
PNGV model	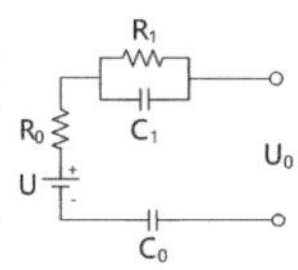	Accurate simulation of dynamic characteristics	Difficulty in parameter identification

response for modeling. Although it does not reflect the internal mechanism of the battery, due to its low model structure complexity and fewer parameters, it is convenient for online application and has strong practicality, thus receiving widespread attention.

Saw et al. [26] established an empirical electrical characteristic model for $LiFePO_4$ batteries, and calculated the battery temperature using a lumped thermal model to correct the parameters of the empirical electrical characteristic model. Lin et al. [27] used an ECM to construct an electrothermal coupling model for cylindrical LIB, and based on this model, they achieved the estimation of the battery's core temperature. In addition, during the low-temperature heating application process, Ruan et al. [28] used a fractional-order model to construct an electrothermal coupling model for high-frequency AC heating design, and the parameters of this model were calibrated through EIS tests at different temperatures.

4.1.3 Models for AC Heating

Researchers have constructed some models for AC heating in low-temperature environments based on above two types of models. Qin et al. [29] introduced a coupling electro-thermal model by simplifying the electrochemical submodel. In reference [30], an electro-thermal coupling model was constructed using the P2D model. While these methods achieve precise voltage and thermal predictions, their computational demands escalate significantly owing to the electrochemical model's intricate nature. Furthermore, parameter calibration typically focuses on low-frequency domains, potentially compromising high-frequency response accuracy. In response to this challenge, several equivalent circuit-thermal coupling models have been proposed for low-temperature AC heating applications, including those by Wu et al. [31], Jiang et al. [32], and Huang et al. [33]. The model parameters were calibrated using the EIS tests. Research by Huang et al. [34] revealed that AC amplitude significantly affects battery impedance at low temperatures. Although Meyers et al. [35] accounted for amplitude variations in their electro-thermal model, the extent of amplitude-dependent modeling errors remains unquantified. Therefore, to develop an accurate electro-thermal coupling model for AC heating strategy design, a thorough investigation into the influence of different AC amplitudes on model accuracy is essential.

Sections 4.2 and 4.3 delineate the implementation of two distinct modeling frameworks for low-temperature AC heating in LIBs: the electrochemical-thermal coupling model and the equivalent circuit thermal coupling model. The electrochemical-thermal model serves to elucidate the underlying reaction kinetics and degradation mechanisms under cryogenic conditions, providing critical insights into battery aging phenomena. Conversely, the ECM is leveraged to systematically evaluate and optimize AC heating strategies, with the dual objectives of maximizing thermal efficiency and ensuring operational safety. Experimental validation corroborates the predictive accuracy and practical applicability of both modeling approaches.

4.2 Model for Mechanism Analysis of AC Heating

From the previous analysis, it is evident that as temperatures decrease, the energy and power characteristics of LIBs severely deteriorate. Through literature review and experimental testing, the key factors constraining the low-temperature performance of LIBs have been identified. The low-temperature characteristics of batteries are constrained by internal material conductivity, reaction rates, diffusion coefficients, and other characteristic parameters. Moreover, LIBs face challenges in charging at low temperatures, as it can lead to lithium plating on the surface of the graphite particles at the anode. As charging duration and the number of charging cycles increase, lithium plating can gradually form dendrites. When lithium dendrites penetrate the battery separator, they can cause internal short circuits, leading to the end of the battery's lifespan and even triggering thermal runaway, significantly reducing the safety of battery.

Due to the numerous and complex internal parameters of batteries, accurately obtaining them through experimental measurements poses certain challenges. The electrochemical model can simulate the internal state of a battery and, under the premise of accurately acquiring internal parameters, intuitively observe changes in the internal parameters of power batteries. The porous electrode theory for LIBs proposed by Thomas et al. [3] has long been recognized as an important foundational theory for describing electrode processes. The internal parameters of this model have clear physical significance. The modeling theory describes the internal electrode processes of batteries from the perspectives of charge balance, mass balance, and electrode reaction kinetics.

In this chapter, by designing coin-type half-cell configurations for single-electrode electrochemical experiments, the conductivity, diffusion coefficient, and chemical reaction activation energy parameters of electrode materials are measured. This allows for the study of the relationship between key internal characteristic parameters and temperature. Subsequently, based on the porous electrode theory, an electrode process model for LFP system power batteries is established. This model accurately reflects the internal state of batteries in both the time and frequency domains. Finally, based on this model, the influence of low-temperature characteristic parameters on the impedance spectrum is investigated through simulation.

4.2.1 Electrochemical Model

From a geometric structure perspective, the most basic reaction unit of a LIB consists of a thin sheet-like negative electrode, a separator, a positive electrode, and an electrolyte. It can be assumed that the physical properties within each sheet-like material are uniform, and only the physical processes along the battery-unit's thickness direction need to be considered. On this foundation, the electrochemical model adopted in this chapter is a single battery reaction unit, which includes the negative electrode current collector, the negative electrode active material, the separator, the positive electrode active material, and the positive electrode current collector from left to right (as shown in Fig. 4.2). The solid black line within the red dashed box in the Fig. 4.2 represents the simplified one-dimensional battery geometric model. Specifically, taking the negative terminal of the LIB as the origin, and the direction perpendicular to the electrode surface along the battery thickness as the x-axis, a one-dimensional coordinate system is established. The input to the model is the terminal current excitation at $x = 0$ (i_{apply}). The quantities to be solved in the model include: the solid and liquid phase lithium-ion concentrations (c_s, c_l), the solid and liquid phase potentials (ϕ_s, ϕ_l) , the solid and liquid phase currents (i_s, i_l), the non-Faradaic current (i_{dl}), and the reaction current (i_F), among other parameters. Based on the electrode processes, modeling is sequentially conducted for the conduction of electrons and ions, the charge transfer process of electrode reactions, as well as the non-Faradaic processes and diffusion processes. To facilitate analysis, the impedance is expressed based on the electrode processes, and this section assumes that the spatial (x) and temporal (t) distributions of parameters are neglected.

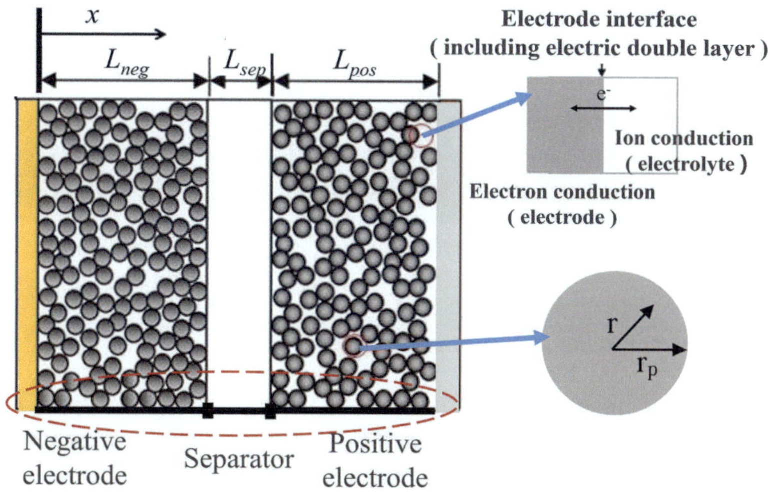

Fig. 4.2 Porous electrode theory-based 1D electrochemical model

4.2.1.1 Electron and Ion Conduction Process

Throughout the battery's charge and discharge cycles, electron transport in the solid-phase electrode occurs only between the current collector and the solid-phase electrode. The relationship between the solid-phase voltage and current follows Ohm's law, expressed as:

$$i_s = \sigma_s^{eff} \frac{\partial}{\partial x} \phi_s, \sigma_s^{eff} = \varepsilon_s^{1.5} \sigma_s \tag{4.8}$$

where i_s represents the solid-phase current, δ_s^{eff} represents the effective conductivity of the solid-phase porous electrode, ϕ_s represents the potential in the solid-phase region of the electrode, ε_s represents the porosity of the solid-phase porous electrode, indicating the volume fraction of pores in the solid-phase electrode, σ_s represents the conductivity of the solid-phase electrode material.

The transport process of charged ions in the liquid phase is manifested as the transfer of charged ions in the liquid electrolyte, which is influenced by ion mobility and liquid-phase conductivity. The corresponding modified formula is given as:

$$i_l = -\delta_l^{eff} \nabla \phi_l + \delta_D^{eff} \nabla (\ln c_l), \delta_l^{eff} = \varepsilon_l^{1.5} \sigma_l \tag{4.9}$$

where i_l represents the liquid-phase current; δ_l^{eff} represents the effective conductivity of the liquid phase, which is a function of the lithium-ion concentration in the liquid phase; ϕ_l represents the potential in the liquid-phase region of the electrode; δ_D^{eff} represents the effective ion mobility; c_l represents the concentration of lithium ions in the liquid phase; ε_l represents the porosity of the separator or porous electrode. And the specific expression for δ_D^{eff} is:

$$\delta_D^{eff} = \frac{2RT\delta_l^{eff}}{F}\left[1+\frac{\partial lnf_1}{\partial lnc_1}\right][1-t_+] \tag{4.10}$$

where R denotes the universal gas constant, T is the absolute temperature, and F stands for Faraday's constant. The mean molar activity coefficient of the electrolyte is represented by f_l, while c_l indicates the Li^+ concentration in the liquid phase. The transference number of Li^+ in the liquid phase is denoted by t_+, which remains constant in a homogeneous electrolyte solution.

4.2.1.2 Electrochemical Reaction Process

Due to the porous characteristics of the solid electrodes in LIBs, both two electrode coatings are regarded as be porous electrodes filled with electrolyte. At every point within the porous electrode, there exists a hypothetical reaction interface. The solid active material generally exhibits a relatively loose lattice structure, and this solid lattice is assumed to be an isotropic spherical particle. At every coordinate within the solid electrode, there exists a hypothetical solid active particle. The spherical solid active particle surface serves as the reaction site for the electrode process.

(1) Charge Transfer Process

At the interface where the active particles contact the electrolyte, charge transfer occurs due to the insertion and extraction of lithium ions. This process, representing the current of the electrode reaction at that location, can be described by the Butler–Volmer electrochemical equation [36–38]. To facilitate subsequent derivation and analysis, the linear calculation formula of this theory is selected as follows:

$$J_{loc} = \frac{i_0(\alpha_a+\alpha_c)F}{RT}[\phi_s-\phi_l-E_{eq}(SOC)] \tag{4.11}$$

where J_{loc} denotes the current passing through the unit area of the reaction interface; α_a denotes the cathode reaction conversion coefficient; α_c denotes the anode reaction conversion coefficient. ϕ_s denotes the potential of the solid phase region of the electrode, ϕ_l denotes the potential of the liquid phase region of the electrode, which is related to position and time; $E_{eq}(SOC)$ denotes the equilibrium potential of the active material relative to the reference electrode during lithium utilization.

The parameter i_0 denotes the exchange current density, which characterizes the absolute rate of oxidation and reduction reactions at the equilibrium potential. It is determined by the concentration of the reacting species and the activation free energy at the equilibrium potential. This parameter reflects the intrinsic reaction capability of the electrode at the equilibrium potential. It can be considered a constant when the electrode material, surface state, solution concentration, and temperature are determined.

In addition, in the formula, i_0 can also be expressed as a function of the concentration of lithium ions in the electrolyte and the electrode, calculated through the following formula [39]:

$$i_0 = Fk_0\sqrt{c_l(c_{s,max} - c_{s,surf})c_{s,surf}} \tag{4.12}$$

where k_0 represents the rate constant of the electrochemical reaction; c_l represents the concentration of Li^+ in the liquid phase; $c_{s,max}$ represents the maximum theoretical concentration of lithium ions in the active material of the solid electrode; $c_{s,surf}$ represents the concentration of Li^+ at the solid–liquid phase reaction interface.

In the J_{loc} formula, $E_{eq}(SOC)$ can be obtained by the open circuit voltage (OCV) method. SOC represents the state of charge, which indicates the degree of saturation of Li^+ (active material) in the electrode. It is calculated through the following formula:

$$SOC = \frac{C_{s,surf}}{C_{s,max}} \tag{4.13}$$

(2) Electrical Double Layer Process (Non-Faradaic Process)

The interface between the solid and liquid phases of the electrode is the site where electrochemical reactions occur. Its structure and properties significantly influence the reaction rate and mechanism. At this interface, the electrode process encompasses not only charge transfer, but also the electric double layer's charging between the electrolyte and the solid matrix. This electric double layer capacitance originates from the non-uniform distribution of charged particles at the solid–liquid interface. This distribution disrupts the electrical neutrality of the isolated phases. The charging process of the electric double layer is a non-Faradaic process. The current passing through it is the non-Faradaic current. The factor causing the non-Faradaic current is the change in the relative potential across the reaction interface between the solid and liquid phases. Furthermore, from a general electrode system's impedance characteristics, it can be concluded that:

$$i_{dl} = C_{dl}\left(\frac{\partial \phi_s}{\partial t} - \frac{\partial \phi_l}{\partial t}\right). \tag{4.14}$$

where i_{dl} represents the non-Faradaic current at the electrode interface; and C_{dl} represents the electric double layer capacitance; ϕ_s represents the solid-phase potential; ϕ_l represents the liquid-phase potential.

(3) Diffusion Process

The diffusion process inside the battery includes the diffusion of Li^+ in the liquid phase and the diffusion within the lattice of the solid active material. First, the diffusion process of Li^+ in the liquid phase is analyzed. In porous electrode theory, the Li^+'s mass transfer process in the electrolyte mainly includes three modes: ionic conduction, diffusion, and convection. In the interior of LIBs, the convective effect of the electrolyte can be neglected. The process of ionic conduction has been discussed and analyzed in the previous text. Then, the diffusion process of lithium ions in the electrolyte will be discussed. The diffusion process is primarily related to changes in concentration. Considering each point in the electrolyte as a micro system, the change in liquid-phase concentration is equal to the lithium ion reaction source minus the inflow and outflow concentration. The formula is expressed as follows:

$$\varepsilon_l \frac{\partial c_l}{\partial t} = -\frac{i_F}{F} - \nabla\left(-D_l^{eff}\nabla c_l + \frac{i_l t_+}{F}\right) \tag{4.15}$$

In the formula, the primary term located on the right-hand side represents the source of lithium ion reactions. The first term inside the parentheses denotes the Li^+ outflow caused by diffusion. The following term represents the outflow of Li^+ caused by charge migration. Among them, i_F represents the reaction current at the electrode interface; F is Faraday's constant. And D_l^{eff} is the equivalent diffusion coefficient of Li^+ in the liquid phase, which is expressed as:

$$D_l^{eff} = \varepsilon_l^{1.5} D_l \tag{4.16}$$

In electrode process studies, liquid-phase diffusion is generally considered faster than solid-state diffusion and is not a rate-limiting step. Thus, the impact of liquid-phase mass transfer on lithium-ion intercalation kinetics is often negligible in research. The following analysis focuses on discussing the solid-state diffusion process.

In the theory of porous electrodes, it is assumed that the active material in the solid electrode is composed of spherical particles with a radius of r_p (Fig. 4.2). Throughout the charge–discharge cycles of batteries, lithium ion diffusion processes occur within and on the surface of the active particles.

In the solid electrode, the concentration distribution of Li^+ follows non-steady-state diffusion and obeys Fick's second law:

$$\frac{dc_s}{dt} = D_s \frac{\partial^2 c_s}{\partial r^2} \tag{4.17}$$

where c_s denotes the concentration of lithium ions in the solid active particles; D_s represents the lithium ions' diffusion coefficient in the solid electrode particles; r represents the radial distance between the Li^+ and the active particles (Fig. 4.2).

If $r = 0$, it indicates that the Li^+ are at the center of the particle, where there is no diffusion occurring. It can be derived that:

$$-D_s \left.\frac{\partial c_s}{\partial r}\right|_{r=0} = 0 \tag{4.18}$$

When $r = r_p$, lithium ions are located at the particle surface, i.e., the solid–liquid interface where electrochemical reactions occur. At this point, the generation and consumption rate of lithium ions is:

$$-FD_s\frac{\partial c_s}{\partial r}|_{r=r_p} = \frac{i_F}{A_F} = J_{loc} \tag{4.19}$$

where r_p represents the particle radius; i_F represents the reaction current at the electrode interface; A_F represents the area of the reaction interface of the spherical particle; J_{loc} represents the current passing through the unit area of the reaction interface.

(4) Charge Conservation

In the three sections of the previous text, a model analysis was conducted on three processes within the lithium-ion battery: electron and ion conduction, electrochemical reaction processes, and lithium ion solid–liquid phase diffusion. Six relationships were derived among eight unknowns: solid–liquid phase lithium ion concentrations (c_s, c_l), solid–liquid phase potentials (ϕ_s, ϕ_l), solid–liquid phase currents (i_s, i_l), non-Faradaic currents (i_{dl}), and reaction currents (i_F), see Eqs. (4.8), (4.9), (4.11), (4.14), (4.15), (4.19). Two more relationships are needed to solve for the above unknowns.

According to Kirchhoff's law, the total current through each interface in the thickness direction of the LIB is equal and matches the terminal current excitation of the battery. The current passing through the collector section is i_s; The current passing through the porous electrode section is i_s, i_l; The current passing through the separator section is i_j. If the solid phase current in the separator area is set to $i_s = 0$ and the liquid phase currents in the copper and aluminum areas are $i_l = 0$, then the current conservation equation can be obtained:

$$i_s + i_l = i_{apply} \tag{4.20}$$

Equation (4.19) merely describes the transfer process of lithium ions during the electrode reaction, where electrons should enter or leave the reaction interface synchronously with lithium ions. The entire process of solid-phase current reaction involves electron conduction. At the reaction interface, the solid-phase current gradually transforms into Faradaic current (reaction current) and non-Faradaic current. Thus, the following charge conservation relationship exists:

$$\nabla i_s = i_F + i_{dl} \tag{4.21}$$

Through the analysis above, a set of PDEs can be established. These equations describe the concentrations of lithium ions in the solid and liquid phases (c_s, c_l), the potentials in the solid and liquid phases (ϕ_s, ϕ_l), the currents in the solid and liquid phases (i_s, i_l), the non-Faradaic current (i_{dl}), and the reaction current (i_F). This set of equations holds at every point on the x-axis.

4.2.1.3 Formulation of AC Impedance Spectrum

In EIS, ohmic impedance R_0 is mainly reflected in the processes of electron transport in the solid electrode and ion transport in the liquid electrolyte. The current in the electrode reaction adheres to Ohm's law as it flows through the solid electrode and the liquid electrolyte. The impedance of electrons in the solid electrode is defined as the solid-phase ohmic internal resistance R_s, the impedance of charged ions in the liquid electrolyte is termed the liquid-phase ohmic internal resistance R_l. Consequently, the ohmic impedance R_0 is expressed as $R_0 = R_s + R_l$.

From Eq. (4.8), the ohmic impedance arising from electron conduction in the solid phase is simplified and represented as:

$$R_s = \frac{1}{\varepsilon_s^{1.5} \sigma_s(T)} \tag{4.22}$$

From Eq. (4.9), the ohmic impedance generated during the transport of lithium ions in the liquid phase can be simplified and expressed as:

$$R_l = \frac{1}{\varepsilon_l^{1.5} \delta_l(c_l, T)}. \tag{4.23}$$

Finally, electron transport in the solid electrode and lithium-ion transport in the liquid electrolyte are two distinct processes. Since the current passing through both processes is identical, the total voltage across the ohmic internal resistance is the sum of the voltages associated with these two processes. These processes are connected in series. Combining Eqs. (4.22) and (4.23), the expression for ohmic impedance in the electrochemical impedance spectrum is obtained as:

$$R_0 = R_s + R_l = \frac{1}{\varepsilon_s^{1.5} \sigma_s(T)} + \frac{1}{\varepsilon_l^{1.5} \delta_l(c_l, T)} \tag{4.24}$$

Analysis of Eq. (4.24) reveals that ohmic impedance is determined by the separator's and porous electrode's porosity (ε_l and ε_s), the conductivity of the solid electrode material σ_s, and the conductivity of the liquid phase δ_l. Among these, the porosity of the separator and porous electrode is defined by the battery itself. The conductivity of the solid electrode varies with temperature. The conductivity of the liquid phase is affected by the concentration of lithium ions and temperature in the liquid electrolyte. Generally, the concentration of lithium ions in the liquid electrolyte is difficult to measure and is closely related to SOC of the battery [39, 40]. It can be considered that the factors affecting ohmic impedance can be summarized as T and SOC of the battery.

From the analysis of the equivalent impedance model (Fig. 4.3), it is found that the EIS includes impedances of electrochemical reactions and diffusion processes. These include internal resistance R_{ct} of the charge transfer process in electrochemical reactions, capacitance C_{dl} of the electric double layer charging process, and

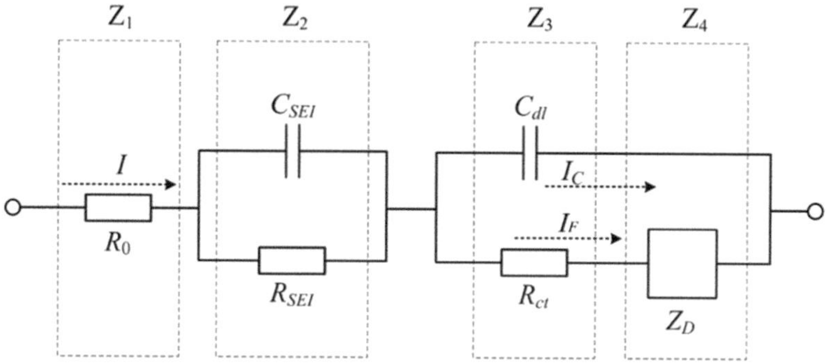

Fig. 4.3 Equivalent impedance model based on EIS analysis

impedance $Z_D(f)$ of the diffusion mass transfer process. The diffusion mass transfer process was previously described using Z_D. In this chapter, considering the factor of frequency, it is described as $Z_D(f)$. The series and parallel relationships between the impedances of each process and the capacitance are determined by the corresponding current and voltage relationships in each process.

The charge transfer process in electrochemical reactions and the diffusion mass transfer process occur continuously, both impeding the Faradaic potential. The sum of these is referred to as the Faradaic impedance $Z_F(f)$. The charging process of the electric double layer is a non-Faradaic process. The capacitance C_{dl} is referred to as the non-Faradaic impedance. The polarization current generated by the electrode reaction process is partly used to charge the electric double layer. The other part flows through the Faradaic impedance. Moreover, the polarization overpotential on the solid–liquid interface is equal to the voltage across the electric double layer capacitor. Therefore, the Faradaic impedance $Z_F(f)$ and the electric double layer capacitance C_{dl} are in parallel, which leads to the following expression:

$$Z_i^{-1}(f) = Z_{dl}^{-1}(f) + Z_F^{-1}(f) \tag{4.25}$$

where $Z_i(f)$ represents the impedance of the electrochemical reaction/diffusion process [40]; f represents the excitation frequency; $Z_{dl}(f)$ represents the impedance corresponding to the electric double layer capacitance; $Z_F(f)$ represents the impedance in the Faradaic process.

First, the impedance associated with the double-layer capacitor is analyzed. According to Eq. (4.14) and the basic theory of electrical circuits, the double-layer capacitor's impedance can be simplified and expressed as:

$$Z_{dl}(f) = (j2\pi f C_{dl})^{-1}, j = \sqrt{-1} \tag{4.26}$$

The impedance in the Faradaic process $Z_F(f)$, which includes the impedance of the electrochemical reaction charge transfer process R_{ct} and the diffusion mass transfer process $Z_D(f)$. Next, a detailed analysis of $Z_F(f)$ will be conducted.

According to the analysis of the equivalent impedance model, the internal resistance R_{ct} of the electrochemical reaction charge transfer process and the impedance $Z_D(f)$ of the diffusion mass transfer process are in series. That is,

$$Z_F(f) = R_{ct} + Z_D(f) \tag{4.27}$$

Combining Eqs. (4.11), (4.12), (4.19) and through analysis [35, 41, 42], the impedance expression corresponding to the Faradaic process is obtained as:

$$R_{ct} = RTc_{s,\max}^{-1}F^{-2}k_0^{-1}(\alpha_a + \alpha_c)^{-1}c_l^{-\frac{1}{2}}(1 - SOC)^{-\frac{1}{2}}SOC^{-\frac{1}{2}} \tag{4.28}$$

$$Z_D(f) = r_p(3D_sC_D)^{-1}(j2\pi f/D_s)^{-1/2}(\cot\sqrt{j2\pi fr_p^2/D_s} - 1)^{-1} \tag{4.29}$$

where C_D represents the boundary value of the electrode particle capacitance characteristic in the low-frequency region.

By analyzing Eqs. (4.26), (4.28), and (4.29), the impedance $Z_i(f)$ is primarily influenced by the temperature, SOC, the concentration gradient of Li^+ in the solid phase, the concentration of Li^+ in the liquid phase c_l, the diffusion coefficient of lithium ions in the solid phase D_s. Among these, the changes of concentration in the solid and liquid phases and the diffusion coefficient in the solid phase of lithium ions are also influenced by T and SOC. Therefore, the influencing factors of $Z_i(f)$ can ultimately be summarized as T and SOC of LIBs.

Finally, for LIBs, disregarding the internal electrode SEI film impedance [43, 44], and combining the analysis of EIS along with the series and parallel relationships of impedances, the model's total impedance can be denoted as:

$$Z(f) = R_0 + Z_i(f) = R_0 + \frac{R_{ct} + Z_D(f)}{1 + j2\pi fC_{dl}[R_{ct} + Z_D(f)]} \tag{4.30}$$

In summary, the ohmic internal resistance R_0, which includes the processes of electron transport in the solid electrode and Li^+ transfer in the liquid electrolyte, is largely influenced by T and SOC. The impedance Z_f corresponding to the electrochemical reaction/diffusion process includes the internal resistance R_{ct} of the charge transfer process, the capacitance C_{dl} of the electric double layer charging process, and the impedance $Z_D(f)$ of the diffusion mass transfer process. If not considering the impact of the SOH, it's also mainly influenced by the temperature T and SOC.

EIS is composed of the impedance of the battery measured at multiple frequency points of excitation, corresponding to the AC impedance at different excitation frequencies. Furthermore, the values of EIS exhibit a discrete characteristic.To refine its mathematical description, this section employs a matrix-based analytical approach to the EIS data. According to Ohm's law, the derivation is:

$$ZI = U \tag{4.31}$$

where I denotes the excitation current vector matrix; U denotes the response voltage vector matrix; Z denotes the impedance vector matrix.

As the excitation frequency varies $f_x, x \in [1, 2, \ldots, n]$, the parameter matrices can be represented as:

$$I = \begin{pmatrix} I(f_1) & & \\ & \ddots & \\ & & I(f_n) \end{pmatrix}; U = \begin{pmatrix} U(f_1) & & \\ & \ddots & \\ & & U(f_n) \end{pmatrix}; Z = \begin{pmatrix} Z(f_1) & & \\ & \ddots & \\ & & Z(f_n) \end{pmatrix} \tag{4.32}$$

After performing matrix operations $Z = U \cdot I^{-1}$, and representing the current vector as $I(f_x)$, the voltage vector as $U(f_x)$, the impedance vector as $Z(f_x)$, the matrix formula can be transformed with the variation of the excitation frequency into:

$$Z(f_x) = U(f_x) \cdot I(f_x)^{-1}, x \in [1, 2, \cdots, n] \tag{4.33}$$

The voltage vector matrix can be expressed using trigonometric functions as: $U(f_x) = |U(f_x)| \cos(2\pi f_x t)$, so

$$U(f_x) = \mathrm{Re}\{|U(f_x)| \exp(j2\pi f_x t)\}, x \in [1, 2, \cdots, n] \tag{4.34}$$

The current vector matrix can be expressed using trigonometric functions as:

$$I(f_x) = |I(f_x)| \cos\left[2\pi f_x t + \varphi(f_x)\right] \tag{4.35}$$

$$I(f_x) = \mathrm{Re}\{|I(f_x)| \exp[j\varphi(f_x)] \exp(j2\pi f_x t)\}, x \in [1, 2, \cdots, n] \tag{4.36}$$

Thus, the impedance matrix vector can be defined as:

$$Z(f_x) = \frac{U(f_x)}{I(f_x)} \exp[j\varphi(f_x)]^{-1} \tag{4.37}$$

$$Z(f_x) = |Z(f_x)| \{\cos\left[\varphi(f_x)\right] - j \sin\left[\varphi(f_x)\right]\}, x \in [1, 2, \cdots, n] \tag{4.38}$$

where

$$\mathrm{Re}[Z(f_x)] = |Z(f_x)| \cos[\varphi(f_x)] \tag{4.39}$$

$$\mathrm{Im}[Z(f_x)] = |Z(f_x)| \sin\left[\varphi(f_x)\right] \tag{4.40}$$

For electrochemical impedance spectroscopy, the imaginary part is represented on the y-axis as $\mathrm{Im}[Z(f_x)]$, so $Z(f_x)$ can be expressed as [37]

$$Z(f_x) = |Z(f_x)|\{\cos[\varphi(f_x)] + j\sin[\varphi(f_x)]\} \tag{4.41}$$

This can be further expressed as:

$$Z(f_x) = |Z(f_x)|\exp[j\varphi(f_x)], x \in [1, 2, \cdots, n] \tag{4.42}$$

From this, the phase angle and magnitude of the AC impedance vector corresponding to Fig. 4.4 can be expressed as:

$$\varphi(f_x) = \tan^{-1}\left\{\frac{-\mathrm{Im}\{Z(f_x)\}}{\mathrm{Re}\{Z(f_x)\}}\right\}, x \in [1, 2, \cdots, n] \tag{4.43}$$

$$|Z(f_x)| = \sqrt{\mathrm{Re}\{Z(f_x)\}^2 + \mathrm{Im}\{Z(f_x)\}^2}, x \in [1, 2, \cdots n] \tag{4.44}$$

Finally, according to Eq. (4.44), as the excitation frequency varies, for $f_x \in [1, 2, \ldots, n]$, the AC impedance, phase angle and magnitude vector matrix can be expressed as:

$$Z = \begin{pmatrix} |Z(f_1)|\exp[j\varphi(f_1)] & & 0 \\ & \ddots & \\ 0 & & |Z(f_n)|\exp[j\varphi(f_n)] \end{pmatrix} \tag{4.45}$$

$$\varphi = \begin{pmatrix} \tan^{-1}\left\{\frac{-\mathrm{Im}\{Z(f_1)\}}{\mathrm{Re}\{Z(f_1)\}}\right\} & & 0 \\ & \ddots & \\ 0 & & \tan^{-1}\left\{\frac{-\mathrm{Im}\{Z(f_n)\}}{\mathrm{Re}\{Z(f_n)\}}\right\} \end{pmatrix} \tag{4.46}$$

$$|Z| = \begin{pmatrix} \sqrt{\mathrm{Re}\{Z(f_1)\}^2 + \mathrm{Im}\{Z(f_1)\}^2} & & 0 \\ & \ddots & \\ & & \sqrt{\mathrm{Re}\{Z(f_n)\}^2 + \mathrm{Im}\{Z(f_n)\}^2} \end{pmatrix} \tag{4.47}$$

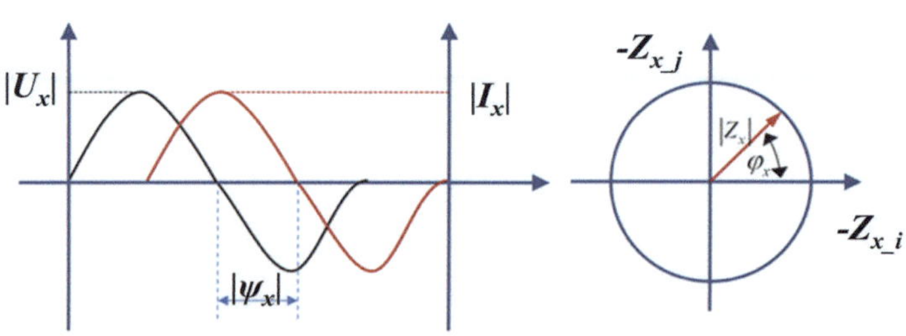

Fig. 4.4 AC Impedance Vector Relationship Diagram

4.2.2 *Electrochemical Model Establishment and Verification*

Due to the time-varying and nonlinear nature of LIBs, their internal parameters often change with the state of the battery. Therefore, accurately obtaining the physical parameters of the positive and negative active materials and considering the coupling of these parameters with other battery state variables is fundamental to accurately establishing a battery model. Typically, the LIBs' electrochemical performance can be analyzed using electrochemical cells in the form of full-cells or half-cells. A full-cell refers to the complete form of battery discussed in this paper, where electrochemical reactions occur at both the positive and negative electrodes. It allows for direct measurement of the battery's characteristics and performance. If an additional reference electrode is used, the full-cell can obtain the electrode potentials of the positive and negative electrodes through separate measurements. The reference electrode in the half-cell simplifies measurement and analysis of reactions. It helps to understand the basic performance of each electrode material.

In this section, we take the sample 8Ah $LiFePO_4$/C battery (Sample B) as the research subject, design coin-type half-cells for its positive and negative electrode materials, and conduct single-electrode experimental tests and analysis. On the one hand, this serves to verify the impedance analysis based on the electrode process from the previous section. On the other hand, it aims to obtain the coupling patterns of some relevant internal characteristics of battery with temperature.

4.2.2.1 Design of Coin-Type Half-Cell Single Electrode Electrochemical Experiments

In this part, LFP and graphite coin-type half-cells are prepared, with the specific manufacturing dimensions shown in Table 4.2.

The positive electrode material for the coin-type half-cell is selected from the same batch of 8Ah LFP battery's positive electrode material, specifically $LiFePO_4$ active material. Lithium metal foil is employed as the negative electrode material. The negative electrode material of the battery (graphite active material) is selected as the positive electrode material for the coin-type half-cell, with lithium metal foil also used as the negative electrode material. Additionally, the electrolyte and separator

Table 4.2 Manufacturing dimensions of coin-type half-cells

LFP coin-type half-cell: LFP active material			
LFP Electrode thickness	Separator thickness	Lithium metal thickness	Diameter
250 μ m	40 μ m	25 μ m	1.3 cm
Graphite coin-type half-cell: graphite active material			
Graphite electrode thickness	Separator thickness	Lithium metal thickness	Diameter
200 μ m	40 μ m	25 μ m	1.3 cm

of above coin-type half-cells are identical to those used in sample battery., ensuring the feasibility of comparative analysis.

4.2.2.2 Parameter Measurement and Analysis

This paper conducts single-electrode electrochemical experimental tests on coin-type half-cells. The main focus is on determining the parameters of the solid active materials (LFP, graphite) in LIBs, including the equilibrium electrode potential E_{eq}, the solid phase diffusion coefficient, and the conductivity of the active materials. The equilibrium electrode potentials of the positive and negative electrode materials can be measured using the OCV method. The conductivity of the active materials is estimated through EIS. The solid phase lithium ion diffusion coefficient is measured using the galvanostatic intermittent titration technique (GITT) [45–47].

(1) Equilibrium Electrode Potential E_{eq}

The equilibrium electrode potential is defined as the electrode potential achieved when the rates of the microscopic redox reactions on the electrode are equal. Under this condition, the macroscopic current density on the electrode becomes zero, and no current passes through the electrode. This is the potential generated by the reaction between the electrode and the reference electrode. The measurement of the electrode potential is essentially the measurement of electromotive force (EMF) of the cell. In this paper, the equilibrium electrode potentials of the electrodes are all relative to the counter electrode lithium (vs. Li^+/Li). The measurement method involves using the OCV method to measure the charge–discharge equilibrium potentials of the coin-type half-cells, with the lithium metal foil serving as the reference electrode. The discharge equilibrium potential is measured using the following method. First, the battery is fully charged employing a CC–CV charging protocol. Subsequently, the battery is discharged to a specific SOC point at 25 °C with a 0.1C rate. After reaching the desired SOC, the battery is allowed to rest sufficiently. At intervals of $\Delta SOC = 0.1$, the battery's terminal voltage is recorded. The measurement of the equilibrium potential of the electrode during charging is similar, and will not be repeated here. This method is an approximate calculation method, and the obtained data serve as input values for the electrode process model. The equilibrium electrode potential data for the coin-type half-cells are obtained as shown in Fig. 4.5.

Figure 4.5a shows the measured equilibrium potential of the positive electrode LFP, and Fig. 4.5b shows the measured equilibrium potential of the negative electrode graphite. The variation pattern of the equilibrium potential is related to the thermodynamic principles of the battery. During different stages of the charge and discharge process, the concentration of lithium ions in electrode material. And the gradual change of reactants and products affect the change in free energy and voltage. The OCV of an individual battery can be expressed as the difference between the equilibrium potentials of the positive and negative electrode materials. It is an important indicator for measuring the performance of the battery, and it is related to the remaining capacity. Research has shown that for LIBs with different temperatures,

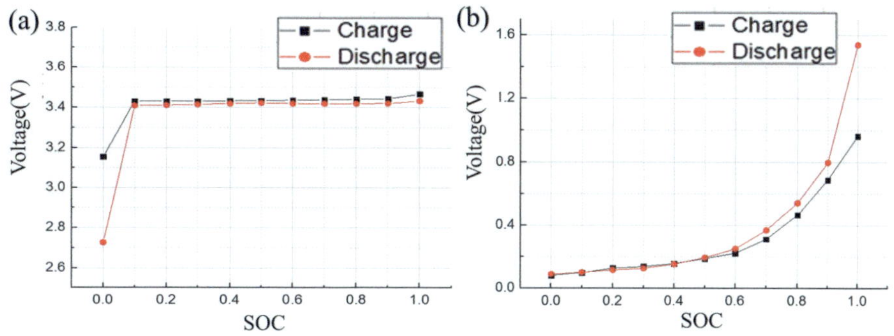

Fig. 4.5 The equilibrium potential of LIB active material and graphite active material. **a** LFP/Li. **b** Graphite/Li

different currents (the same capacity), and different degrees of aging, the OCV characteristics are almost unaffected [48]. There are numerous studies that use the OCV of battery to determine the SOC [49, 50]. This approach essentially based on the functional relationship between the equilibrium potentials of the internal positive and negative electrode materials and the concentration of Li^+.

(2) Conductivity of the Active Materials

This part employs EIS to measure the active materials' conductivity in batteries, which includes both the solid electrode materials and the liquid electrolyte. As shown by Fig. 4.6, these are the AC impedance spectra obtained from the experimental setup under 20 °C ambient conditions for LFP/graphite coin-type half-cells at different SOC.

Conductivity is a manifestation of resistance. Derived from the analysis of EIS characteristics mentioned above, the electrode materials' conductivity can be estimated by the intersection of high-frequency impedance with the real axis. According to the method for calculating conductivity in physics and the measurement of ionic

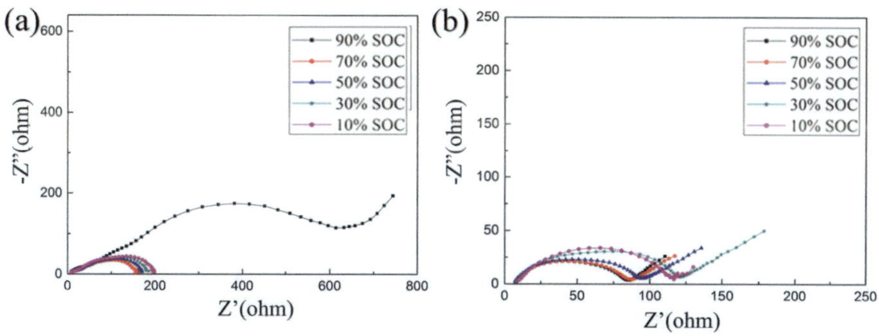

Fig. 4.6 Relationship between AC impedance spectrum and SOC of LFP/graphite coin-type half-cell under 20 °C environment. **a** LFP/Li. **b** Graphite/Li

conductivity in pure liquid electrolytes in reference [51], this section introduces Eq. (4.47) for calculating the conductivity of active materials in coin-type half-cells:

$$\sigma = \frac{L}{S_{half} R_0} \tag{4.48}$$

where L represents the thickness of the LFP or graphite electrode in coin-type half-cell (in meters); S_{half} represents the area of electrode in coin-type half-cell (in square meters). R_0 is the real part value of the impedance at high frequencies (in ohms), which corresponds to the ohmic internal resistance of the battery.

According to Eq. (4.47) and the EIS at different T and SOC, the R_0 values can be identified by calculating the effective conductivity of LFP and graphite materials under the corresponding conditions.

In Fig. 4.6, for both types of coin-type half-cells, the mid-to-low frequency impedance characteristics change significantly with SOC. However, in the high-frequency region, the position of the high-frequency starting point does not change with the variation of SOC. It suggests that the ohmic internal resistance of the coin-type half-cell is less nfluenced by changes in SOC. The theoretical analysis in the previous section concluded that the ohmic internal resistance represents the transport of electrons and ions through the electrode active materials and electrolyte. In this section, the ohmic internal resistance of coin-type half-cells is identified separately, and the values of R_0 at different SOCs are obtained as shown in Fig. 4.8a.

According to the impedance Eq. (4.24), the concentration of liquid-phase lithium ions is associated with SOC. In this test, the variation pattern of concentration is not significantly pronounced. This result indicates that the effective conductivity of the liquid phase within the active materials is minimally affected by changes in SOC across the tested range. Figure 4.7 shows the AC impedance spectra of LFP/graphite coin-type half-cell at different temperature test points when the battery is at a 50% SOC.

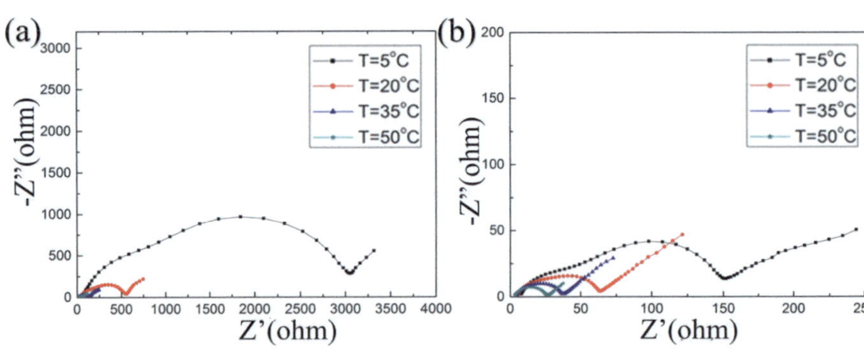

Fig. 4.7 Relationship between AC impedance spectrum and temperature of LFP (graphite) type coin half-cell at 50% SOC. **a** LFP/Li. **b** Graphite/Li

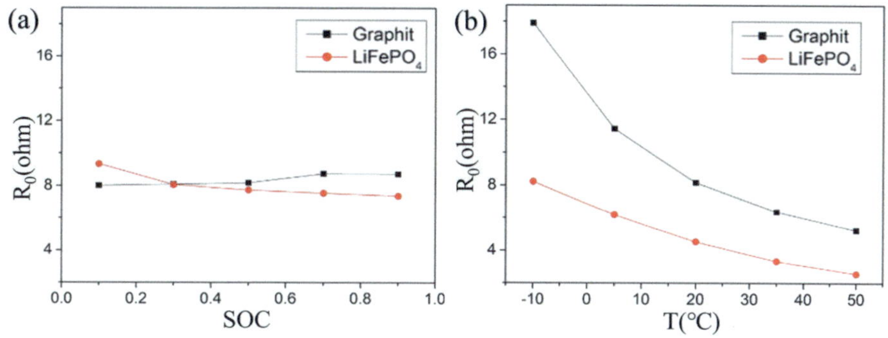

Fig. 4.8 The coupling relationship between R_0 of LFP (graphite) electrode materials and SOC and temperature. **a** 20 °C temperature, different SOCs. **b** 50% SOC, different temperatures

As shown in Fig. 4.7, the high-frequency starting point of EIS curve gradually shifts to the right as the temperature decreases. It indicates that the ohmic internal resistance of battery increases with the lowering of temperature, and the effective conductivity of the electrode active materials is significantly affected by temperature. From Eq. (4.47), it is known that the conductivity is inversely proportional to the impedance R_0. Thus from Fig. 4.8b, it can be preliminarily inferred that the effective conductivity σ increases with the rise in temperature.

The LIBs' ohmic internal resistance is due to the impedance of Li^+ and electrons passing through the electrolyte, porous separator, collector, and active material particles. The collector is generally made of metal (copper or aluminum) and can be considered essentially constant within the operating temperature range of battery (−30 to 50 °C). This section mainly explores the coupling relationship between the electrode active materials (solid and liquid phases), temperature and SOC. Therefore, the calculated values from Eq. (4.47) represent the conductivity of the electrode active materials. The solution concentration has a significant impact on the measured conductivity, and variations in sample preparation and experimental equipment measurement requirements further complicate the analysis. Given these factors, this section focuses solely on a qualitative analysis of the conductivity trends of the electrode active materials. Specifically, the solid and liquid phases' conductivities are not explicitly distinguished in this analysis. After calculation, the conductivities of the active materials for the two types of materials in the coin-type half-cells at different SOCs and T are obtained, as shown in Tables 4.3 and 4.4. Drawing from the data in Table 4.3, it can be concluded that the effective conductivity of LFP and graphite active materials under the combined effects of T and SOC and, it is minimally influenced by SOC. And the conductivity of these materials exhibits a significant decrease as the temperature declines. The conductivity of the active materials is mainly related to the ohmic internal resistance, which is primarily controlled by temperature through the electron transport in the solid electrode and the transfer of lithium ions in the liquid phase. Reference [52] indicates that the ionic conductivity of the liquid phase is affected by the solution concentration under different SOCs. In

Table 4.3 Effective conductivity ($s \cdot m^{-1}$) of LFP active material at different temperatures and different SOCs

SOC	T(°C)			
	5	20	35	50
0.1	0.190268	0.217703	0.340527	0.429127
0.3	0.192938	0.260717	0.339912	0.412392
0.5	0.191076	0.263832	0.339643	0.416147
0.7	0.190503	0.271081	0.338361	0.425713
0.9	0.190090	0.27593	0.338647	0.425540

Table 4.4 Effective conductivity ($s \cdot m^{-1}$) of graphite active material at different temperatures and different SOCs

SOC	T(°C)			
	5	20	35	50
0.1	0.273813	0.391982	0.875396	0.921207
0.3	0.278355	0.397756	0.787941	0.897824
0.5	0.274426	0.396615	0.784107	0.946910
0.7	0.279771	0.380904	0.779753	0.885660
0.9	0.275208	0.396500	0.793355	0.895689

this section, due to the high ohmic resistance of the positive and negative electrode active materials of the sample half-cell, the changes in the ionic conductivity of the liquid phase are not evident.

(3) The Activation Energy for the Reaction of Electrode Active Materials

From the kinetic perspective, whether oxidation or reduction reactions occur at the electrode surface, they must overcome a certain energy barrier. As shown in Fig. 4.9, using statistical theory, the rate constant of elementary reactions at different temperatures can be derived from parameters such as microscopic energy barriers and threshold energies. Then, by plotting an Arrhenius graph, the activation energy and other Arrhenius parameters can be estimated. Similarly, after calculating the rate constants of elementary reactions at different temperatures, an Arrhenius plot can be made to obtain the activation energy and other Arrhenius parameters.

Equation (4.48) provides the Arrhenius empirical formula that describes the general form of the rate constant for an electrochemical reaction:

$$k = A_3 \cdot \exp(\frac{-Ea}{RT}) \tag{4.49}$$

where A_3 represents the pre-exponential factor describing the rate constant of the electrochemical reaction; E_a is the activation energy for the reaction.

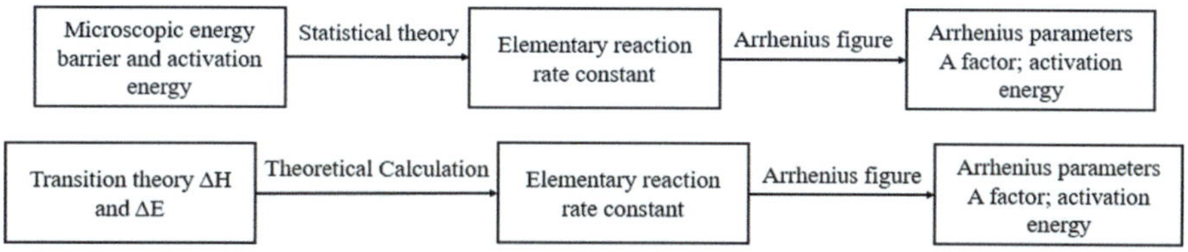

Fig. 4.9 Arrhenius parameters theoretical calculation process

Both the charge transfer internal resistance and the rate constant of the electrochemical reaction describe the electrochemical reactionl process. Equation (4.28) describes the functional relationship between the two, and they have the same activation energy value. Subsequent descriptions of the electrochemical reaction process will use E_a to denote the battery reaction activation energy. After mathematical transformation, a linear relationship can be obtained:

$$\ln k = \ln A_3 - \frac{Ea}{RT} \tag{4.50}$$

In this section, parameter identification is conducted on the EIS of both LFP and graphite coin-type half-cells. The interrelationship between charge transfer internal resistance and temperature is examined. The activation energy of the reactions is obtained using the method described in reference [53]. As shown in Fig. 4.10, the charge transfer internal resistance satisfies an exponential relationship with temperature. The activation energy parameters can be calculated through Arrhenius plot. And it can be concluded that the chemical reaction activation energy of graphite is higher than that of the positive electrode lithium iron phosphate.

In chemical kinetics, an Arrhenius plot can be constructed to determine the slope at a specific temperature T for any reaction whose rate is temperature-dependent. This slope value is the activation energy E_a of the battery. E_a is located in the exponential term of the formula, hence changes in the activation energy value have a

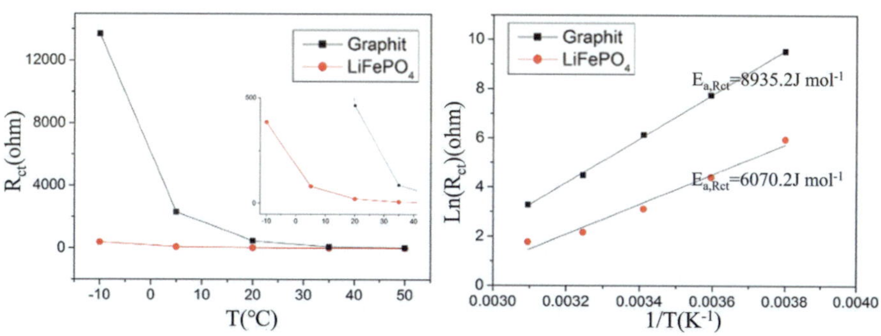

Fig. 4.10 Arrhenius figure of lithium iron phosphate and graphite coin-type half-cell

significant impact on the rate constant k of the chemical reaction. As analyzed in the previous text, the charge transfer internal resistance R_{ct} is a direct reflection of the electrochemical reactions in batteries. In contemporary LIB research, the application of the Arrhenius empirical formula to determine the activation energy of electrode processes via AC impedance spectra is widely recognized and utilized [53, 54]. This provides a correction for the rate constant of the electrochemical reaction when modeling the electrode process.

(4) Solid Phase Diffusion Coefficient

In the field of electrochemical research, methods for measuring the solid-state diffusion coefficient of lithium ions include Current Pulse Relaxation (CPR), Potential Step Chronoamperometry (PSCA), EIS, and GITT. The method used in this study to determine the solid-state diffusion coefficient of Li^+ is the GITT. In Reference [55], a comparative analysis of various calculation methods for the solid-state diffusion coefficient of Li^+ in the negative electrode graphite is conducted, revealing a high degree of regular consistency in the test results.

According to the literature [45–47], the solid-state diffusion coefficient D_s of the active material in the coin-type half-cell can be obtained from the following Eq. (4.50).

$$D_{s-Li} = \frac{4}{\pi\tau}\left(\frac{m_B V_M}{M_B S_{half}}\right)^2\left(\frac{\Delta E_s}{\Delta E_t}\right)^2 \tag{4.51}$$

The schematic diagram of the GITT testing method is shown in Fig. 4.11. In Eq. (4.50), τ denotes the duration during which a small constant current is applied; m_B is the mass of the measured material; M_B is the molar mass of the measured active material; V_M is the molar volume of the measured active material; S_{half} is the cross-sectional area of the active material coating in the coin-type half-cell, ΔE_s is the equilibrium potential difference value of the battery when the coin-type half-cell reaches a steady state before and after charge and discharge; ΔE_t is the voltage drop caused by polarization during the charge and discharge process. Through GITT testing experiments and formula calculations, the diffusion coefficients of the positive and negative electrode active materials at different temperatures and SOCs are obtained, as shown in Fig. 4.12.

Based on the test results, analysis was conducted on both the LFP and graphite coin-type half-cells. It was found that at 20 °C, the LFP's solid-state diffusion coefficient is lower than that of graphite. Thus, within the individual LFP-based batteries, the solid-state diffusion coefficient of LFP plays a dominant role in the diffusion of lithium ions. That is, the diffusion coefficient of the LFP active material is the main factor affecting the charge–discharge polarization internal resistance of batteries.

For LFP electrode materials, the temperature effect is minimal, with significant variations primarily observed in the low and high SOC regions. The primary cause of this phenomenon is attributed at low and high SOC, concentration gradients are more likely to form, thus facilitating the diffusion process in the battery. For graphite electrode materials, it is evident that the solid-state diffusion coefficient of the active

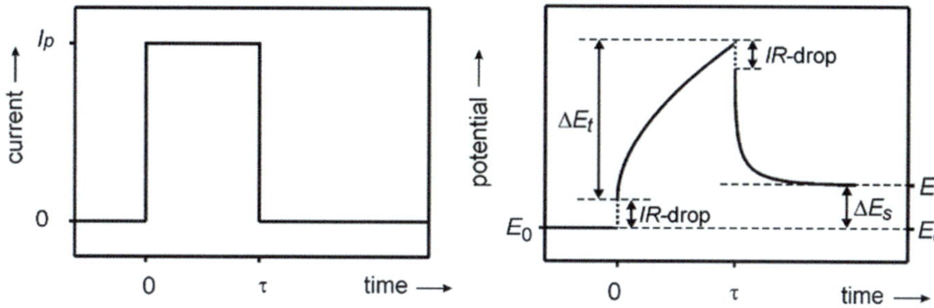

Fig. 4.11 GITT testing method schematic diagram

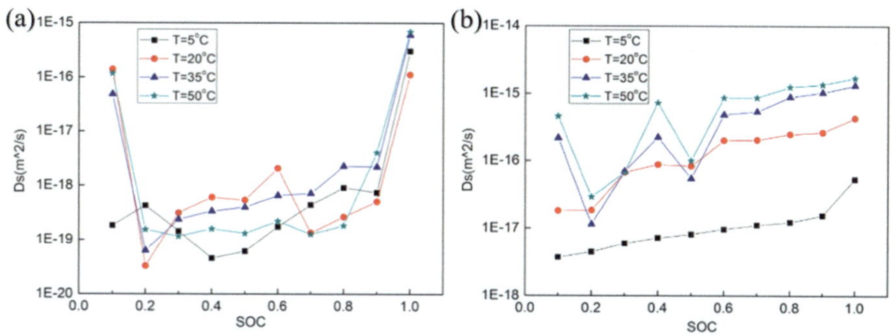

Fig. 4.12 Variation curves of the diffusion coefficient of coin-type battery active materials at different temperatures and SOCs **a** LFP **b** Graphite

material decreases with the reduction in temperature. It indicates that the diffusion resistance of ions within graphite particles increases at low temperatures. Due to the limitations of the experimental equipment's range, the lowest test temperature was 5 °C, and for even lower environmental temperatures, extrapolation or reasonable prediction is necessary. The coupling relationships of the solid-state diffusion coefficients corresponding to different temperatures and different SOCs in this section are used for the simulation analysis and research in the following text.

4.2.2.3 Electrochemical Model Validation

(1) Verification of LIBs Electrode Process Model

Based on the analysis of LIBs electrode processes and experimental parameter acquisition, this section validates the electrode process model for the 8Ah LFP-based battery sample.

In this section, some parameters of the simulation model are obtained by referring to literature; some parameters are calculated theoretically based on the materials and structure of the battery; some parameters are estimated. Detailed parameters are shown in Table 4.5.

Figure 4.13 shows the comparison of the model and experimental test discharge voltage at an ambient temperature of 20 °C under different discharge rates. Specifically, there is a good fit at discharge rates of 0.5 and 1 C. At a discharge rate of 2 C, there is a slight discrepancy between the experimental and simulation values towards the end of the discharge. Overall, it can be concluded that the model simulation has a good fit with the experimental results. To some extent, the model can reflect the internal state during continuous discharge [59].

Figure 4.14 presents a comparison of experimental test and impedance frequency domain simulation results at environmental temperatures of 20 and 30 °C. The SOC of batteries is set to 50%. The sweeping method and range used in simulation and experiment are the same. It is evident that the AC impedance spectra of the simulation model closely resemble those of the experiments, with the results in the mid-to-high frequency range almost overlapping. However, there is some discrepancy between the simulation and experimental results in the low-frequency range. Overall, the impedance characteristics of the simulation model closely match those of the experimental model, indicating that the simulation model can effectively reflect the impedance characteristics of LFP-based batteries across various frequency bands. Further simulation studies will be conducted using the model to further verify its effectiveness.

(2) Impact of Low-Temperature Characteristic Parameters on EIS

The analysis confirms that the porous electrode theory-based model accurately captures the voltage and impedance characteristics of batteries during charge and discharge, establishing it as an effective tool for LIBs research. Other parameters that support the porous electrode theory, such as electrode thickness dimensions, solid–liquid phase conductivities, and theoretical capacities of electrode materials, are all directly obtainable and have their unique physical laws. Some parameters are difficult to accurately obtain even through experimental testing, such as the rate of electrochemical reactions and solid–liquid phase diffusion coefficients. Furthermore, the reliability of the obtained values is limited, particularly when considering temperature variations. For instance, the estimation of conductivity and diffusion coefficients in this section is constrained by the half-cell lifespan and the impedance tester's range, with experiments conducted solely at 5 °C.

In this chapter, the model analysis approach is utilized to examine the model parameters, with particular focus on the internal characteristic parameters. Additionally, their corresponding relationship with the temperature and EIS is examined. It is known that at low temperatures, the main characteristic parameters that limit the batteries' low-temperature characteristics are the liquid-phase ionic conductivity, solid–liquid phase diffusion coefficients, and electrochemical reaction rates. In this study, a method of artificially defining the trend of parameter changes is adopted. A proportional parameter k is used to proportionally increase or decrease the actual

Table 4.5 Battery electrode process model simulation parameters

Parameter	Parameter value (Unit)	Description
r_{p_pos}	$5.2e{-}6(\mathrm{m})^{b}$	Positive electrode particle radius
k_{neg}	$2e{-}11(\mathrm{m}\cdot\mathrm{s}^{-1})$ [56, 57]	Negative electrode reaction rate constant
k_{pos}	$2e{-}11(\mathrm{m}\cdot\mathrm{s}^{-1})$ [56, 57]	Positive electrode reaction rate constant
D_{s_neg}	$8.3e{-}17\ (\mathrm{m}^2\cdot\mathrm{s}^{-1})^{b}$	Negative electrode solid-phase diffusion coefficient
D_{s_pos}	$5.4e{-}19\ (\mathrm{m}^2\cdot\mathrm{s}^{-1})^{b}$	Positive electrode solid-phase diffusion coefficient
$C_{s\_neg,max}$	$26{,}390(\mathrm{mol}\cdot\mathrm{m}^{-3})$ [56, 57]	Maximum concentration of negative electrode solid-phase lithium ions
$C_{s\_pos,max}$	$22{,}800(\mathrm{mol}\cdot\mathrm{m}^{-3})$ [56, 57]	Maximum concentration of positive electrode solid-phase lithium ions
C_0	$2000(\mathrm{mol}\cdot\mathrm{m}^{-3})$ [40]	Initial electrolyte concentration
C_{s0_pos}	$11{,}400(\mathrm{mol}\cdot\mathrm{m}^{-3})^{b}$	Initial concentration of positive electrode solid-phase lithium ions
C_{s0_neg}	$14{,}863(\mathrm{mol}\cdot\mathrm{m}^{-3})^{b}$	Initial concentration of negative electrode solid-phase lithium ions
D_l	$3e{-}11(\mathrm{m}^2\cdot\mathrm{s}^{-1})^{a}$	Liquid-phase lithium ion diffusion coefficient
	$4.5e{-}11(\mathrm{m}^2\cdot\mathrm{s}^{-1})^{a}$ (30 °C)	
$\alpha_{a.c\_pos}$	0.5^{a}	Positive electrode anode/cathode reaction coefficient
$\alpha_{a.c\_neg}$	0.5^{a}	Negative electrode anode/cathode reaction coefficient
γ	1.5^{a}	Bruggeman porosity coefficient
ε_{s_pos}	0.38^{b}	Effective volume fraction of positive electrode active material
ε_{s_neg}	0.65^{b}	Effective volume fraction of negative electrode active material
ε_{l_pos}	0.41^{b}	Positive electrode porosity
ε_{l_neg}	0.15^{b}	Negative electrode porosity
σ_{s_pos}	$0.005(\mathrm{s}\cdot\mathrm{m}^{-1})$ [57]	Positive electrode solid-phase conductivity
σ_{s_neg}	$100(\mathrm{s}\cdot\mathrm{m}^{-1})$ [58]	Negative electrode solid-phase conductivity
L_{neg}	$40e{-}6(\mathrm{m})^{b}$	Thickness of negative electrode active material
L_{pos}	$40e{-}6(\mathrm{m})^{b}$	Separator thickness
L_{pos}	$35e{-}6(\mathrm{m})^{b}$	Thickness of positive electrode active material
C_{dl}	$4\ (\mathrm{F}\cdot\mathrm{m}^{-2})^{a}$	Double-layer capacitor
δ_l	δ_l [39]	Ionic conductivity of the electrolyte
T	$293(\mathrm{K})^{a}$	Temperature
t_+	0.363 [40]	Li^+ transference number

a: Estimated parameters; b: Parameters calculated theoretically based on battery materials and structure

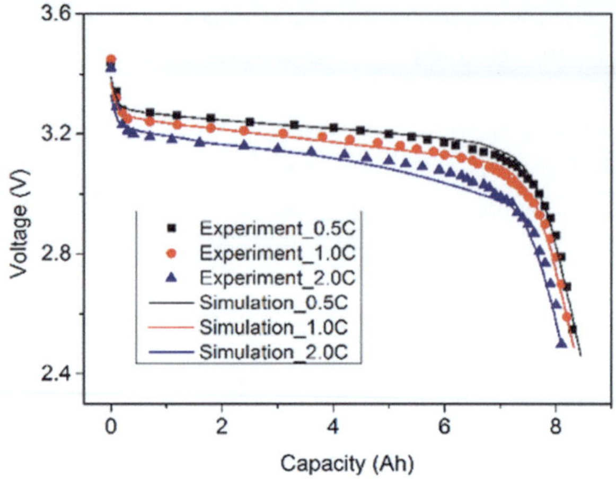

Fig. 4.13 Comparison between model simulation and experimental discharge voltage under different discharge rates

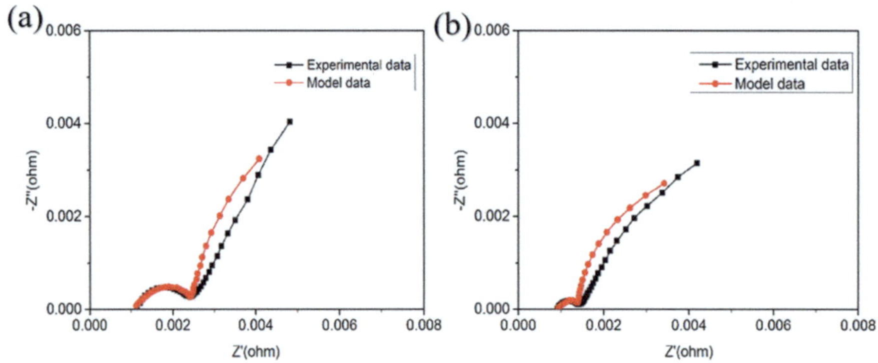

Fig. 4.14 Comparison of model simulation and experimental EIS at **a** 20 °C and **b** 30 °C

parameter values to modify model's parameter, whose aim is to obtain the characteristics of EIS under different parameter values through model simulation, discuss the impedance features, and thus analyze the coupling patterns.

Figure 4.15 shows the simulation of EIS under different reaction rate constants. In the model, the reaction rates for the positive and negative electrodes are defined as k_{pos} and k_{neg}, respectively. Here, we take a proportionality coefficient $k = \{0.2, 0.5, 1.0, 2.0\}$ and $k' = k \cdot k_{pos/neg}$. This enables the comparison of how the EIS of the model is altered when different values of k are assigned. From the analysis of the graph, it can be concluded that: changes in the reaction rate constant do not affect the ohmic resistance R_0 of the battery. The impedance of the electrochemical reaction process is inversely related to the reaction rate, significantly increasing as the

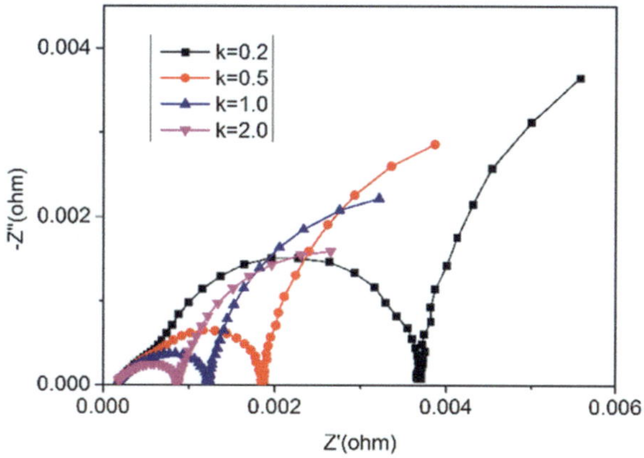

Fig. 4.15 Simulation of the variation pattern of the model's impedance spectrum under different reaction rates

reaction rate decreases. According to the Arrhenius empirical equation, the reaction rate constant is described by Eq. (4.48), and the reaction rate is exponentially related to temperature. As the temperature decreases, the reaction rate decreases, leading to an increase in the mid-frequency impedance arc.

Figure 4.16 displays the simulation of impedance spectrum variation patterns under different liquid-phase ionic conductivities. The ionic conductivity of the electrolyte is an interpolation function of temperature [39, 40] $k_l{'}(T)$, where k takes values {0.25, 0.5, 1, 2}, allowing $k_l'(T) = k \cdot k_l(T)$. Here, we select a proportionality coefficient to compare the changes in the model's EIS when k takes on different values. From the graph, it can be observed that as the liquid-phase ionic conductivity decreases, the EIS shifts to the right overall, indicating an increase in ohmic internal resistance. The ohmic internal resistance includes both the solid-phase electron conduction resistance and the liquid-phase ion transfer resistance. A decrease in liquid-phase ionic conductivity leads to an increase in the liquid-phase ion transfer resistance. At the same time, an increase in k_l accelerates the transport speed of ions, thereby speeding up the reaction rate of lithium ions. It is reflected in the EIS as a decrease in the mid-frequency impedance arc. When lithium plating occurs, the reaction between lithium metal and the electrolyte, on one hand, forms an interface resistance that leads to an increase in the battery's internal resistance. Concurrently, due to the irreversible consumption of the electrolyte solute, the liquid-phase conductivity decreases [60].

Figure 4.17 shows the simulation of EIS variation patterns under different solid-phase diffusion coefficients at 20 °C. In this model, the positive electrode's solid-phase diffusion coefficient is a measured parameter. At a temperature of 20°C and 50% SOC, its value is determined by actual experimental test results as $D_{s,pos} = 5.4 \times 10^{-12}$m²/s. The negative electrode's solid-phase diffusion coefficient is $D_{s,neg} = 8.3 \times 10^{-12}$m²/s. In this section, a proportionality coefficient $k = 10^{\{-5,0,2,5\}}$ is selected, making $D_{s,pos}{'} = k \cdot D_{s,pos}$ and $D_{s,neg}{'} = k \cdot D_{s,neg}$. This allows for comparison of the changes in the frequency characteristics of the model EIS when k takes on different values.

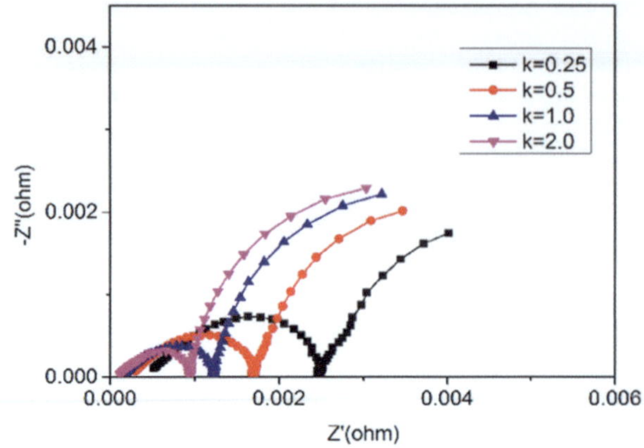

Fig. 4.16 Simulation of the impact of different liquid-phase ionic conductivities on impedance spectrum

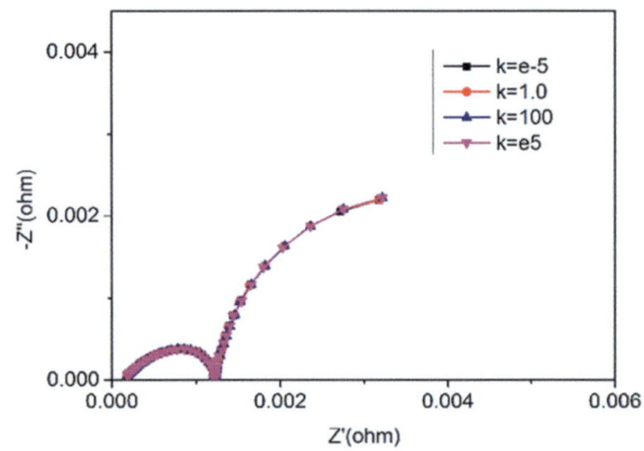

Fig. 4.17 Simulation of the impact of different solid-phase diffusion coefficients on EIS

The solid-phase diffusion process is predominantly observed in the low-frequency region of EIS. However, the simulation revealed that despite considerable variations in the solid-phase lithium ion diffusion coefficient, the impedance characteristics of the model remained almost unchanged. The impedance of solid-phase diffusion in batteries is primarily reflected in Eq. (4.17), which means that the impedance of solid-phase diffusion becomes more apparent when the concentration gradient within the solid-phase active particles is smaller. In this simulation, the minimal variation in low-frequency impedance may be attributed to the short excitation time and small amplitude of the EIS current. Although the concentration at the particle surface changes significantly during the simulation, the concentration gradient inside the particles changes little. Thus showing that although the solid-phase diffusion coefficient changes noticeably, its impedance remains roughly the same.

Selecting the lowest excitation frequency in the simulation frequency scanning range (0.01 Hz) for analysis. Figure 4.18 shows the concentration distribution diagrams of the surface (Fig. 4.18a) and center (Fig. 4.18b) of the battery particles

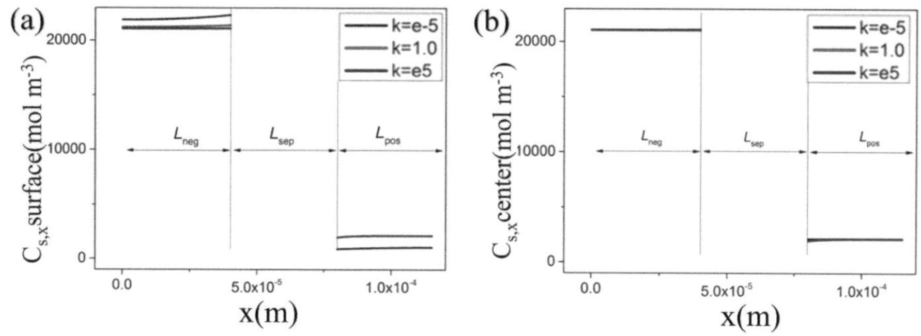

Fig. 4.18 Under 0.01 Hz excitation, the concentration distribution diagram of the particles' **a** Surface, **b** Center under diverse solid-phase diffusion coefficients

at a time of 50 s, which corresponds to the peak point of the battery during charging. The X-axis depicts the structure of a one-dimensional geometric model, with the negative electrode graphite, separator, and positive electrode material arranged from left to right. The vertical axis indicates the concentration value. From the plot, it can be inferred that changes in the diffusion coefficient are primarily reflected in the changes at the particle surface concentration. The smaller the diffusion coefficient, the greater the diffusion resistance, leading to larger fluctuations in the battery particles' surface concentration. However, with changes in the diffusion coefficient, the concentration at the center of the particles remains largely unaffected. It suggests that the diffusion process either has not occurred or the depth of diffusion is low. Based on literature analysis and experimental results, the solid-phase diffusion coefficient of the negative electrode graphite exhibits a significant temperature dependence, decreasing notably as temperature declines.

Figure 4.19 illustrates the solid–liquid phase potential difference (ϕ_{delta}) at the interface between the negative electrode graphite and the separator (this interface can be used for the analysis of lithium plating). It can be observed that as the diffusion coefficient decreases, the potential tends towards zero, indicating an increased likelihood of lithium plating occurring. The low-frequency impedance arc includes both liquid-phase diffusion impedance and solid-phase diffusion impedance. And the solid-phase diffusion time constant generally being greater than that of the liquid phase. In this section's simulation of the solid-phase diffusion coefficient, the lowest frequency is 0.01 Hz, the significant region of solid-phase diffusion may not yet be reflected.

Figure 4.20 presents the variation pattern of EIS under different liquid-phase diffusion coefficients. In the original model, the diffusion coefficient of lithium ions in the electrolyte is defined as D_l. Here, we take a proportionality coefficient $10^{\{1.0,0,-1,-2\}}$, making $D_l' = k{\cdot}D_l$. This enables the comparison of impedance characteristic variations in the model under different proportionality coefficient k values. The liquid-phase diffusion of lithium ions is mainly related to the concentration polarization

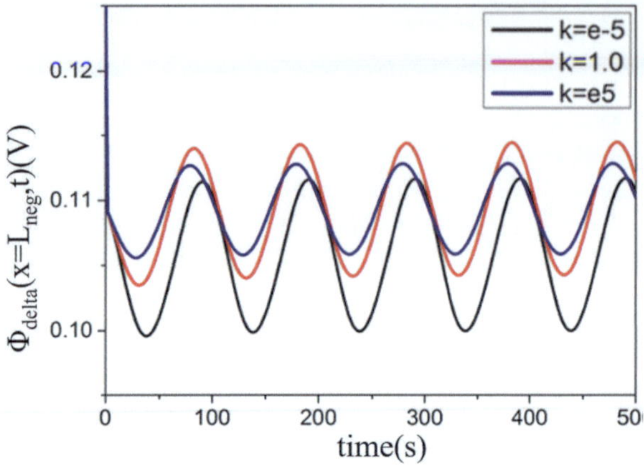

Fig. 4.19 The solid–liquid phase potential difference at the interface between negative electrode graphite and separator under different solid phase diffusion coefficients excited by 0.01 Hz

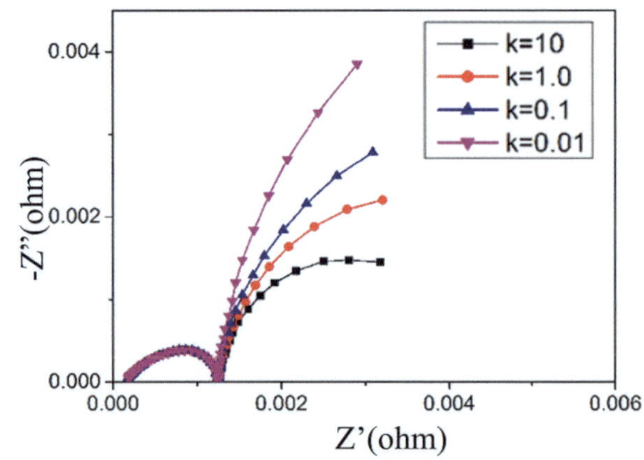

Fig. 4.20 Influence of various liquid phase diffusion coefficients on impedance

within the battery. Upon battery energization, the liquid phase transitions into a non-equilibrium state, where a larger concentration gradient leads to insufficient replenishment of lithium ions at the reaction interface, resulting in increased impedance. And a larger liquid-phase diffusion coefficient is conducive to the lithium ions' transport.

From the EIS analysis, it can be concluded that the liquid-phase diffusion coefficient significantly affects the low-frequency diffusion region of the battery. As the liquid-phase diffusion coefficient increases, the low-frequency diffusion impedance decreases, and the impedance gradually takes on a circular arc shape. It indicates that the diffusion time constant is gradually decreasing. The liquid-phase diffusion coefficient hardly changes the impedance's frequency characteristics. And the the electrochemical reaction process's ohmic internal resistance and the charge transfer internal resistance are unrelated to the liquid-phase diffusion coefficient.

In summary, the fundamental material properties governing battery performance at cryogenic temperatures are predominantly determined by three aspects: (1) the rate of electrochemical reactions, (2) liquid-phase ionic conductivit, and (3) diffusion coefficients. They all affect the impedance characteristics of LIBs models to varying degrees. Among these, in the simulation study of this section, variations in the solid-phase diffusion coefficient exert a relatively subtle influence on EIS. However, under varying simulation coefficients, significant changes in particle surface concentration occur, substantially influencing the internal electrode potential, particularly the the negative graphite electrode's potential distribution. According to the reference [61], the surface concentration of the negative electrode particles can lead to changes in the diffusion concentration ratio iSOC, which is one of the factors causing lithium plating in LIBs at low temperatures. The research in this chapter provides a certain theoretical basis for establishing a low-temperature electrochemical analysis model and for subsequent studies on the impact of AC heating on health of LIBs.

4.3 Models for AC Heating Strategy Design

The electrochemical model established in Sect. 4.2 can reveal the polarization characteristics of the battery from the perspective of microscopic reaction mechanisms. However, the complexity of parameters and calculations in electrochemical models makes them difficult to apply in the study of heating strategies. To address this issue, this section couples ECM with the thermal model. By simulating the external characteristics of the battery through electrical components and integrating them with the heat transfer equation, a coupling model is constructed. This model achieves a balance between computational efficiency and physical representation capabilities.

The thermal model is analyzed first because its theory is relatively well-established. The impedance in the electrode process can lead to heat generation within the battery during charge–discharge cycles. The heat generation during this process mainly includes electrochemical reaction heat, Joule heat, phase transformation heat, mixing heat. Electrochemical reaction heat pertains to the heat produced by the chemical reactions inside the LIBs, including the main reactions of cathode and anode. Additionally, the heat generated by side reactions is also included in reaction heat, such as the decomposition of the electrolyte, the formation and decomposition of the SEI layer, and lithium plating. Joule heat is caused by the resistance of the electrolyte, electrode materials, and current collectors, which is a major source of heat generation in batteries. Phase transformation heat refers to the heat released or absorbed when the cathode or anode materials undergo phase changes during cycles. Mixing heat originates from the concentration gradient formed in reactions due to the inconsistent reaction rates in different parts of the battery, leading to non-uniform mixing of substances. Since phase transformation heat and mixing heat constitute a minor fraction of the total heat generation and are only present in specific material systems, they are usually not considered in the calculation of heat generation in batteries. Instead, only the reaction heat and Joule heat are typically taken into account. For AC heating, which involves both charging and discharging processes, the reaction heat associated with these two mechanisms tends to neutralize one another.

Therefore, the heat generation module for this study usually only considers Joule heat.

The establishment of the ECM is a major challenge in the development of the electro-thermal coupling model. This section initially introduces an ECM based on a three-electrode battery, analyzing the current amplitude's influence on heating efficiency at various temperatures under 20% SOC. The rationale for employing a three-electrode battery model is to establish the criterion for lithium plating. However, the implantation of a reference electrode not only increases manufacturing costs but also involves invasive modifications to the mechanical structure, making this method difficult to generalize. Consequently, an ECM based on full cell is proposed, whose parameters incorporate not only the effects of temperature and heating frequency but also the impact of SOC. This provides a foundational framework for the study of AC heating strategies in low-temperature environments.

4.3.1 Model Structure Design

4.3.1.1 Modeling Based on Three-Electrode Battery

Figure 4.21 presenteds the electro-thermal coupling model of the three-electrode battery in this study.

Notably, the lithium plating criterion's derivation, also its application for determining the plating boundary, can be achieved solely using a graphite anode model. Subsequently, the temperature increase of the battery resulting from BPC heating can be determined using the full-cell electro-thermal coupling model. Nevertheless, utilizing a three-electrode battery configuration is crucial for isolating the

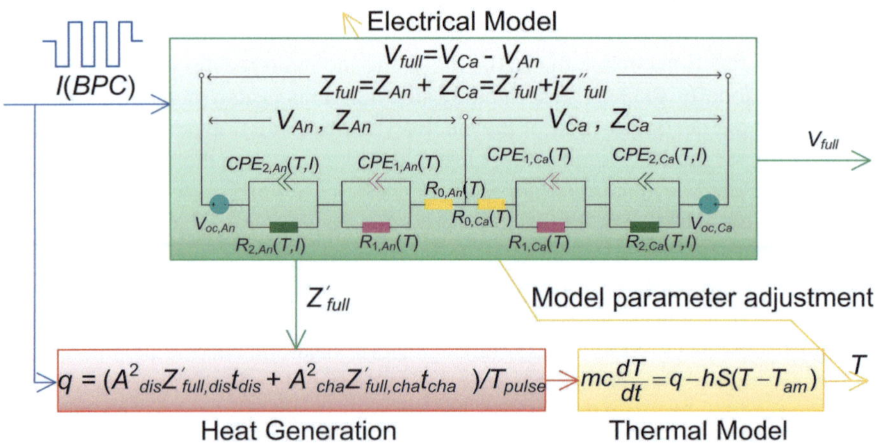

Fig. 4.21 The electro-thermal coupling model [62]

anode. Furthermore, the temperature change of the battery can also be determined using the three-electrode configuration. And the three-electrode electro-thermal coupling model enables simultaneous determination of lithium plating boundaries and temperature chan without increasing computational complexity relative to existing methods.

Consequently, this study employs the three-electrode electro-thermal coupling model in conducting computational analysis The coupling framework consists of two distinct components: an electrical sub-model and a thermal sub-model. Electrical sub-model parameters are influenced by temperature [63]. A bidirectional coupling effect exists between the electrochemical model and the thermal dynamics model due to the significant influence of battery internal resistance on thermal energy dissipation. The electrochemical reactions occurring at the electrodes of LIBs involve complex multi-step processes that can be effectively modeled using an equivalent circuit representation consisting of a series resistor connected to two parallel circuit branches. The parallel structure is a resistor connected in parallel with a constant phase element (CPE) [39, 64]. The representation of this circuit is categorized as a fractional-order equivalent circuit model. Among the context of the fractional-order circuit model, the component R_0 is conventionally employed to signify the ohmic resistance. The configuration involving the parallel combination of R_1 and CPE_1 serves to characterize the SEI/CEI process. Furthermore, the arrangement of R_2 and CPE_2 is utilized to depict the charge transfer process. In this study, for the 1 Ah commercial LIB under investigation, a lumped thermal model is adopted to approximate the battery temperature, primarily attributed to its relatively compact dimensions. This thermal model can be expressed as [30]

$$mc\frac{dT}{dt} = q - hS(T - T_{am}) \tag{4.52}$$

the symbols are defined as follows: m denotes the mass; c represents the specific heat capacity; T and T_{am} correspond to the battery's and ambient temperature, respectively; S signifies the surface area of battery; h denotes the equivalent heat transfer coefficient; and q indicates the heat generation rate. For BPC heating, the rate of heat production q can be formulated in the following:

$$q = \frac{A_{dis}^2 Z'_{full} t_{dis} + A_{cha}^2 Z'_{full} t_{cha}}{T_{pulse}} \tag{4.53}$$

In this formula, $Z'_{full,dis}$ and $Z'_{full,cha}$ denote the full cell impedance's real components, which associated with the discharge pulse and charge pulse, respectively. And impedance Z'_{full} of the full cell can be mathematically represented as

$$Z'_{full} = Z'_{An} + Z'_{Ca} \tag{4.54}$$

Herein, Z'_{An} and Z'_{Ca} correspond to the impedance's real components associated with the anode and cathode. According to ECM depicted in Fig. 4.21, the aforementioned parameters can be formulated as follows [65]

$$Z_i' = R_{0,i} + \sum_{k=1}^{2} \frac{R_{k,i} + \omega^{n_{k,i}} R_{k,i}^2 Q_{k,i} \cos\frac{n_{k,i}\pi}{2}}{1 + 2\omega^{n_{k,i}} R_{k,i} Q_{k,i} \cos\frac{n_{k,i}\pi}{2} + \omega^{n_{k,i}} R_{k,i}^2 Q_{k,i}} (i = An,\ Ca) \tag{4.55}$$

Herein, ω represents the angular frequency of the current, while Q and n correspond to the generalized capacitance and the depression factor associated with the CPE, respectively.

Within the framework of the electrical model, the terminal voltage V_{full} can be mathematically described as

$$V_{full} = V_{Ca} - V_{An} = (V_{oc,Ca} + V_{p,Ca}) - (V_{oc,An} + V_{p,An}) \tag{4.56}$$

Herein, V_{Ca} and V_{An} represent the voltages across the cathode and anode, $V_{oc,Ca}$ and $V_{oc,An}$ denote the OCV of the cathode and anode. $V_{p,Ca}$ and $V_{p,An}$ correspond to the polarization voltages of the cathode and anode, respectively. The polarization impedance is described by Ruan et al. [66]

$$Z_i(s) = \frac{V_{p,i}(s)}{I(s)} = \frac{\sum_{k=1}^{4} a_{k,i} s^{\alpha_{k,i}}}{\sum_{k-1}^{4} b_{k,i} s^{\beta_{k,i}}} + R_{0,i} = \sum_{k=1}^{2} \frac{R_{k,i}}{R_{k,i} Q_{k,i} s^{n_{k,i}} + 1} + R_{0,i} (i = An,\ Ca) \tag{4.57}$$

The coefficients are represented as

$$\begin{cases} [a_{1,i}\ a_{2,i}\ a_{3,i}\ a_{4,i}] = [R_{1,i} R_{2,i} Q_{2,i}\ R_{1,i}\ R_{1,i} R_{2,i} Q_{1,i}\ R_{2,i}] \\ [\alpha_{1,i}\ \alpha_{2,i}\ \alpha_{3,i}\ \alpha_{4,i}] = [n_{2,i}\ 0\ n_{1,i}\ 0] \\ [b_{1,i}\ b_{2,i}\ b_{3,i}\ b_{4,i}] = [R_{1,i} R_{2,i} Q_{1,i} Q_{2,i}\ R_{1,i} Q_{1,i}\ R_{2,i} Q_{2,i}\ 1] \\ [\beta_{1,i}\ \beta_{2,i}\ \beta_{3,i}\ \beta_{4,i}] = [n_{1,i} + n_{2,i}\ n_{1,i}\ n_{2,i}\ 0] \end{cases} \quad (i = An,\ Ca) \tag{4.58}$$

Based on the Grünwald-Letnikov (GL) definition [67], the polarization voltage is described by

$$V_{p,i}(t) = \frac{1}{\sum_{k=1}^{4} \frac{b_{k,i}}{\Delta t^{\beta_{k,i}}}} \left[\sum_{k=1}^{4} \sum_{g=0}^{[t/\Delta t]} \frac{a_{k,i}}{\Delta t^{\alpha_{k,i}}} (-1)^g \binom{\alpha_{k,i}}{g} I(t - g\Delta t) + \sum_{k=1}^{4} \sum_{g=0}^{[t/\Delta t]} \frac{b_{k,i}}{\Delta t^{\beta_{k,i}}} (-1)^g \binom{\beta_{k,i}}{g} I(t - g\Delta t) R_{0,i} - \sum_{k=1}^{4} \sum_{g=0}^{[t/\Delta t]} \frac{b_{k,i}}{\Delta t^{\beta_{k,i}}} (-1)^g \binom{\beta_{k,i}}{g} V_{p,i}(t - g\Delta t) \right] \quad (i = An,\ Ca) \tag{4.59}$$

Herein, Δt represents the sampling time interval. $[t/\Delta t]$ denotes the integer part of $t/\Delta t$. The Newton binomial term $\binom{\mu}{g}$ is described by

$$\binom{\mu}{g} = \begin{cases} \frac{\mu(\mu-1)\cdots(\mu-g+1)}{g!} & (g > 0) \\ 1 & (g = 0) \end{cases} \tag{4.60}$$

4.3.1.2 Modeling Based on Full Battery

The fractional-order ECM uesd in this part is shown in Fig. 4.22. The thermal-electrical coupling model, which is based on a three-electrode battery configuration, is capable of precisely determining the lithium plating boundary. This capability ensures that lithium plating within the battery can be effectively avoided during the development of heating strategies. However, the fabrication process of the three-electrode battery is cumbersome, with poor consistency and short service life. Moreover, commercial batteries only contain two electrodes. Thus, it is necessary to develop models based on full batteries and derive lithium plating criteria.

The full cell model's thermal part is similar to that of three-electrode battery and will not be described further. For the electrical part, the modeling based on the full cell also employs a fractional-order model, thermal model, and heat generation module, which are the same as shown in Fig. 4.21. The fractional-order model's order needs to be evaluated based on actual battery experiments, considering both precision and computational complexity., the fractional-order model for the full cell is of order 3 in this study, as shown in Fig. 4.22. the impedance Z' can be mathematically represented as

$$Z' = R_0 + \sum\nolimits_{i=1}^{3} \frac{R_i + \omega^{n_i} R_i^2 Q_i \cos \frac{n_i \pi}{2}}{1 + 2\omega^{n_i} R_i Q_i \cos \frac{n_i \pi}{2} + \omega^{2n_i} R_i^2 Q_i} \tag{4.61}$$

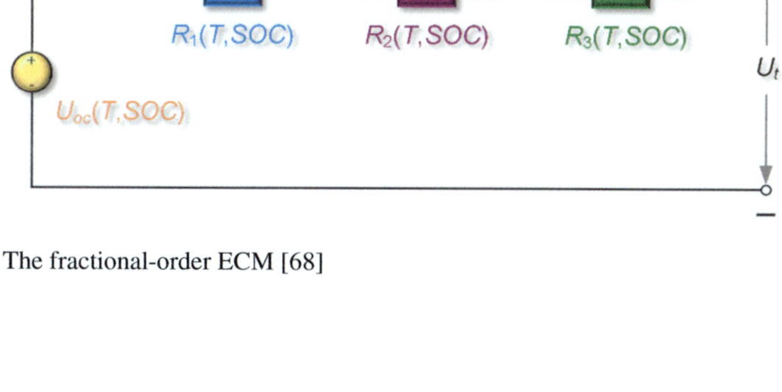

Fig. 4.22 The fractional-order ECM [68]

Herein, ω denotes the applied current's angular frequency. Meanwhile, Q and n correspond to the generalized capacitance and the dispersion coefficient associated with the CPE, respectively.

Within this sub-model, the terminal voltage U_t can be formulated as

$$U_t = U_{oc} + U_p \tag{4.62}$$

Herein, U_{oc} represents the OCV of the battery. The polarization voltage U_p can be mathematically described as [65]

$$U_p(s) = Z(s)I(s) = \left(\frac{\sum_{k=1}^{7} a_k s^{\alpha_k}}{\sum_{k=1}^{8} b_k s^{\beta_k}} + R_0\right)I(s) = \left(\sum_{i=1}^{3} \frac{R_i}{R_i Q_i s^{n_i} + 1} + R_0\right)I(s) \tag{4.63}$$

Then, the parameters in Eq. (4.62) can be expressed as

$$\begin{cases} [\, a_1\ a_2\ a_3\ a_4\ a_5\ a_6\ a_7 \,] = \Big[R_1R_2R_3Q_1Q_3\ R_1R_2R_3Q_1Q_2\ R_1R_2R_3Q_2Q_3\ R_1R_2Q_1 \\ \quad + R_1R_3Q_1R_1R_2Q_2 + R_2R_3Q_2\ R_1R_3Q_3 + R_2R_3Q_3\ R_1 + R_2 + R_3 \Big] \\ [\, \alpha_1\ \alpha_2\ \alpha_3\ \alpha_4\ \alpha_5\ \alpha_6\ \alpha_7 \,] = \Big[n_1 + n_3\ n_1 + n_2\ n_2 + n_3\ n_1\ n_2\ n_3\ 0 \Big] \\ [\, b_1\ b_2\ b_3\ b_4\ b_5\ b_6\ b_7\ b_8 \,] = \Big[R_1R_2R_3Q_1Q_2Q_3\ R_1R_2Q_1Q_2\ R_1R_3Q_1Q_3 \\ \quad R_2R_3Q_2Q_3\ R_1Q_1\ R_2Q_2\ R_3Q_3\ 1 \Big] \\ [\, \beta_1\ \beta_2\ \beta_3\ \beta_4\ \beta_5\ \beta_6\ \beta_7\ \beta_8 \,] = [\, n_1 + n_2 + n_3\ n_1 + n_2\ n_1 + n_3\ n_2 + n_3\ n_1\ n_2\ n_3\ 0 \,] \end{cases} \tag{4.64}$$

The solution of fractional calculus poses a significant challenge due to its inherent complexity. However, the Grünwald-Letnikov (GL) definition offers a valuable approach by discretizing fractional calculus, thereby enabling the solution of fractional-order equations [67]. Utilizing this method, Eq. (4.62) can be effectively resolved as

$$\begin{aligned} U_p(t) = \frac{1}{\sum_{k=1}^{8} \frac{b_k}{\Delta t^{\beta_k}}} \Bigg[& \sum_{k=1}^{7} \sum_{g=0}^{[\frac{t}{\Delta t}]} \frac{a_k}{\Delta t^{\alpha_k}} (-1)^g \binom{\alpha_k}{g} i(t - g\Delta t) \\ & + R_0 \sum_{k=1}^{8} \sum_{g=0}^{[\frac{t}{\Delta t}]} \frac{b_k}{\Delta t^{\beta_k}} (-1)^g \binom{\beta_k}{g} i(t - g\Delta t) - \sum_{k=1}^{8} \sum_{g=1}^{[\frac{t}{\Delta t}]} \frac{b_k}{\Delta t^{\beta_k}} (-1)^g \binom{\beta_k}{g} v(t - g\Delta t) \Bigg] \end{aligned} \tag{4.65}$$

Herein, Δt represents the sampling time interval, while $[t/\Delta t]$ denotes the integer part of $t/\Delta t$. $\binom{\mu}{g}$ is the Newton binomial term, and it can be expressed as Eq. (4.59), where μ denotes the fractional-order exponent, which governs the dynamics of α_k and β_k in accordance with fractional calculus principles.

4.3.2 Experiment Design

There are many parameters in the aforementioned models, which can be obtained through battery testing. The experimental design is as follows.

In this study, 1 Ah batteries (referred to as Sample I) were employed. Table 4.6 summarizes the battery parameters. The experimental procedures included the assembly of a three-electrode battery, AC impedance testing, heating experiments, and evaluations of battery performance degradation. The experimental platform, as depicted in Fig. 4.23a, consists of several key components: a host computer, a temperature acquisition recorder (Hongrun OHR-XH710), an electrochemical workstation (EnergyLab XM), a temperature-controlled chamber (Binder MK56), and a power booster (Solartron analytical, rated at 24 V/10 A),. Additionally, using the Maccor MC16 device, battery testing was performed.

The host computer is utilized to regulate and oversee the equipment's operation and to collect data. The EnergyLab XM is employed for conducting AC impedance tests on batteries and generating BPC wave, with a maximum current of 2 A. And due to the high amplitude of the high-frequency AC heating, the Solartron analytical power booster is utilized to amplify the current output from the EnergyLab XM. The power amplifier operates across an extended voltage range up to 24 V and can support a maximum current of 10 A. Additionally, the Hongrun OHR-XH710 temperature acquisition recorder is used to monitor battery's temperature. The sampling interval is set at 1 s, with a temperature measurement range from 50 to 100 °C (the measurement error is maintained within 0.1 °C). The thermocouple type employed for temperature sensing is T-type. The temperature is determined by averaging the

Table 4.6 Specifications of the sample I

Item	Specification
Cathode material	$LiNi_{0.8}Co_{0.1}Mn_{0.1}O_2$, single crystal
Anode material	Graphite
Electrolyte	Solution of lithium hexafluorophosphate ($LiPF_6$)
Nominal capacity	1 Ah
Nominal voltage	3.7 V
Charge cut-off voltage	4.2 V
Discharge cut-off voltage	3 V
Mass	22 g
Surface area	73.05 cm^2
Specific heat capacity	1.1 $Jg^{-1}\,K^{-1}$

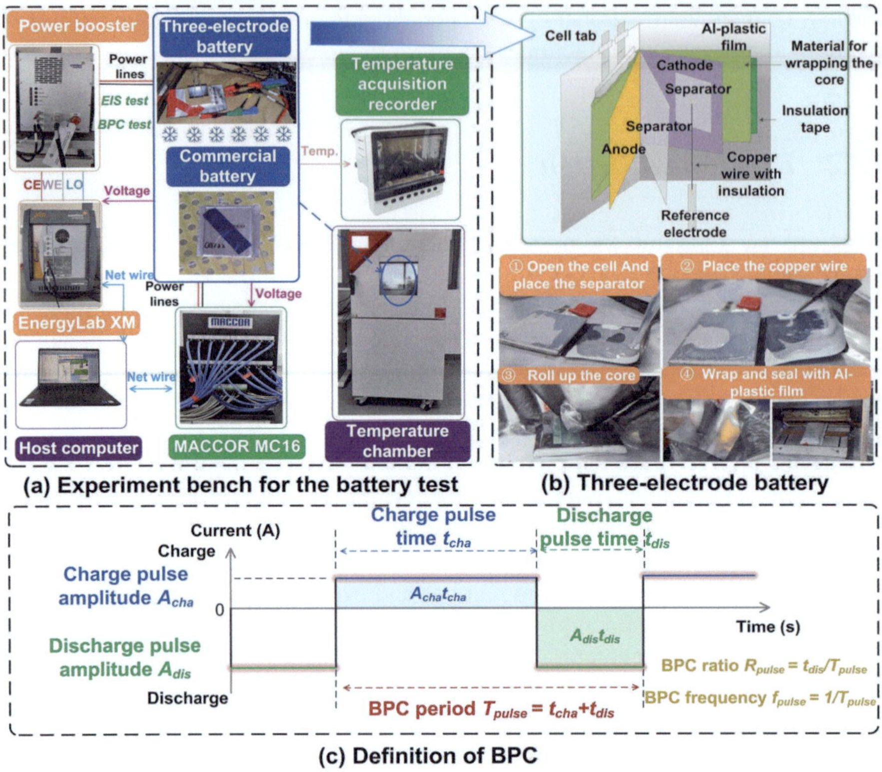

Fig. 4.23 Experiment platform [68]

measurements from six thermocouples positioned on the battery surface. Moreover, the Binder MK56 temperature chamber is utilized to regulate the ambient temperature surrounding the battery, with a temperature span from −40 to 180 °C and a stability of ±0.3 °C. Furthermore, the Maccor MC16 battery cycler is employed for charge and discharge, capable of measuring a maximum voltage of 5 V and offering four current ranges: 150 μA, 5 mA, 150 mA, and 5 A.

To calibrate the parameters and identify the suitable current range, the three-electrode battery model was evaluated exclusively at 20% SOC across different temperatures and current amplitudes. Conversely, the full battery model was subjected to a broader range of conditions, including temperatures of −10, 0, and 10 °C, as well as SOC levels of 20, 50, and 80%. This comprehensive investigation aimed to analyze the variations in parameters and provide valuable insights for the creation of AC heating strategies.

4.3.2.1 Experiment Design for Three-Electrode Battery

(1) Three-electrode Battery Preparation

The three-electrode battery used in this study was fabricated by altering a commercial battery, and the process consists of two main steps: acquisition of the reference electrode and its subsequent implantation. The implantation step was conducted in a glovebox filled with Argon gas (MBRAUN, with water content less than 0.1 ppm and O_2 content less than 0.1 ppm), as illustrated in Fig. 4.23b.

The reference electrode was constructed using a copper wire with a diameter of 20 μm, coated with an insulating layer. To remove the insulation, both ends of the wire were immersed in concentrated sulfuric acid for approximately 20 min, with an immersion depth of 2 cm. Following acid treatment, the wire segments were extensively rinsed with deionized water until a neutral pH was achieved. Subsequently, the wire was soaked in absolute ethanol for 10 min to remove any residual impurities and then dried in a vacuum desiccator. One end of the processed copper wire was then welded to a conductive tab, facilitating connection to the measurement fixture of the test equipment.

To integrate the reference electrode into a commercial battery, the aluminum-plastic film was carefully peeled back to expose the inner core. A portion of the core was unrolled, and a piece of separator material was inserted between the existing separator and the positive electrode. The bare end of the copper wire was positioned between the separators, and a few drops of electrolyte were added to ensure proper contact. The inner core was then re-rolled and sealed with a new piece of aluminum-plastic film. To secure the reference electrode and ensure good internal contact, the modified battery was clamped using plate clamps (as shown in Fig. 4.23a). The three-electrode battery was allowed to rest for 12 h after this modification.

Lithium plating was performed on the copper wire to further enhance its functionality. The process involved plating the copper wire with lithium first through the positive electrode and then through the negative electrode, with the copper wire acting as the anode. The lithium plating was conducted at a constant current of 2 μA for a duration of 4 h. This step effectively deposited a layer of lithium on the copper wire, preparing it for subsequent electrochemical measurements.

(2) AC Impedance Test for Three-electrode Battery

AC impedance tests were performed under two distinct conditions for the three-electrode battery. The primary objectives were to assess the performance of the three-electrode battery and to measure its impedance characteristics across various temperatures and current amplitudes.

In order to evaluate the performance of the three-electrode battery, a comparative investigation was carried out in parallel with a standard two-electrode battery. This assessment involved a series of electrochemical tests to conduct a comprehensive analysis of the characteristics of both battery configurations. AC impedance measurements were implemented at 25 °C by utilizing the EnergyLab XM system. Before the testing, both batteries were equilibrated to SOC of 20%. The impedance tests were

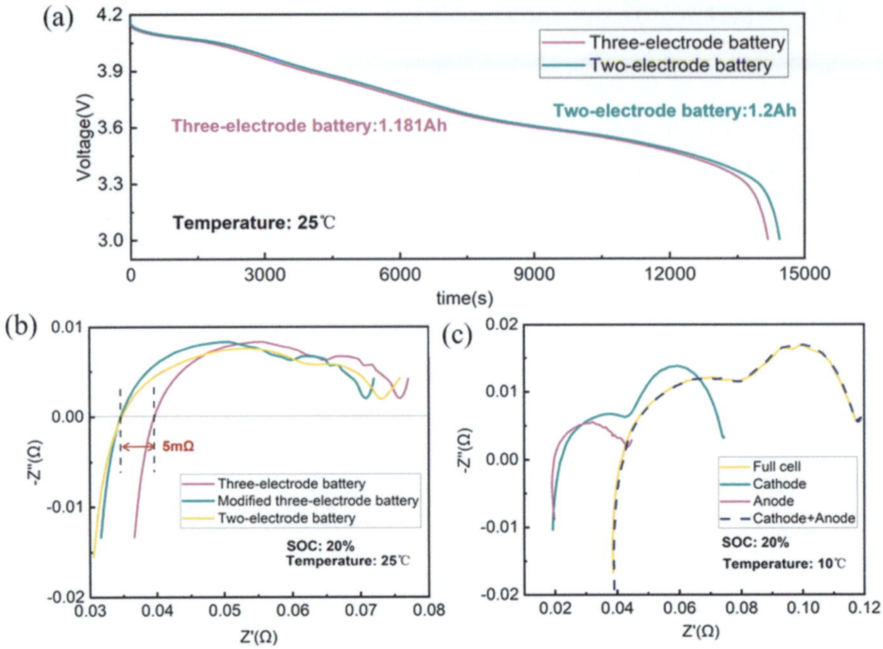

Fig. 4.24 Electrochemical performance validation of the three-electrode battery: **a** Discharge voltage profiles at 0.3 A; **b** Comparative EIS between three-electrode and two-electrode configurations at 25 °C; **c** Individual electrode and full-cell impedance at 10 °C [62]

executed in galvanostatic mode, with a current amplitude of 0.25 A and a frequency range extending from 0.01 to 10^6 Hz. This specific frequency range was chosen to capture the impedance behavior over different time scales, thus offering insights into the electrochemical processes occurring within the batteries. The discharge capacities of the two battery configurations were evaluated according to a constant current–constant voltage (CCCV) protocol. The charging process commenced at a current of 0.3 A until the voltage reached 4.2 V. Subsequently, a constant-voltage stage was initiated and continued until the current decreased below 0.01 A. After that, the batteries were discharged at a current of 0.3 A, and the discharge process was terminated when the voltage dropped to the cutoff voltage of 3 V. The discharge capacity, which was recorded as the effective battery capacity, served as a crucial performance indicator.

The experimental results, including voltage profiles, impedance comparisons, and electrode—specific impedance distributions, are depicted in Fig. 4.24. These results present a detailed comparison of the electrochemical performance between the three-electrode and two-electrode batteries, highlighting the advantages and potential applications of the three-electrode design.

Subsequently, AC impedance experiments were conducted to characterize the three-electrode battery's impedance behavior across a range of temperatures and amplitudes. The temperatures investigated spanned from −15 to 10 °C, with increments of 5 °C. For each temperature, AC impedance measurements were performed using a sequence of current amplitudes: 0.25, 0.65, 1, 1.5, 2, and 2.5 A. The frequency range was maintained from 0.1 Hz to 10^6 Hz. Prior to initiating the tests, SOC of the three-electrode battery was calibrated to 20%, and the battery was stabilized at 10 °C for 2 h. The AC impedance tests were then conducted in a sequential manner: starting with an amplitude of 0.25 A, followed by a 1-h rest period to ensure stability, and subsequently proceeding with tests at increasing amplitudes (0.65, 1, 1.5, 2, and 2.5 A), each separated by a 1-h rest interval. Once the tests at 2.5 A were completed, the battery was subjected to the same procedure at the next lower temperature (5 °C), with a 2-h stabilization period preceding the tests. This protocol was consistently applied across all specified temperatures, ensuring a systematic evaluation of the battery's impedance characteristics under varying thermal and current conditions.

(3) BPC Test for Three-electrode Battery

To validate the accuracy of the thermal model, it is necessary to conduct temperature rise experiments on the battery. The two-electrode battery is initially placed at −13 °C for a duration of 2 h. Subsequently, the battery is subjected to heating for 30 min using various types of BPCs. After each heating test, the battery is allowed to rest for 1 h before proceeding to the next heating test. The BPCs are generated using a combination of the EnergyLab XM and a power booster. The temperature of the battery is measured by averaging the readings from six thermocouples attached to different locations on the battery surface. The specific BPC waveform is illustrated in Fig. 4.23c. The parameters defining the BPC include the period (T_{pulse}), frequency (f_{pulse}), duty cycle ratio (R_{pulse}), charge pulse time (t_{cha}), discharge pulse time (t_{dis}), charge pulse amplitude (A_{cha}), and discharge pulse amplitude (A_{dis}). The specific types of BPCs utilized in the heating tests are detailed in Table 4.7.

Furthermore, to assess the electrical model's accuracy, the two-electrode battery is tested with BPCs across various temperatures and frequencies. Table 4.8 presents the details of those wave, and the tests are conducted at temperatures of 10, −5, and −15 °C. Using the EnergyLab XM, the high-frequency current and voltage data are collected.

The data presented in Tables 4.7 and 4.8 serve to validate the accuracy of the electro-thermal coupling model, and the selection of these data adheres to the principle of randomization. Regarding the electrical model's validation, precise current and voltage waveforms cannot be reliably obtained when the frequency exceeds 100 Hz due to the limitations of the platform's sampling frequency. Consequently, the maximum frequency specified in Table 4.8 is capped at 100 Hz. In contrast, for the heating tests, the focus is solely on measuring the temperature change rather than the current and voltage of the battery. Therefore, the maximum frequency for these tests is determined by the highest frequency that the platform can generate, which is set at 500 Hz as indicated in Table 4.7.

Table 4.7 The types of BPCs for heating tests [62]

Frequency f_{pulse} (Hz)	Ratio R_{pulse}	Discharge amplitude A_{dis} (A)	Charge amplitude A_{cha} (A)	Abbreviation
1	0.5	1.5	1.5	1.5 A@1 Hz
	0.5	2.5	2.5	2.5 A@1 Hz
	0.7	1.5	3.5	A@1 Hz
	0.4	2.5	1.5	B@1 Hz
10	0.5	1.5	1.5	1.5 A@10 Hz
	0.5	2.5	2.5	2.5 A@10 Hz
	0.7	1.5	3.5	A@10 Hz
	0.4	2.5	1.5	B@10 Hz
100	0.5	1.5	1.5	1.5 A@100 Hz
	0.5	2.5	2.5	2.5 A@100 Hz
	0.7	1.5	3.5	A@100 Hz
	0.4	2.5	1.5	B@100 Hz
500	0.5	4.9	4.9	4.9 A@500 Hz

Table 4.8 The types of BPCs for validation of the electrical model [62]

Frequency f_{pulse} (Hz)	Ratio R_{pulse}	Discharge amplitude A_{dis} (A)	Charge amplitude A_{cha} (A)	Abbreviation
1	0.5	1.5	1.5	I@1 Hz
	0.5	3.5	1.5	II@1 Hz
	0.7	2.5	2.5	III@1 Hz
	0.3	2.5	2.5	IV@1 Hz
50	0.5	1.5	3.5	V@50 Hz
	0.5	3.5	1.5	II@50 Hz
	0.7	2.5	2.5	III@50 Hz
	0.3	2.5	2.5	IV@50 Hz
100	0.5	1.5	3.5	V@100 Hz
	0.5	3.5	1.5	II@100 Hz
	0.7	2.5	2.5	III@100 Hz
	0.3	2.5	2.5	IV@100 Hz

(4) Battery Performance Degradation Test

The battery performance degradation test encompasses two primary components: the verification of the lithium plating criterion and the BPC heating test. To validate the accuracy of the lithium plating criterion, two batteries with identical capacities were subjected to aging experiments using 1 Hz symmetrical BPCs with two kinds of amplitudes (1.4 and 1.6 A). In order to minimize the heat generation's influence on the actual temperature, the testing methodology employed was adapted from the approach proposed by Qin et al. [69]. Initially, the batteries were preconditioned at temperatures below −10 °C for 2 h. They were then subjected to BPC stimulation at amplitudes of 1.4 A and 1.6 A. During this phase, thermal equilibrium was achieved, stabilizing the temperature at −10 °C. The choice of 1 Hz and −10 °C is determined by considering the current range of the experimental platform. Achieving a lower equilibrium temperature is challenging given the existing cooling capacity of the temperature chamber. The study also includes intermittent capacity testing at 25 °C, supplementing the lithium plating validation. The capacity evaluation method mirrors that used in the three-electrode and two-electrode analysis, but with a modified protocol where charge and discharge are executed at 1 A. After completing the tests, the batteries are disassembled in a glovebox filled with Argon gas (MBRAUN, with water content less than 0.1 ppm and O_2 content less than 0.1 ppm). The internal components of the batteries are then examined using a scanning electron microscope (SEM, JEOL JSM-7610FPLUS).

The BPC heating protocol, designed within lithium plating and voltage safety limits, examined thermal strategy effects on battery aging. During this test, each heating cycle (−15 to 10 °C) was followed by a 60-min rest period for thermal recovery. Capacity assessments, identical to the lithium plating verification method, were conducted at 25-cycle intervals.

4.3.2.2 Experiment Design for Full Battery

(1) EIS Tests at Different SOCs and Temperatures

Since the electro-thermal coupling model parameters are highly sensitive to SOC and temperature [27], experimental validation through EIS testing at multiple SOC and temperature is essential for establishing their correlations [68]. The impedance characterization employed a 0.25 A galvanostatic perturbation across 0.01–10^6 Hz. Testing conditions encompassed: (1) temperatures from −10 to 10 °C (5 °C increments), (2) SOC at 10% intervals from 0–100%. The experimental protocol commenced with a CCCV charge at 25 °C (1 A constant current, termination at 0.02 A cutoff current). Following charge completion, the battery was equilibrated at the target temperature for 2 h. EIS tests at different SOCs are then conducted. To mitigate temperature-induced capacity variations, SOC adjustments were performed using a fixed 0.1 A discharge current, followed by a 30-min equilibration period preceding each EIS scan.

To determine the parameters for the lithium plating criterion, EIS tests were carried out on a full cell and a three-electrode battery (10% SOC) using the same experimental protocol.

(2) BPC Tests at Different SOCs and Temperatures

As illustrated in Fig. 4.23c, the BPC requires balancing the ampere-hours of charge and discharge within each period T_{pulse}, which guarantees the battery's SOC remains constant under the BPC thermal cycling. Energy-injection and non-energy-injection types represent the two primary categories of current AC heating strategies for LIBs. Energy-injection heating strategies, typically coupled with charging devices [70], do not require SOC maintenance. Unlike energy-injection methods, non-energy-injection schemes (e.g., EV drive system redesign for self-heating) impose strict energy constraints to preserve operational range. In these cases, maintaining a constant SOC is highly desirable. Furthermore, compared to energy-injection AC heating schemes, non-energy-injection schemes are more widely applicable. Therefore, this study focuses primarily on exploring the optimal BPC heating strategy under SOC equilibrium conditions, which can be mathematically fomula as

$$A_{cha}t_{cha} = A_{dis}t_{dis} \tag{4.66}$$

Consequently, the distinct properties of each BPC are determined by the following parameters: the frequency f_{pulse}, the ratio R_{pulse}, and the discharge amplitude A_{dis}.

The validation of the electro-thermal coupling model's accuracy is a crucial aspect of this study. To achieve this, two categories of BPC tests were conducted: terminal voltage tests and temperature tests. The terminal voltage tests were designed to precisely evaluate the model's capability to predict terminal voltage under various operating conditions, while the temperature tests focused on assessing the model's ability to predict temperature variations of the BPC during operation.

In the validation process, three distinct BPCs (labeled as A, B, and C) were employed. The parameters of these BPCs, which are essential for both testing and model simulation, are detailed in Table 4.9. The descriptions of the terminal voltage validation tests are summarized in Table 4.10, and the descriptions of the temperature validation tests are presented in Table 4.11. These tables provide a clear presentation of the validation results, facilitating an intuitive comparison between the model predictions and the actual test data, thereby enabling an accurate assessment of the electro-thermal coupling model's validity.

For the terminal voltage's validation, measurements are conducted at three different SOC levels (20, 50, and 80%) across a range of temperatures: −10, 0,

Table 4.9 The BPCs types for tests [68]

Ratio R_{pulse}	Discharge pulse amplitude A_{dis} (A)	Charge pulse amplitude A_{cha} (A)	Abbreviation
0.50	3	3	A
0.25	6	2	B
0.75	1	3	C

Table 4.10 Verification results of terminal voltage through BPC testing [68]

SOC	Temperature (°C)	Frequency f_{pulse} (Hz)	BPC tests
20%, 50%, 80%	−10, 0, 10	100, 500	A, B, C

Table 4.11 BPC tests for temperature verification [68]

Initial temperature (°C)	SOC	Frequency f_{pulse} (Hz)	BPC tests
−10	20%, 50%, 80%	100, 500	A, B, C

and 10 °C. The test frequencies were set to 100 and 500 Hz, with the EnergyLab XM sampling frequency maintained at 20-fold the BPC pulse frequency f_{pulse} After stabilizing the battery at the target SOC and temperature (2 h), Tests A to C are executed at 100 and 500 Hz. A 10-min rest interval separates consecutive BPC tests to ensure stabilization.

In contrast to terminal voltage validation, temperature validation focuses primarily on monitoring the temperature increase of the battery when subjected to various BPCs over a 30-min period at −10 °C. The battery is tested at three different SOC levels (20%, 50%, and 80%) with BPC frequencies set at 100 and 500 Hz. Following each heating test, a 40-min interval is required before initiating the next heating test.

(3) Battery Degradation Test

In order to assess the proposed heating strategy's impact on battery degradation, this heating strategy, which involves raising the temperature from −10 to 10 °C at a frequency of 100 Hz and an SOC level of 50%, is employed for validation. The degradation test involves two primary components: the calibration test and the cycle test, which are performed at 25 °C and −10 °C, respectively. The calibration test includes a battery capacity assessment and an EIS analysis at 50% SOC, while the cycle test incorporates BPC heating. When heating the battery from −10 to 10 °C, a 40-min rest interval is necessary to allow the battery to cool adequately before repeating the heating process. 60 cycle tests are performed in total, with a calibration test carried out every 20 cycles.

4.3.3 Determination of Model Parameters

4.3.3.1 Parameters and Validation of the Three-Electrode Battery Model

(1) Electrical Model Parameters

Figure 4.25 presents the impedance characteristics of the three-electrode battery at 20% SOC. As the current amplitude increases, the impedance arcs in the medium and low-frequency regions tend to shrink. Moreover, this shrinkage becomes more evidentd at lower temperatures. This phenomenon is primarily attributed to the enhanced sensitivity of the charge transfer process to changes in current amplitude under low-temperature conditions [34]. Additionally, the impedance of the cathode is significantly higher than that of the anode. This difference is mainly due to the

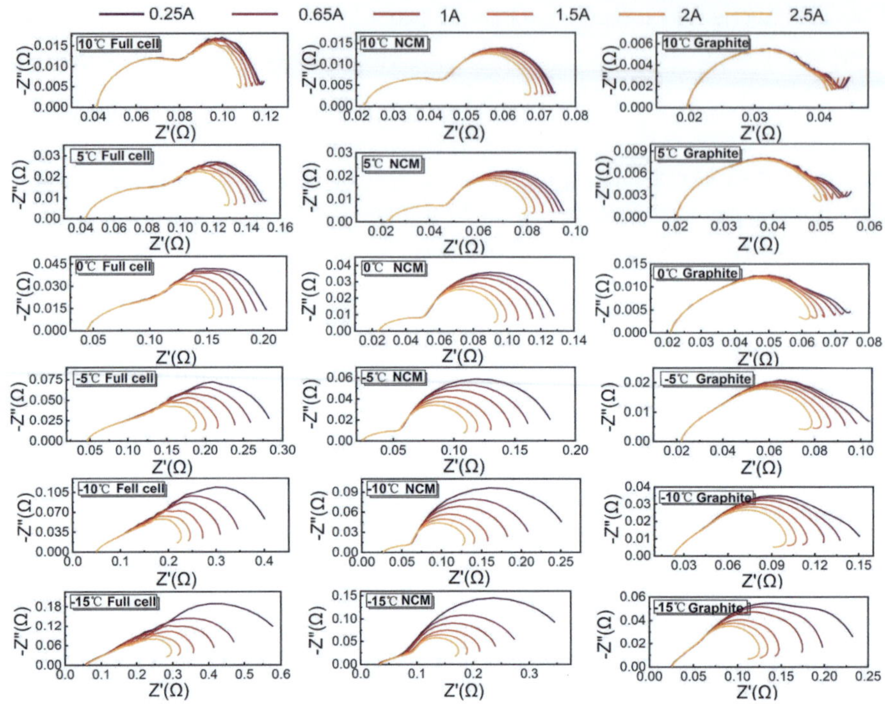

Fig. 4.25 AC impedance characteristics of the three-electrode battery at 20% SOC across different temperatures and current amplitudes [62]

dominant role of the cathode charge transfer process at low SOC levels. These observations are consistent with our prior research findings [34]. To further explore the variations in electrical model parameters with temperature and current amplitude, the equivalent circuit model (ECM) depicted in Fig. 4.21 was constructed using ZView software. The model parameters were identified based on the impedance data obtained from the cathode and anode, as shown in Fig. 4.26.

As depicted in Fig. 4.26, the fractional-order model parameters exhibit a clear dependence on T. Nevertheless, the current's impact amplitude on these parameters is not uniform across all parameters. Based on the analysis presented in references [71, 72], the ohmic resistance R_0 is independent of the current amplitude. In contrast, the charge transfer process is modulated by the current amplitude. The current amplitude primarily influences the charge transfer process. Moreover, as indicated by the findings in reference [35], the resistance associated with the charge transfer process is highly sensitive to the current amplitude. In contrast, the ohmic resistance R_0 and the resistances related to the SEI and CEI processes remain largely unaffected by the current amplitude. These conclusions are consistent with the results presented in Fig. 4.26. Thus, for the three-electrode battery model, the parameters R_0, R_1, Q_1, and n_1 are only adjusted to account for the effects of temperature during fitting. In

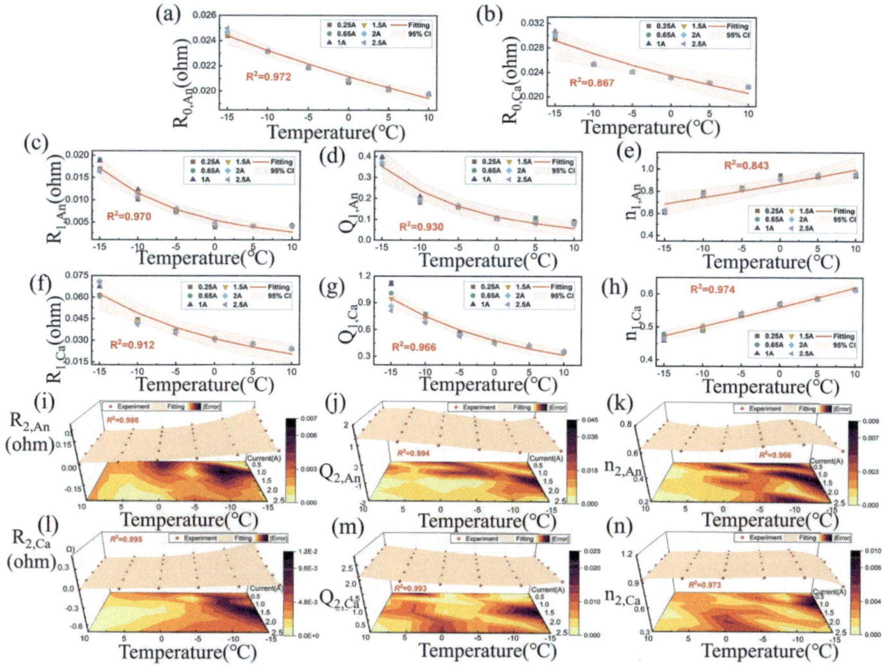

Fig. 4.26 Electrical model parameters and corresponding fitting results across diverse temperatures and current amplitudes at 20% SOC [62]

contrast, the parameters R_2, Q_2, and n_2 are adjusted to account for the combined effects of both temperature and current amplitude.

The Arrhenius equation [27] can be employed to model the correlation between the parameters and temperature, and it is formulated as

$$\frac{1}{X_T} = X_A e^{\frac{-E_{a,X}}{RT}} \tag{4.67}$$

Herein, X_T denotes the parameters R_0, R_1, Q_1, and n_1. $E_{a,X}$ represents the activation energy associated with X_T. X_A denotes the pre-exponential constant, and R denotes the universal gas constant.

Initially, the mean values of each parameter are calculated over a range of current amplitudes and across multiple temperatures. Subsequently, the Arrhenius equation is employed for fitting, and the results are presented in Figs. 4.26a–h. Among the fitted parameters, the coefficient of determination R^2 for $R_{1,An}$ is the smallest, at 0.843, while the R^2 values for the other parameters are greater than 0.91.

The parameters linked to the charge transfer process are affected by both temperature and current amplitude. As documented in prior research [34, 35], the relationship between R_2 and these two factors can be characterized into

$$R_2(I,T) = \frac{\ln\left\{\frac{1}{2S_A X_A' B} e^{\frac{E_a}{RT}} I + \left[\left(\frac{1}{2S_A X_A' B} e^{\frac{E_a}{RT}} I\right)^2 + 1\right]^{0.5}\right\}}{\frac{1}{2S_A X_A' B} X_B' I} \tag{4.68}$$

Herein, S_A represents the active particles' total surface area. B is a function of Li^+ concentration. X_A' and X_B' are the pre-exponential constants, E_a denotes the activation energy.

The fitting results presented in Figs. 4.26i–l illustrate the accuracy of Eq. (4.67). The relationship between the parameters of CPE_2 and temperature or current amplitude has not been thoroughly investigated. Therefore, in this study, a polynomial fitting method is utilized, which can be formulated as

$$\begin{aligned} O(I,T) = d_0 + d_1 T + d_2 I + d_3 T^2 + d_4 TI + d_5 I^2 + d_6 T^3 \\ + d_7 T^2 I + d_8 TI^2 + d_9 T^4 + d_{10} T^3 I + d_{11} T^2 I^2 \end{aligned} \tag{4.69}$$

The fitting results are presented in Figs. 4.26j, k, m, and n, where O represents Q_2 and n_2. The minimum R^2 reaches 0.966, demonstrating the high accuracy of Eq. (4.68).

While the relationship between the parameters of CPE_2 and the current amplitude/temperature can be accurately modeled using polynomial functions, such a correlation is more of an empirical observation than a mechanistic explanation. Due to the potential risk of overfitting in Eq. (4.68), a conservative approach is adopted when extrapolating beyond the experimentally calibrated current and temperature ranges—parameters are constrained by boundary values to ensure reliability.

(2) Thermal Model Parameter

The equivalent heat transfer coefficient h also represents parameter in the thermal model. This parameter can be determined by analyzing the battery temperature variation during resting periods. The solution to Eq. (4.51) provides the basis for calculating h.

$$T = T_{am} + (T_{t=0} - T_{am}) e^{-\frac{hS}{mc} t} \tag{4.70}$$

Herein, $T_{t=0}$ represents the initial battery temperature at the start of the resting period. As depicted in Fig. 4.27, the experimental cooling curve closely aligns with the fitted model, exhibiting a maximum absolute temperature deviation of less than 0.5 °C.

(3) Three-electrode Battery Model Verifications

Figure 4.28 presents a comparison between the simulated voltage obtained from the electro-thermal coupling model and the experimentally measured voltage, highlighting the current amplitude's impact on voltage prediction.

Figures 4.28a–c illustrate the voltage comparison at 10 °C, while Figs. 4.28d–g provide a quantitative evaluation of voltage errors across different BPCs. The results

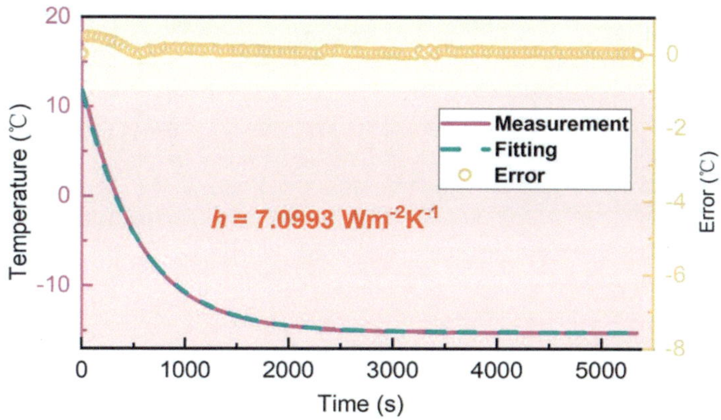

Fig. 4.27 Battery cooling curve fitting result [62]

indicate that the absolute voltage error remains below 0.05 V at 10 °C. Similar trends are observed at −5 °C and −15 °C, though with slightly higher deviations. Statistical analysis reveals that, Generally, with some specific exceptions (I@1 Hz, II@50 Hz, and IV@1H), the mean absolute error (MAE) and root mean square error (RMSE) at 10 °C are generally lower than those at −5 and −15 °C. Notably, the MAE of the voltage absolute error is typically less than 0.03 V, with the exception of the II@50 Hz at −15 °C. No clear correlation between the MAE and BPC frequency is observed. However, the RMSE exhibits higher values at 1 Hz and lower values at 100 Hz. Despite these variations, all error magnitudes remain within an acceptable tolerance range. These results confirm that the electro-thermal coupling model can reliably simulate voltage behavior across different BPC conditions with high accuracy.

Figure 4.28h compares the voltage absolute error between current-dependent and current-independent cases for BPC I (1 Hz) and BPC V (100 Hz) at 10, −5, and −15 °C. The electro-thermal coupling model parameters were calibrated at 0.25 A when current effects were neglected. The results demonstrate that the current-independent case exhibits higher MAE and RMSE values at 1 Hz compared to the current-dependent scenario at identical temperatures. Moreover, these errors tend to escalate as the temperature decreases. Such behavior is consistent with the observations depicted in Fig. 4.25, which reveal that battery impedance shows stronger current amplitude dependence at lower frequencies and temperatures. Consequently, the current-independent assumption introduces greater voltage error at 1 Hz. In contrast, at frequencies beyond a critical threshold (100 Hz), the impedance becomes current-independent, resulting in identical and minimal voltage errors for both modeling approaches.

Figures 4.29a, b validate the battery temperature rise predictions for different BPCs (as specified in Table 4.7). Both symmetrical (Fig. 4.29a) and asymmetrical (Fig. 4.29b) BPC cases exhibit excellent agreement between the modeled and measured temperature change curves. The maximum absolute temperature error

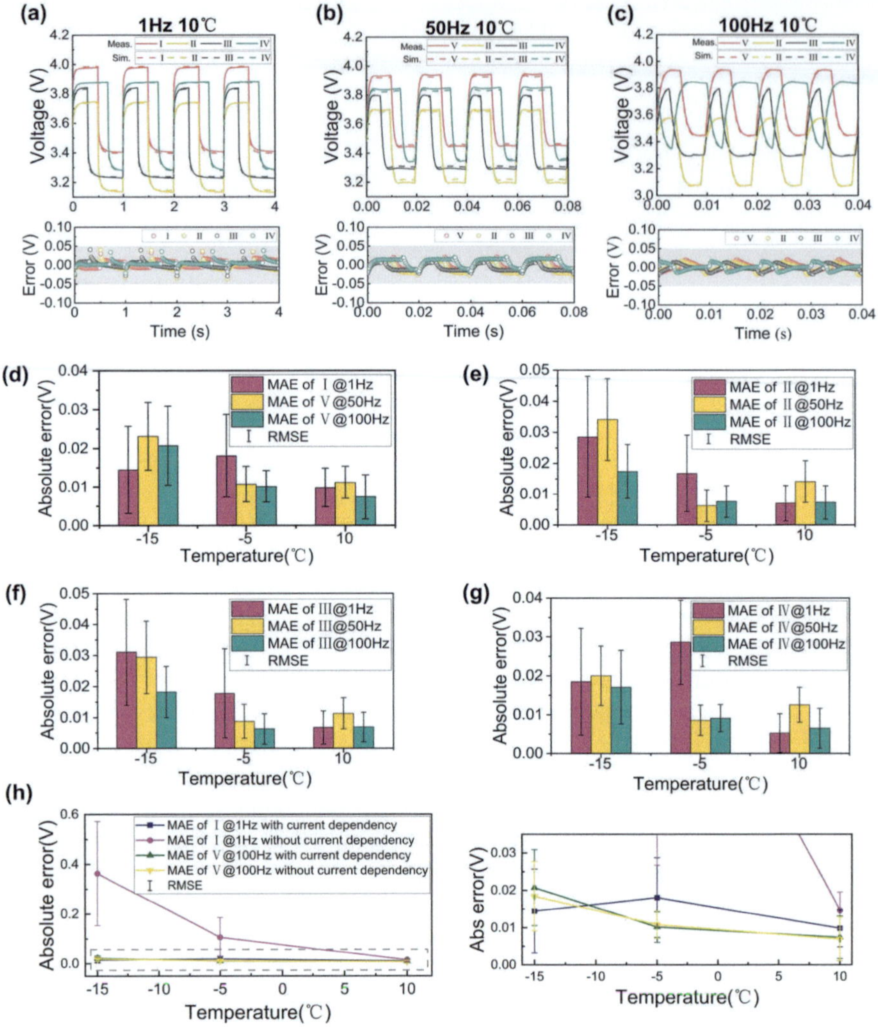

Fig. 4.28 Validation and error analysis of the electro-thermal coupling model. **a–c** Voltage comparison between simulated and experimental results at 10°C under different excitation frequencies: **a** 1 Hz, **b** 50 Hz, and **c** 100 Hz. **d–g** Statistical evaluation of voltage prediction accuracy showing MAE and RMSE for: **d** BPC types I and V, **e** BPC II, **f** BPC III, and **g** BPC IV. **h** Comparative analysis of voltage absolute errors considering current-dependent and current-independent [62]

remains within 0.6 °C, verifying the precision of the electro-thermal coupling model in forecasting thermal performance under diverse BPC type.

Figure 4.29c compares the temperature errors with and without current dependency for BPC 2.5A@1 Hz, 2.5A@10 Hz, 2.5A@100 Hz, A@1 Hz, A@10 Hz, and A@100 Hz. The results indicate that omitting current dependency leads to higher MAE and RMSE values, particularly at 1 Hz, due to the strong influence

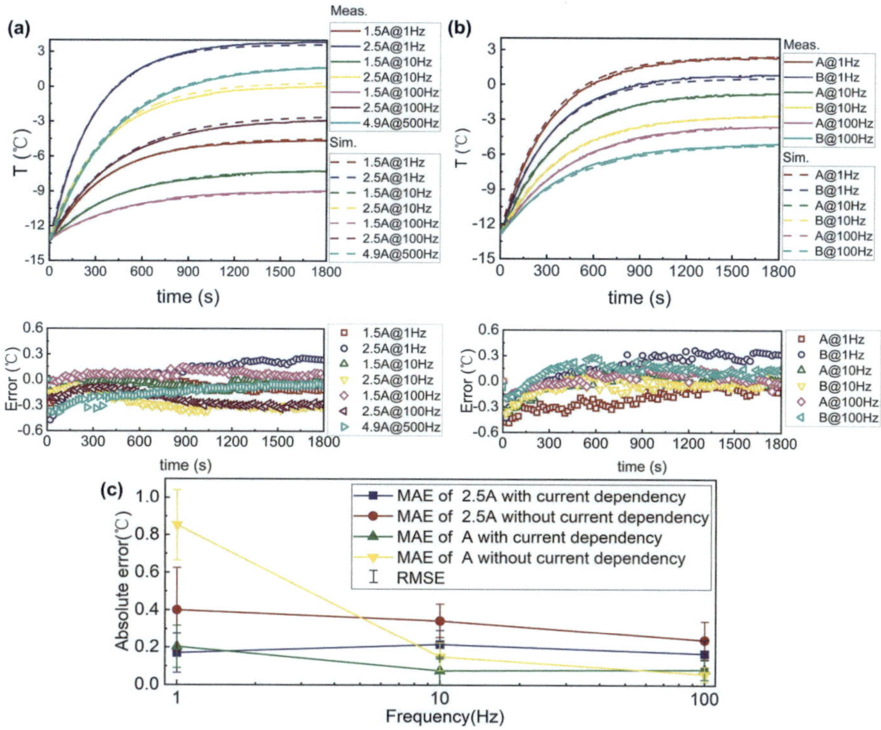

Fig. 4.29 Model validation and temperature error analysis: **a** Symmetrical BPC temperature rise comparison; **b** Asymmetrical BPC temperature rise comparison; **c** Current-dependent versus current-independent temperature absolute error [62]

of low-frequency impedance on current amplitude, which is consistent with the voltage error analysis. Nevertheless, even without current dependency, the maximum MAE remains below 0.9 °C, which falls within an acceptable range compared to prior studies [5, 31, 72]. These findings suggest that current amplitude has a more significant impact on voltage estimation than on temperature prediction.

4.3.3.2 Parameters and Validation of the Full Battery Model

(1) Electrical Model Parameters

Figure 4.30 presents the EIS test results and resistance identification across varying SOCs and temperatures. The experimental data reveal a notable expansion of the battery's impedance arc with decreasing temperature, attributable to the deceleration of internal electrochemical reactions in LIBs under low-temperature conditions [73]. Figure 4.30 additionally reveals that, for a fixed temperature, the impedance arc within the mid-to-low frequency region initially diminishes before rising as the SOC

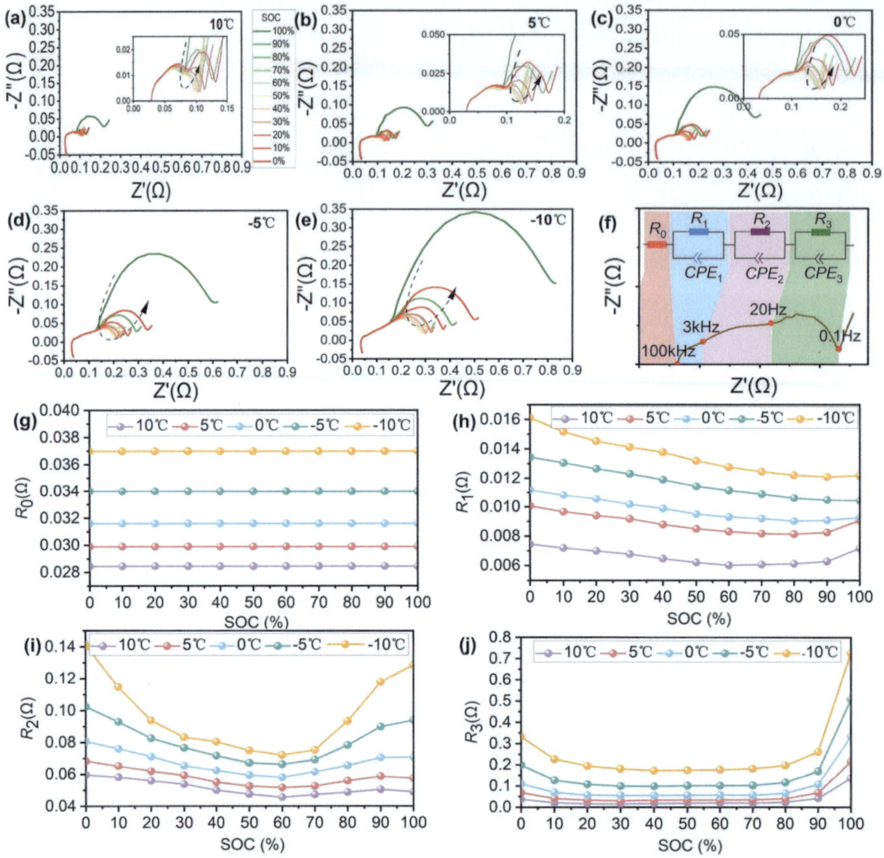

Fig. 4.30 EIS analysis and equivalent circuit parameter identification for the full-cell under varying SOC and temperature conditions: **a**–**e** Nyquist plots of EIS measurements at 10, 5, 0, −5, and −10 °C, respectively; **f** Comparison of the ECM with experimental impedance spectra; **g**–**j** Identified resistance parameters (R_0, R_1, R_2, and R_3) as functions of SOC and temperature [68]

reduces. The peak impedance arc is noted at 100% SOC, attributable to the fluctuating charge transfer resistance and solid-phase diffusion impedance across varying SOC [74]. The high-frequency impedance exhibits minimal sensitivity to SOC variations. Figure 4.30f elucidates the fitting methodology of the impedance arc utilizing the fractional-order model. This figure illustrates the battery impedance at 0°C for a 50% SOC condition.

The inductive characteristics at ultra-high frequencies and the solid-phase diffusion process at very low frequencies are beyond the scope of this study. Consequently, it is evident from the figure that the frequency range analyzed in this study spans from 0.1 Hz to 100 kHz. The parameter R0 denotes the battery's ohmic internal resistance, with its magnitude equating to the real component at the intersection point between the impedance arc and the real axis. The three R//CPE configuration characterize the internal electrochemical reaction processes within the battery. The internal reaction processes of the battery can be categorized into four distinct types according to their characteristic time scales, ranging from the fastest to the slowest: the conduction process, the SEI/CEI processes, the charge transfer process, and the solid-phase diffusion process [75]. The conduction mechanism predominantly originates from the resistances at the interfaces between adjacent active material particles and between particles and current collectors, which are generally characterized by their rapid response [34]. The solid-state diffusion process is typically represented by a linear tail in the Nyquist plot of EIS analysis, mainly appearing in the ultra-low frequency range [76]. Drawing from prior research [34], the three R//CPE equivalent circuits can be ascribed to the conduction mechanism, SEI/CEI phenomena, and charge-transfer kinetics, respectively. Using ZView software, the fractional-order model parameters were extracted through nonlinear least-squares fitting. The fitting resistance values are presented in Fig. 4.30g–j.

Under identical conditions, it is observed that the conduction process resistance R_1 is the smallest, whereas the charge transfer process resistance R_3 is the largest. The ohmic resistance R_0 remains relatively stable across different SOC. Both the SEI/CEI process resistance R_2 and the charge transfer process resistance R_3 initially decrease and then increase with rising SOC. Additionally, the conduction process resistance R_1 tends to be smaller at higher SOC levels and larger at lower SOC. Regarding temperature, each resistance gradually increases as the temperature decreases.

The characterization of ECM parameters as functions of T and SOC can be achieved through diverse mathematical representations, including polynomial approximations and Arrhenius formulations. However, the absence of a universally adopted analytical framework necessitates extensive computational effort to identify optimal relationships. To overcome this limitation, the current investigation adopts an artificial neural network (ANN) methodology, implemented using MATLAB's neural network toolbox, to establish a direct nonlinear mapping between battery model parameters and the influencing variables (T and SOC), as depicted in Fig. 4.31a. The implemented ANN architecture features dual input nodes (SOC and temperature), a ten-layer hidden structure, and a single output layer generating the fractional-order model parameters.

As demonstrated in Fig. 4.31b–f, the network training results exhibit excellent agreement between predicted and experimentally measured impedance values across representative SOC conditions (0, 10, 50, 90, and 100%). While marginally increased discrepancies are observed at elevated SOC levels under low-temperature conditions, the overall predictive performance satisfies the required computational accuracy thresholds. These results substantiate the efficacy of the developed ANN framework in accurately characterizing the complex interdependencies between fractional-order model parameters, SOC, and T.

(2) Thermal Model Parameter

Figure 4.32 reveal that the error of temperature is less than 0.5 °C, the calculated value for h is determined to be 6.9231 $Wm^{-2}\ K^{-1}$ [68].

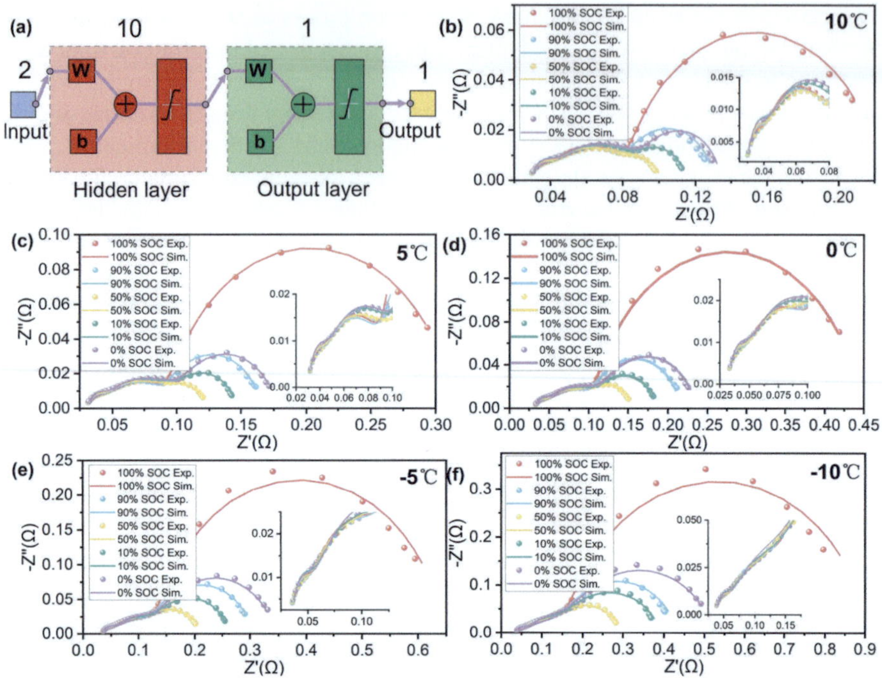

Fig. 4.31 **a** Architecture of the artificial neural network; **b–f** Comparison between experimentally measured impedance spectra and model-predicted values at different temperatures: **b** 10 °C, **c** 5 °C, **d** 0 °C, **e** −5 °C, and **f** −10 °C [68]

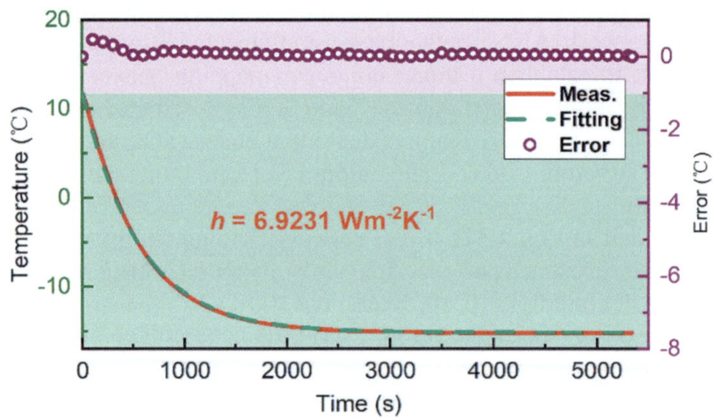

Fig. 4.32 Battery cooling curve fitting result [68]

(3) Full Battery Model verifications

Figures 4.33, 4.34, 4.35 present the validation results of terminal voltage at three distinct SOC levels: 20%, 50%, and 80%. The analysis reveals negligible fluctuations in terminal voltage error under varying SOC conditions. Furthermore, the terminal voltage error demonstrates negligible dependence on temperature and BPC. A notable deviation in error occurs during rapid terminal voltage transitions at specific frequencies, particularly at 100 Hz, where the maximum observed error reaches 50 mV. However, under normal operating conditions, the absolute terminal voltage error remains consistently below 25 mV. Above results show the proposed electro-thermal coupling model's accuracy in terminal voltage prediction.

Figure 4.36 demonstrates the thermal response characteristics under different BPC heating configurations. Experimental results reveal a positive correlation between excitation frequency and heating efficiency, attributable to the increased battery impedance's real component at lower frequencies. Comparative analysis indicates that configuration B achieves optimal heating performance, whereas configuration C demonstrates the lowest thermal response. These findings are consistent with the theoretical heating power predictions from Eq. (4.52). Notably, minimal variations in temperature rise are observed for identical BPC configurations across different SOC levels. The proposed electro-thermal coupling model maintains high computational accuracy, with absolute temperature errors generally confined within 1 °C. Even under challenging conditions (e.g., A@100 Hz at 50% SOC and 80% SOC), the maximum observed deviation remains below 1.5 °C, validating the model's robustness.

4.4 Conclusions

This chapter first provides a comprehensive analysis of two prominent modeling approaches: electrochemical-thermal and equivalent circuit-thermal coupling frameworks. Then, the establishment and parameter identification process of the electrochemical-thermal coupling model are introduced. Through simulations, it is concluded that the performance of LIBs at low temperatures is primarily influenced by the rate of electrochemical reactions, liquid-phase ionic conductivity, and diffusion coefficients. Moreover, the internal state's changes of LIBs under AC heating waveform are also analyzed. The results show that the battery's main failure mode during AC heating is the lithium plating. Given the complexity of applying the electrochemical-thermal coupling model to actual AC heating strategies, the equivalent circuit-thermal coupling models based on three electrode battery and full battery are also given. Three electrode battery modeling is mainly used to monitor lithium plating during AC heating. However, three electrode batteries are difficult to be applied in practice. Most of the batteries used in practice are two electrode batteries. Therefore, the equivalent circuit-thermal coupling model of the full battery is also introduced. Thses models in this chapter provides the basis for the design of AC heating strategy.

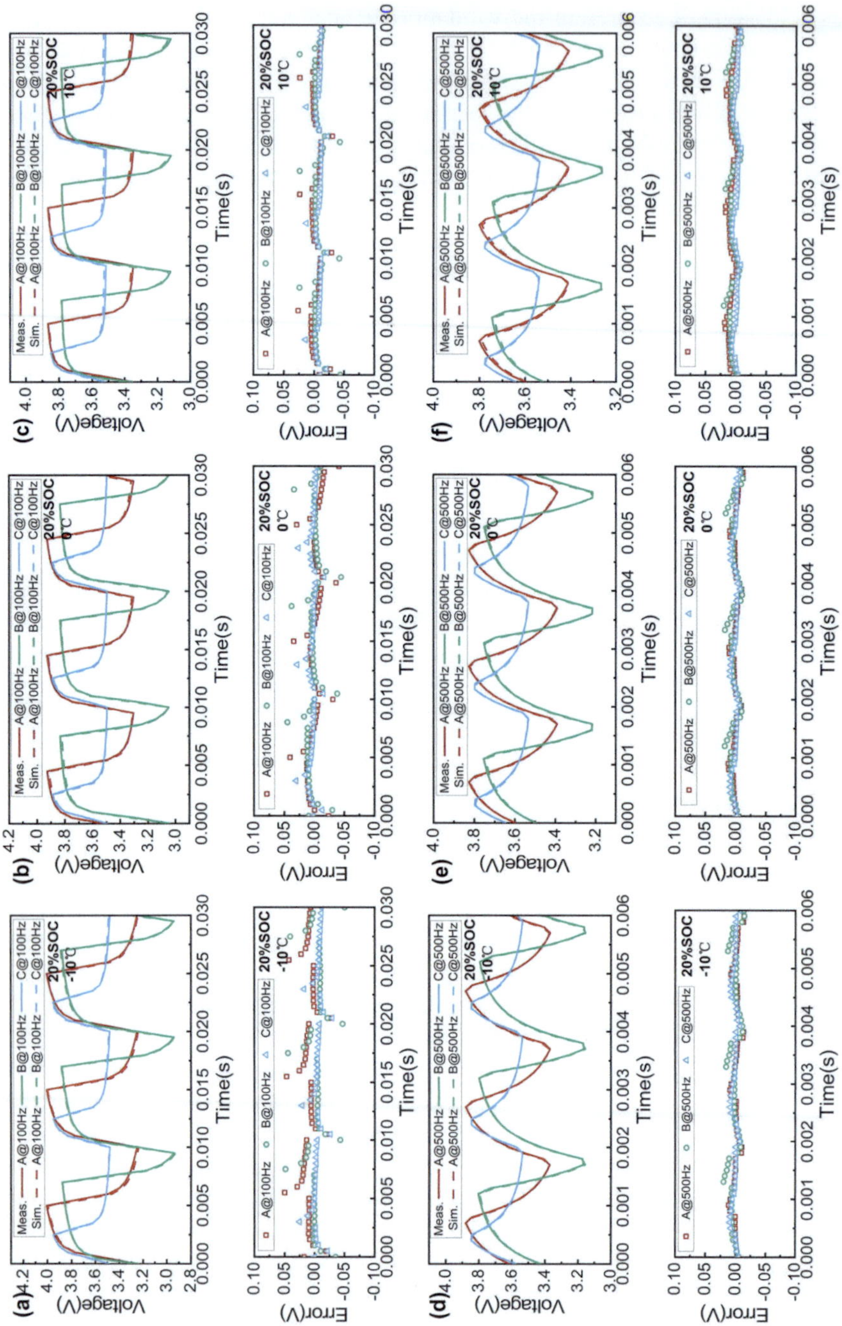

Fig. 4.33 Battery terminal voltage's validation predictions at 20% SOC under varying operating conditions: **a–c** BPC heating frequency of 100 Hz at temperatures of −10, 0, and 10 °C; **d–f** results at 500 Hz heating frequency [68]

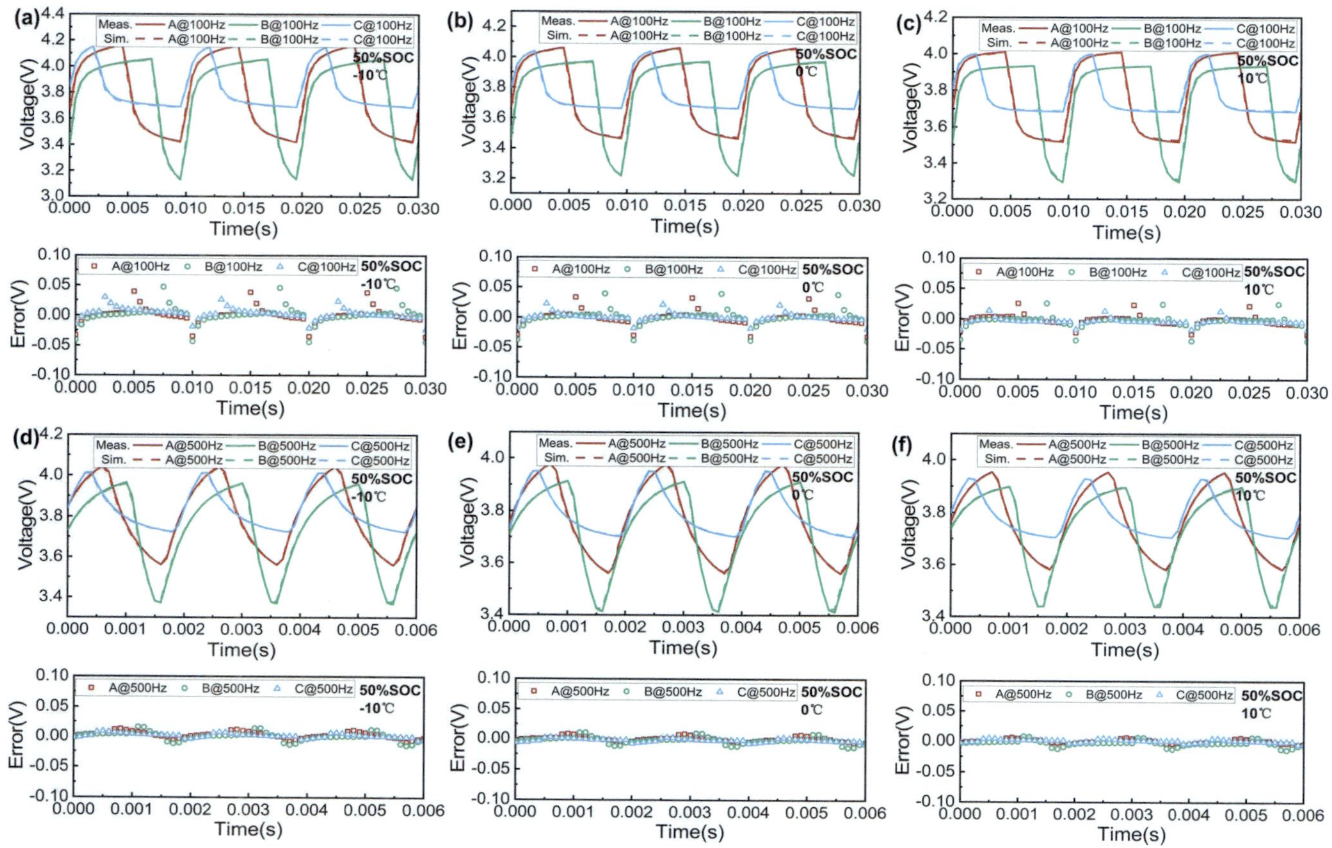

Fig. 4.34 Experimental validation of terminal voltage characteristics at 50% SOC under different thermal-electrical conditions: **a–c** 100 Hz BPC heating frequency at −10, 0, and 10 °C; **d–f** corresponding measurements at 500 Hz [68]

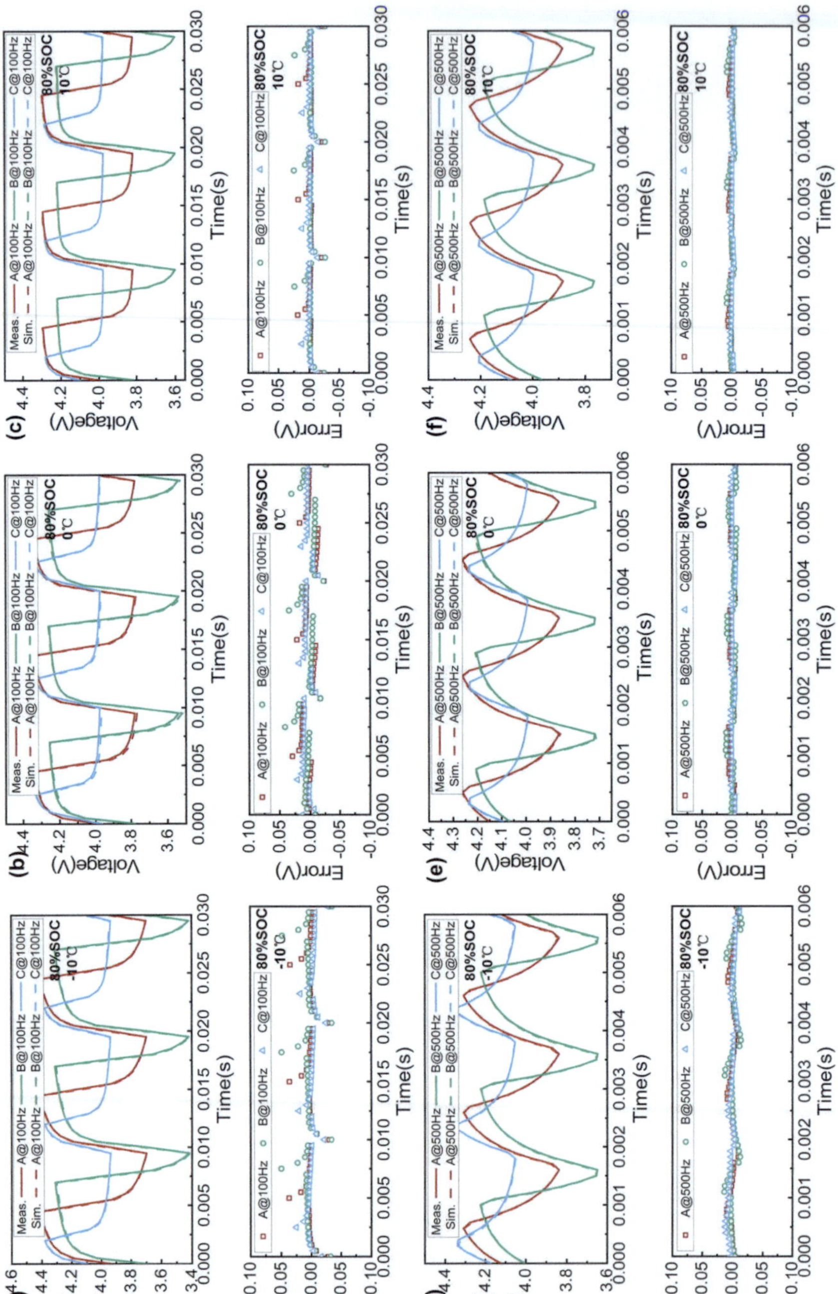

Fig. 4.35 Terminal voltage analysis at 80% SOC under varying operating conditions: **a–c** 100 Hz BPC heating at −10, 0, and 10; **d–f** 500 Hz BPC heating at −10, 0, and 10 °C [68]

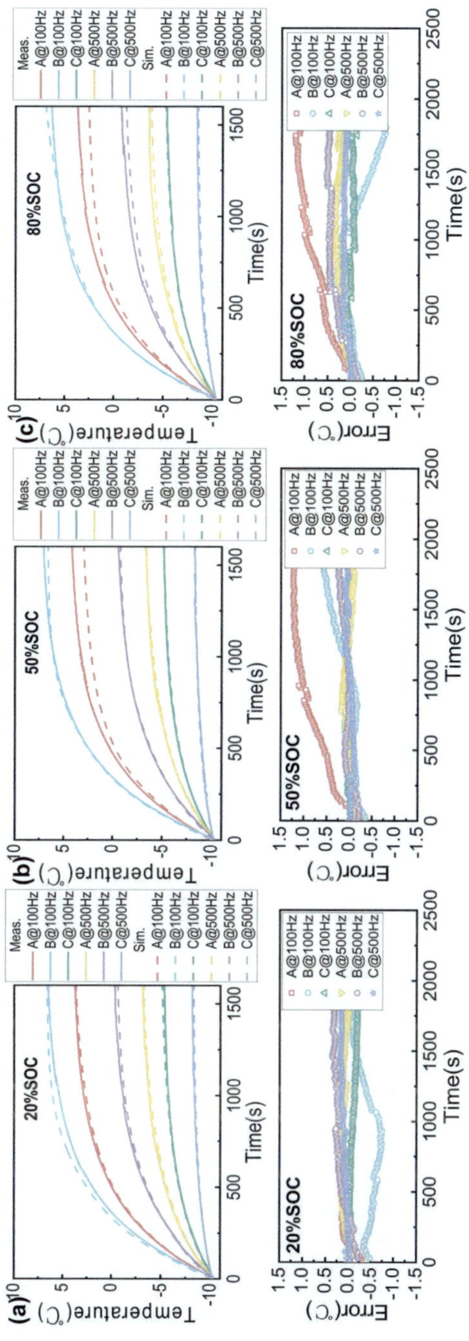

Fig. 4.36 Temperature rise characteristics at varying SOC levels: **a** 20%, **b** 50%, and **c** 80% [68]

References

1. Doyle M, Fuller T F, Newman J. Modeling of Galvanostatic Charge and Discharge of the Lithium/Polymer/Insertion Cell [J]. Journal of The Electrochemical Society, 1993, 140(6): 1526–1533.
2. Cai J, Wei X, Wang X, et al. Revealing effects of pouch Li-ion battery structure on fast charging ability through numerical simulation [J]. Applied Energy, 2025, 377: 124438.
3. Fuller T F, Doyle M, Newman J. Relaxation Phenomena in Lithium-Ion-Insertion Cells[J]. Journal of The Electrochemical Society, 1994, 141(1): 1–10.
4. Li C, Cui N, Wang C, et al. Simplified electrochemical lithium-ion battery model with variable solid-phase diffusion and parameter identification over wide temperature range [J]. Journal of Power Sources, 2021, 497: 229900.
5. Hou M, Hu Y, Zhang J, et al. Development of electrochemical-thermal modelling for large-format Li-ion battery [J]. Electrochimica Acta, 2020, 347: 136280.
6. Wang H, Zhu Y, Kim S C, et al. Underpotential lithium plating on graphite anodes caused by temperature heterogeneity [J]. Proceedings of the National Academy of Sciences, 2020, 117(47): 29453–29461.
7. Mei W, Jiang L, Liang C, et al. Understanding of Li-plating on graphite electrode: detection, quantification and mechanism revelation [J]. Energy Storage Materials, 2021, 41: 209–221.
8. Zhang H, Ren D, Ming H, et al. Digital Twin Enables Rational Design of Ultrahigh-Power Lithium-Ion Batteries [J]. Advanced Energy Materials, 2023, 13(1): 2202660.
9. Gao Y, Zhang X, Guo B, et al. Health-Aware Multiobjective Optimal Charging Strategy With Coupled Electrochemical-Thermal-Aging Model for Lithium-Ion Battery [J]. IEEE Transactions on Industrial Informatics, 2020, 16(5): 3417–3429.
10. Zhao Y, Diaz L B, Patel Y, et al. How to Cool Lithium Ion Batteries: Optimising Cell Design using a Thermally Coupled Model [J]. Journal of The Electrochemical Society, 2019, 166(13): A2849–A2859.
11. Ortiz-Ricardez F, Romero-Becerril A, Alvarez-Icaza L. Residue grouping order reduction method in solid-phase lithium-ion battery models [J]. Journal of Applied Electrochemistry, 2021, 51(12): 1635–1649.
12. Hwang G, Sitapure N, Moon J, et al. Model predictive control of Lithium-ion batteries: Development of optimal charging profile for reduced intracycle capacity fade using an enhanced single particle model (SPM) with first-principled chemical/mechanical degradation mechanisms [J]. Chemical Engineering Journal, 2022, 435: 134768.
13. Zhao Y, Wei M, Dan D, et al. Enhancing battery electrochemical-thermal model accuracy through a hybrid parameter estimation framework [J]. Energy Storage Materials, 2024, 72: 103720.
14. Li J, Wang D, Deng L, et al. Aging modes analysis and physical parameter identification based on a simplified electrochemical model for lithium-ion batteries [J]. Journal of Energy Storage, 2020, 31: 101538.
15. Deng Z, Yang L, Deng H, et al. Polynomial approximation pseudo-two-dimensional battery model for online application in embedded battery management system [J]. 2018, 142(1): 838–850.
16. Han X. Simplification of physics-based electrochemical model for lithium ion battery on electric vehicle. Part I: Diffusion simplification and single particle model [J]. Journal of Power Sources, 2015, 278: 814–825.
17. He H, Xiong R, Fan J. Evaluation of Lithium-Ion Battery Equivalent Circuit Models for State of Charge Estimation by an Experimental Approach [J]. Energies, 2011, 4(4): 582–598.
18. Johnson V H. Battery performance models in ADVISOR [J]. Journal of Power Sources, 2002, 110(2): 321–329.
19. Nejad S, Gladwin D T, Stone D A. A systematic review of lumped-parameter equivalent circuit models for real-time estimation of lithium-ion battery states [J]. Journal of Power Sources, 2016, 316: 183–196.

20. Hongwen He, Rui Xiong, Xiaowei Zhang, et al. State-of-Charge Estimation of the Lithium-Ion Battery Using an Adaptive Extended Kalman Filter Based on an Improved Thevenin Model [J]. IEEE Transactions on Vehicular Technology, 2011, 60(4): 1461–1469.
21. Rizzello A, Scavuzzo S, Ferraris A, et al. Temperature-Dependent Thévenin Model of a Li-Ion Battery for Automotive Management and Control[C]//2020 IEEE International Conference on Environment and Electrical Engineering and 2020 IEEE Industrial and Commercial Power Systems Europe (EEEIC/I&CPS Europe). Madrid, Spain: IEEE, 2020: 1–6.
22. Ding X, Zhang D, Cheng J, et al. An improved Thevenin model of lithium-ion battery with high accuracy for electric vehicles [J]. Applied Energy, 2019, 254: 113615.
23. Kalogiannis T, Hosen M, Sokkeh M, et al. Comparative Study on Parameter Identification Methods for Dual-Polarization Lithium-Ion Equivalent Circuit Model [J]. Energies, 2019, 12(21): 4031.
24. Geng Y, Pang H, Liu X. State-of-charge estimation for lithium-ion battery based on PNGV model and particle filter algorithm [J]. Journal of Power Electronics, 2022, 22(7): 1154–1164.
25. Gao W, Jiang M, Hou Y. Research on PNGV model parameter identification of LiFePO4 Li-ion battery based on FMRLS[C]//2011 6th IEEE Conference on Industrial Electronics and Applications. Beijing, China: IEEE, 2011: 2294–2297.
26. Saw L H, Somasundaram K, Ye Y, et al. Electro-thermal analysis of Lithium Iron Phosphate battery for electric vehicles [J]. Journal of Power Sources, 2014, 249: 231–238.
27. Lin X, Perez H E, Mohan S, et al. A lumped-parameter electro-thermal model for cylindrical batteries [J]. Journal of Power Sources, 2014, 257: 1–11.
28. Ruan H, Jiang J, Sun B, et al. A rapid low-temperature internal heating strategy with optimal frequency based on constant polarization voltage for lithium-ion batteries [J]. Applied Energy, 2016, 177: 771–782.
29. Qin Y, Zuo P, Chen X, et al. An ultra-fast charging strategy for lithium-ion battery at low temperature without lithium plating [J]. Journal of Energy Chemistry, 2022, 72: 442–452.
30. Zhang J, Ge H, Li Z, et al. Internal heating of lithium-ion batteries using alternating current based on the heat generation model in frequency domain [J]. Journal of Power Sources, 2015, 273: 1030–1037.
31. Wu X, Cui Z, Chen E, et al. Capacity degradation minimization oriented optimization for the pulse preheating of lithium-ion batteries under low temperature [J]. Journal of Energy Storage, 2020, 31.
32. Jiang J, Ruan H, Sun B, et al. A low-temperature internal heating strategy without lifetime reduction for large-size automotive lithium-ion battery pack [J]. Applied Energy, 2018, 230: 257–266.
33. Huang R, Wang X, Jiang B, et al. Revealing the electrochemical impedance characteristics of lithium-ion battery (nickel-cobalt-aluminum vs. graphite) under various alternating current amplitudes [J]. Journal of Power Sources, 2023, 566: 232929.
34. Zhu J, Sun Z, Wei X, et al. An alternating current heating method for lithium-ion batteries from subzero temperatures: Heating method for lithium-ion batteries [J]. International Journal of Energy Research, 2016, 40(13): 1869–1883.
35. Meyers J P, Doyle M, Darling R M, et al. The Impedance Response of a Porous Electrode Composed of Intercalation Particles [J]. Journal of The Electrochemical Society, 2000, 147(8): 2930.
36. Chang B Y, Park S M. Electrochemical Impedance Spectroscopy [J]. Annual Review of Analytical Chemistry, 2010, 3(1): 207–229.
37. Li S E, Wang B, Peng H, et al. An electrochemistry-based impedance model for lithium-ion batteries [J]. Journal of Power Sources, 2014, 258: 9–18.
38. Doyle M, Newman J, Gozdz A S, et al. Comparison of Modeling Predictions with Experimental Data from Plastic Lithium Ion Cells [J]. Journal of The Electrochemical Society, 1996, 143(6): 1890–1903.
39. Tippmann S, Walper D, Balboa L, et al. Low-temperature charging of lithium-ion cells part I: Electrochemical modeling and experimental investigation of degradation behavior [J]. Journal of Power Sources, 2014, 252: 305–316.

40. Valøen L O, Reimers J N. Transport Properties of LiPF6-Based Li-Ion Battery Electrolytes [J]. Journal of The Electrochemical Society, 2005, 152(5): A882.
41. Castro E B, Cuscueta D J, Milocco R H, et al. An EIS based study of a Ni–MH battery prototype. Modeling and identification analysis [J]. International Journal of Hydrogen Energy, 2010, 35(11): 5991–5998.
42. Levi M D, Aurbach D. Impedance of a Single Intercalation Particle and of Non-Homogeneous, Multilayered Porous Composite Electrodes for Li-ion Batteries [J]. The Journal of Physical Chemistry B, 2004, 108(31): 11693–11703.
43. Randviir E P, Banks C E. Electrochemical impedance spectroscopy: an overview of bioanalytical applications [J]. Analytical Methods, 2013, 5(5): 1098.
44. Macdonald D D. Reflections on the history of electrochemical impedance spectroscopy [J]. Electrochimica Acta, 2006, 51(8-9): 1376–1388.
45. Dees D W, Kawauchi S, Abraham D P, et al. Analysis of the Galvanostatic Intermittent Titration Technique (GITT) as applied to a lithium-ion porous electrode [J]. Journal of Power Sources, 2009, 189(1): 263–268.
46. Shaju K M, Subba RAO G V, Chowdari B V R. EIS and GITT studies on oxide cathodes, O2-Li(2/3)+x(Co0.15Mn0.85)O2 (x=0 and 1/3) [J]. Electrochimica Acta, 2003, 48(18): 2691–2703.
47. Shaju K, Subbarao G, Chowdari B. Li-ion kinetics and polarization effect on the electrochemical performance of Li(Ni1/2Mn1/2)O2 [J]. Electrochimica Acta, 2004, 49(9-10): 1565–1576.
48. Dai H, Wei X, Sun Z, et al. Online cell SOC estimation of Li-ion battery packs using a dual time-scale Kalman filtering for EV applications [J]. Applied Energy, 2012, 95: 227–237.
49. Dai H, Guo P, Wei X, et al. ANFIS (adaptive neuro-fuzzy inference system) based online SOC (State of Charge) correction considering cell divergence for the EV (electric vehicle) traction batteries [J]. Energy, 2015, 80: 350–360.
50. Zhu L, Sun Z, Dai H, et al. A novel modeling methodology of open circuit voltage hysteresis for LiFePO 4 batteries based on an adaptive discrete Preisach model [J]. Applied Energy, 2015, 155: 91–109.
51. Wang F, Chen J, Tan Z, et al. Low-temperature electrochemical performances of LiFePO4 cathode materials for lithium ion batteries [J]. Journal of the Taiwan Institute of Chemical Engineers, 2014, 45(4): 1321–1330.
52. Bernardi D M, Go J Y. Analysis of pulse and relaxation behavior in lithium-ion batteries [J]. Journal of Power Sources, 2011, 196(1): 412–427.
53. Abraham D P, Heaton J R, Kang S, et al. Investigating the Low-Temperature Impedance Increase of Lithium-Ion Cells [J]. Journal of The Electrochemical Society, 2008, 155(1): A41
54. Yang C C, Jang J H, Jiang J R. Study of electrochemical performances of lithium titanium oxide–coated LiFePO4/C cathode composite at low and high temperatures [J]. Applied Energy, 2016, 162: 1419–1427.
55. Levi M D, Aurbach D. Diffusion Coefficients of Lithium Ions during Intercalation into Graphite Derived from the Simultaneous Measurements and Modeling of Electrochemical Impedance and Potentiostatic Intermittent Titration Characteristics of Thin Graphite Electrodes [J]. The Journal of Physical Chemistry B, 1997, 101(23): 4641–4647.
56. Cai L, White R E. Mathematical modeling of a lithium ion battery with thermal effects in COMSOL Inc. Multiphysics (MP) software [J]. Journal of Power Sources, 2011, 196(14): 5985–5989.
57. Srinivasan V, Newman J. Discharge Model for the Lithium Iron-Phosphate Electrode [J]. Journal of The Electrochemical Society, 2004, 151(10): A1517.
58. Song L, Evans J W. Electrochemical-Thermal Model of Lithium Polymer Batteries [J]. Journal of The Electrochemical Society, 2000, 147(6): 2086.
59. Zhu J, Sun Z, Wei X, et al. A new electrochemical impedance spectroscopy model of a high-power lithium-ion battery [J]. RSC Advances, 2014, 4(57): 29988–29998.
60. Petzl M, Kasper M, Danzer M A. Lithium plating in a commercial lithium-ion battery – A low-temperature aging study [J]. Journal of Power Sources, 2015, 275: 799–807.

61. Chen S, Wei X, Zhang G, et al. All-temperature area battery application mechanism, performance, and strategies [J]. The Innovation, 2023, 4(4): 100465.
62. Huang R, Wei G, Zhou X, et al. Targeting the low-temperature performance degradation of lithium-ion batteries: A non-destructive bidirectional pulse current heating framework [J]. Energy Storage Materials, 2024, 65: 103173.
63. Vashisht S, Rakshit D, Panchal S, et al. Thermal behaviour of Li-ion battery: An improved electrothermal model considering the effects of depth of discharge and temperature [J]. Journal of Energy Storage, 2023, 70: 107797.
64. Guo S, Xiong R, Shen W, et al. Aging investigation of an echelon internal heating method on a three-electrode lithium ion cell at low temperatures [J]. Journal of Energy Storage, 2019, 25: 100878.
65. Zhang L, Fan W, Wang Z, et al. Battery heating for lithium-ion batteries based on multi-stage alternative currents [J]. Journal of Energy Storage, 2020, 32: 101885.
66. Ruan H, Sun B, Jiang J, et al. A modified-electrochemical impedance spectroscopy-based multi-time-scale fractional-order model for lithium-ion batteries [J]. Electrochimica Acta, 2021, 394: 139066.
67. Zou C, Hu X, Dey S, et al. Nonlinear Fractional-Order Estimator With Guaranteed Robustness and Stability for Lithium-Ion Batteries [J]. IEEE Transactions on Industrial Electronics, 2017, 65(7): 5951–5961.
68. Huang R, Wei G, Wang X, et al. A non-destructive heating method for lithium-ion batteries at low temperatures [J]. Renewable and Sustainable Energy Reviews, 2024, 205: 114868.
69. Qin Y, Du J, Lu L, et al. A rapid lithium-ion battery heating method based on bidirectional pulsed current: Heating effect and impact on battery life [J]. Applied Energy, 2020, 280(1): 115957.
70. Montenegro L E G, Villegas Pico H N. AC-Heating and Fast-Charging Power Requirements of EV Battery Packs in Subzero Temperature [J]. IEEE Transactions on Transportation Electrification, 2023, 9(3): 3936–3946.
71. Fleischer C, Waag W, Heyn H M, et al. On-line adaptive battery impedance parameter and state estimation considering physical principles in reduced order equivalent circuit battery models part 2. Parameter and state estimation [J]. Journal of Power Sources, 2014, 262: 457–482.
72. Zhu J, Knapp M, Darma M S D, et al. An improved electro-thermal battery model complemented by current dependent parameters for vehicular low temperature application [J]. Applied Energy, 2019, 248: 149–161.
73. Huang R, Wei G, Jiang B, et al. Investigating the Effect of Different Bidirectional Pulsed Current Parameters on the Heat Generation of Lithium-Ion Battery at Low Temperatures [J]. Batteries, 2023, 9(9): 457.
74. Feng F, Yang R, Meng J, et al. Electrochemical impedance characteristics at various conditions for commercial solid-liquid electrolyte lithium-ion batteries: Part 1. experiment investigation and regression analysis [J]. Energy, 2022, 242(1): 122880.
75. Zhu J, Knapp M, Liu X, et al. Low-Temperature Separating Lithium-Ion Battery Interfacial Polarization Based on Distribution of Relaxation Times (DRT) of Impedance [J]. IEEE Transactions on Transportation Electrification, 2021, 7(2): 410–421.
76. Lu Y, Zhao C Z, Huang J Q, et al. The timescale identification decoupling complicated kinetic processes in lithium batteries [J]. Joule, 2022, 6(6): 1172–1198.

Chapter 5
AC Heating Methods

Abstract The performance of lithium-ion batteries (LIBs) significantly declines in low-temperature environments, manifesting in reduced discharge capacity, diminished cycle life, and challenges with charging. To maintain optimal functionality in such conditions, preheating is often necessary. LIBs typically employ two fundamental approaches for low-temperature heating: external and internal heating mechanisms. For external heating, a Battery Thermal Management System (BTMS) is commonly utilized. This system is vital for managing the thermal conditions of the battery pack. It ensures that the temperature is evenly maintained across the entire pack, avoiding the formation of thermal hotspots and keeping the temperature within a range that supports the best possible performance and durability of the battery. Instead of relying on external sources, internal heating techniques make use of the intrinsic heat generated by the battery during its functioning, particularly while it is being charged or discharged. Compared to external heating, internal heating has several notable merits: it achieves more consistent temperature regulation, consumes less energy, and enhances the overall efficiency of the heating process. Another prominent approach to internal heating is through the utilization of Alternating current (AC) excitation. This method involves applying AC signals with specific frequency and amplitude parameters to produce thermal energy within the battery. Among diverse internal heating options, AC heating is particularly highlighted for its well-balanced relationship between energy expenditure and heating performance. Several studies have preliminarily suggested that this method is benign to the battery, given that the current excitation is precisely optimized. Importantly, AC heating does not require any modifications to the battery's architecture, which makes it a highly desirable and extensively studied approach. It effectively enhances battery performance without jeopardizing its structural integrity. This chapter introduces the AC heating methods. Section 5.1 reviews current heating methods for LIBs in low-temperature environments, categorizing them into external heating (e.g., air, liquid, and phase change materials (PCM) preheating) and internal heating (e.g., self-heating, direct current (DC), and AC preheating). Each method is evaluated for its efficiency, energy consumption, and impact on battery performance, highlighting the trade-offs between heating speed, uniformity, and system complexity. Section 5.2 delve into the critical constraints and strategies for low-temperature AC heating of

J. Zhu et al., *Alternating Current (AC) Heating for Lithium-Ion Batteries in Electric Vehicles*, https://doi.org/10.1007/978-981-96-9071-8_5

LIBs. The primary constraints identified include lithium plating, overcharge, and over-discharge, which can significantly impact battery lifespan and safety. Two new lithium plating standards for three-electrode and two-electrode full cells were developed to prevent irreversible damage during AC heating. The study highlights that the onset potential for lithium plating is still controversial, and an accurate criterion is essential for non-destructive heating strategies. Section 5.3 introduces a comprehensive framework for designing a non-destructive bidirectional pulsed current (BPC) heating strategy. The framework integrates an electro-thermal coupled model with constraints on lithium plating and terminal voltage. The BPC heating strategy is optimized to maximize heating efficiency while maintaining battery health. Experimental validations demonstrate that the proposed strategy achieves significant improvements in heating rates and minimal impact on battery capacity even after multiple heating cycles. The findings suggest that optimizing BPC parameters, such as frequency and pulse ratio, can enhance heating power, especially at higher battery states of charge (SOC). This work provides a robust solution for enhancing the low-temperature performance of LIBs, crucial for applications in electric vehicles (EVs) and energy storage systems (ESSs) in cold climates. Section 5.4 summarizes the work in this chapter.

Keywords Alternating current heating · Lithium-ion battery · Lithium plating · Three-electrode battery · Bidirectional pulsed current

5.1 Classification of Heating Methods

5.1.1 External Heating

In the realm of external heating strategies, the primary techniques involve thermal transfer through conduction and convection [1]. For conduction-based heating, several methods are widely adopted, including resistive heating, thermoelectric (Peltier-effect) heating, heating via heat pipes, heating using burners, and leveraging phase change materials for thermal regulation [2–5]. In addition to conduction, another category of heating techniques involves the transfer of heat through fluid motion, which includes warming air, heating liquids, and utilizing heat pumps [6–8]. Irrespective of the chosen method—be it conduction or convection—relying on external heat sources often demands extra apparatus. This additional machinery tends to make the system more intricate and leads to higher financial outlays [9]. Additionally, the requirement for external heat sources in these heating approaches typically results in high energy demands. A major obstacle lies in the fact that, due to the inherent design restrictions of battery modules, attaining consistent heat dispersion through external heating is quite problematic [10]. This often means that individual cells within the module may encounter differing thermal conditions, thereby causing the emergence of temperature imbalances throughout the module.

5.1.1.1 Air Preheating

Air preheating is one of the most widely adopted methods for battery thermal management in EVs due to its simplicity, low cost, and reliability. As shown in Fig. 5.1, this method involves heating air using an external heat source or an onboard air conditioner and then circulating the hot air around the battery pack to preheat the cells. The heated air transfers heat to the battery cells through convection, raising their temperature and improving their performance in cold climates. The fundamental principle of air preheating is to utilize the convective heat transfer properties of air to warm up the battery cells. The system typically consists of several key components: a heat source (such as an electric heater or an onboard air conditioner), a fan to circulate the hot air, an airflow channel to direct the air flow, and control components to manage the heating process [6, 11]. The heat source generates hot air, which is then forced through the airflow channel by the fan. The hot air flows over the battery cells, transferring heat to them through convection.

The design of the airflow channel is crucial for efficient heat transfer. The channel can be configured in parallel or series, with parallel configurations generally preferred due to lower wind resistance and independent airflow in each channel. In a parallel configuration, each battery cell is exposed to the hot air simultaneously, which can reduce the overall heating time and improve temperature uniformity. However, the design must also consider the thermal resistance of the air and the battery pack, as well as the potential for temperature gradients within the pack. The performance

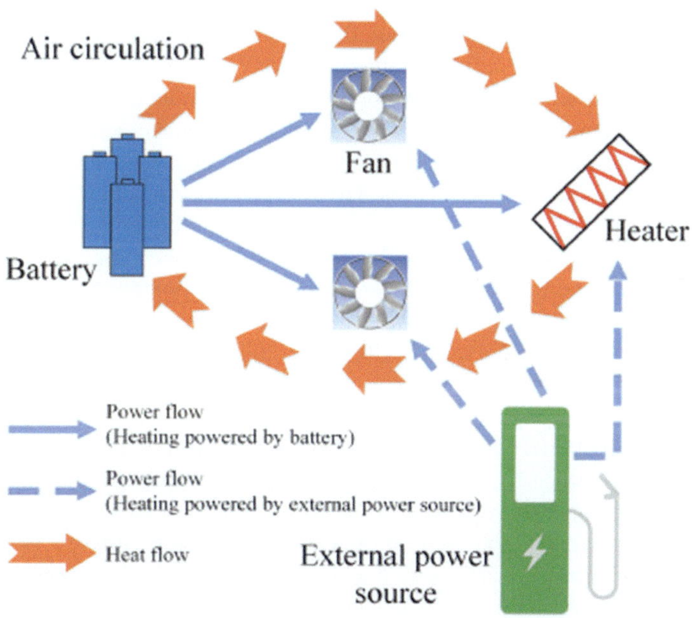

Fig. 5.1 Schematic of an air heating system [12]

of air preheating systems is characterized by several key metrics, including the rate of temperature rise (RTR), energy consumption, and temperature uniformity. The RTR is typically in the range of 0.5–3 °C/min, depending on the system design and operating conditions. For example, Ji et al. demonstrated that an 18,650 cell could be heated from −20 to 20 °C in approximately 196 s using air preheating, achieving an RTR of 0.69 °C/min [6]. This rate can be improved by optimizing the airflow velocity and temperature, but increasing the air temperature can also lead to larger temperature differences within the battery, potentially causing uneven electrochemical reactions and reducing battery life.

Energy consumption is another critical factor. Air preheating systems typically consume a significant amount of energy, especially when heating large battery packs. The energy consumption can be reduced by optimizing the heating process, such as using variable-speed fans and temperature controllers. However, the low thermal conductivity and heat capacity of air mean that the heating process is inherently less efficient compared to liquid-based systems. Temperature uniformity is also a significant concern in air preheating systems. The convective nature of air heating can lead to temperature gradients within the battery pack, especially in larger packs. These gradients can cause uneven aging and performance degradation of the battery cells [13]. To address this issue, researchers have explored various strategies, such as optimizing the arrangement of the battery cells and the shape of the airflow channels. For example, serial ventilation with higher flow rates can reduce temperature differences but may increase the heating time.

Air preheating systems are relatively simple and inexpensive to implement, making them attractive for early-generation EVs and low-cost applications. This method is suitable for all types of batteries, including cylindrical, prismatic, and pouch cells, making it a versatile solution for various EV designs. Air preheating can be easily integrated with existing air-cooling systems, reducing the need for additional components and simplifying the overall thermal management system. The low thermal properties of air result in a low rate of temperature rise and longer heating times compared to liquid-based systems. The convective nature of air heating can lead to temperature gradients within the battery pack, potentially causing uneven aging and performance degradation of the battery cells. The need for fans and airflow loops increases the complexity and cost of the system, potentially affecting reliability and maintenance. Fans and blowers used in air preheating systems can generate significant noise, which may compromise the comfort of drivers and passengers.

Recent research has focused on improving the efficiency and performance of air preheating systems through various innovations. One approach is to optimize the arrangement of the battery cells and the shape of the airflow channels to enhance heat transfer and reduce temperature gradients. For example, studies have shown that increasing the airflow velocity can improve the rate of temperature rise, but it must be balanced with the potential for increased temperature differences within the battery. Another innovation is the use of advanced materials and coatings to enhance the thermal conductivity of the air and the battery pack. For instance, researchers have explored the use of nanofluids and other thermally conductive materials to improve heat transfer efficiency. These materials can be integrated into the airflow channels or applied to the surface of the battery cells to enhance heat absorption and distribution. Furthermore, the development of smart control systems has enabled more precise management of the heating process. These systems can dynamically adjust the airflow velocity and temperature based on real-time monitoring of the battery's thermal state, optimizing energy consumption and temperature uniformity.

Air preheating systems have been widely adopted in various EV models due to their simplicity and low cost. For example, the Honda Insight and Toyota Prius both utilize air preheating systems to manage the temperature of their battery packs. These systems have demonstrated effective performance in maintaining battery temperature within an optimal range, improving the overall efficiency and reliability of the vehicles. In a study by Ji et al., an air preheating system was designed and tested for a battery pack consisting of 18,650 cells. The system achieved an RTR of 0.69 °C/min, heating the battery from −20 to 20 °C in approximately 196 s [6]. The study also investigated the effects of airflow velocity and temperature on the heating efficiency, finding that increasing the airflow velocity was more effective than increasing the air temperature in reducing the heating time. Another study by Li et al. [14] explored the impact of serial and parallel airflow configurations on the preheating performance of a battery pack. The results showed that a parallel configuration achieved a higher RTR and better temperature uniformity compared to a serial configuration. This study highlights the importance of optimizing the airflow design to enhance the performance of air preheating systems.

5.1.1.2 Liquid Preheating

Liquid preheating is a highly effective method for managing the thermal performance of LIBs in EVs, particularly in cold climates. This method leverages the superior thermal properties of liquids to transfer heat to the battery cells, resulting in faster and more uniform heating compared to air preheating. The use of liquids as a heat transfer medium offers several advantages, including higher thermal conductivity and heat capacity, which enable more efficient heat transfer and better temperature control. However, liquid preheating systems are also more complex and costly to implement, requiring careful design and optimization to maximize their benefits.

As shown in Fig. 5.2, the fundamental principle of liquid preheating involves circulating a heated liquid through channels or jackets in direct contact with the battery cells to transfer heat through conduction. The liquid medium, which can be water, glycol, oil, or a specialized coolant, is heated using an external heat source, such as an electric heater or a heat pump. The heated liquid then flows through the battery pack, transferring heat to the cells and raising their temperature [15, 16].

The design of a liquid preheating system typically includes several key components: a heater to warm the liquid, a pump to circulate the liquid, a heat exchanger to transfer heat to the battery cells, and control components to manage the heating process. The liquid can be circulated in either direct contact with the cells (immersion heating) or through jackets or channels that are in close proximity to the cells (non-contact heating). Immersion heating generally provides higher heat transfer

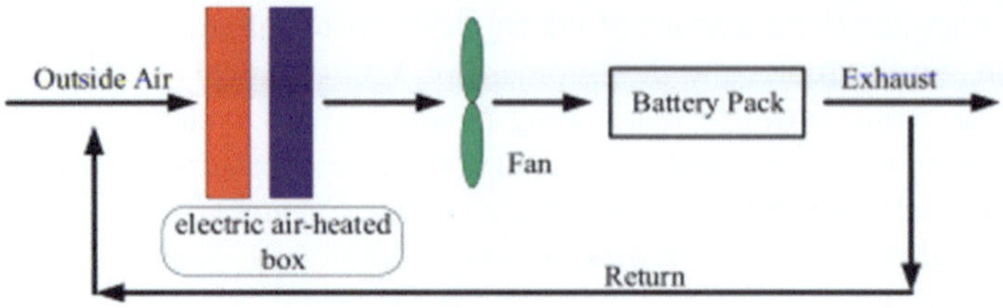

Fig. 5.2 The schematic of air-based BTMS [17]

coefficients and more uniform temperature distribution compared to non-contact heating methods [18, 19].

Liquid preheating systems typically achieve higher RTRs compared to air preheating, often in the range of 1–4 °C/min. For example, a study by Wang et al. [20] demonstrated that a liquid preheating system could heat a battery pack from −30 to 10 °C within 1 h, achieving an RTR of 0.67 °C/min. This higher RTR is attributed to the superior thermal properties of liquids, which facilitate faster heat transfer to the battery cells. Energy consumption is another critical factor in liquid preheating systems. While liquids offer higher heat transfer efficiency, the need for pumps and heaters can increase energy consumption compared to air preheating systems. However, the overall energy efficiency can be optimized by selecting appropriate liquid flow rates and temperatures. For instance, using a lower flow rate can reduce energy consumption while maintaining effective heat transfer [15]. Temperature uniformity is a significant advantage of liquid preheating systems. The high thermal conductivity of liquids ensures that heat is distributed more evenly across the battery pack, reducing the risk of temperature gradients and uneven aging of the battery cells. This uniformity is crucial for maintaining the performance and lifespan of the battery pack, especially in large-format batteries where temperature gradients can be more pronounced.

Liquids have higher thermal properties compared to air, resulting in faster and more efficient heat transfer. The high thermal conductivity of liquids ensures that heat is distributed more evenly across the battery pack, reducing temperature gradients and improving battery performance and lifespan. Liquid preheating is particularly effective for large-format batteries, where uniform temperature distribution is critical [15]. Liquid preheating systems can be designed to provide both heating and cooling functions, making them versatile for year-round thermal management. Liquid preheating systems require more complex designs, including pumps, heaters, and heat exchangers, which increase system cost and complexity. The use of liquid mediums poses risks of leaks and corrosion, which can damage the battery pack and other components. The liquid medium and associated components add weight to the vehicle, potentially reducing its range and efficiency. The need for pumps and heaters can increase energy consumption compared to air preheating systems.

One approach is to optimize the design of the heat exchangers and channels to enhance heat transfer efficiency. For example, studies have explored the use of minichannel heat exchangers, which offer higher heat transfer coefficients and better temperature uniformity [21, 22]. These designs can significantly reduce the time required to heat the battery pack while maintaining uniform temperature distribution. Another innovation is the use of advanced materials and nanofluids to enhance the thermal properties of the liquid medium. Nanofluids, which contain nanoparticles dispersed in a base fluid, have shown promise in improving heat transfer efficiency and reducing temperature gradients within the battery pack [23]. These materials can be integrated into the liquid medium to enhance its thermal conductivity and overall performance. Furthermore, the development of smart control systems has enabled more precise management of the heating process. These systems can dynamically adjust the flow rate and temperature of the liquid based on real-time monitoring of the battery's thermal state, optimizing energy consumption and temperature uniformity.

Liquid preheating systems have been widely adopted in various EV models due to their superior thermal management capabilities. For example, Tesla and Volt have both implemented liquid preheating systems to manage the temperature of their battery packs [24]. These systems have demonstrated effective performance in maintaining battery temperature within an optimal range, improving the overall efficiency and reliability of the vehicles. In a study by Wang et al. [20], a liquid preheating system was designed and tested for a battery pack consisting of prismatic cells. The system achieved an RTR of 0.67 °C/min, heating the battery from −30 to 10 °C within 1 h. The study also investigated the effects of liquid flow rate and temperature on the heating efficiency, finding that optimizing these parameters could significantly reduce the heating time and energy consumption. Another study by Zhao et al. [25] explored the performance of a mini-channel liquid-cooled system for cylindrical batteries. The results showed that the system achieved an RTR of 4.18 °C/min, with a temperature difference of less than 4 °C across the battery pack. This study highlights the importance of optimizing the channel design to enhance the performance of liquid preheating systems.

5.1.1.3 Phase Change Materials (PCMs) Preheating

PCMs have emerged as a promising solution for preheating LIBs in EVs, particularly in cold climates. PCMs are materials that absorb and release a large amount of heat during phase transitions, such as melting and solidifying, at a relatively constant temperature. This unique property allows PCMs to provide a stable and uniform temperature environment for the battery cells, enhancing their performance and lifespan.

As shown in Fig. 5.3, the fundamental principle of PCM preheating is to utilize the latent heat of phase change to maintain the battery temperature within an optimal range. Conversely, when the battery temperature rises above the PCM's melting point, the PCM melts, absorbing heat and preventing the battery from overheating [26, 27]. A typical PCM preheating system consists of several key components: the PCM itself, a container or encapsulation to hold the PCM, and a thermal interface to ensure effective heat transfer between the PCM and the battery cells. This design ensures

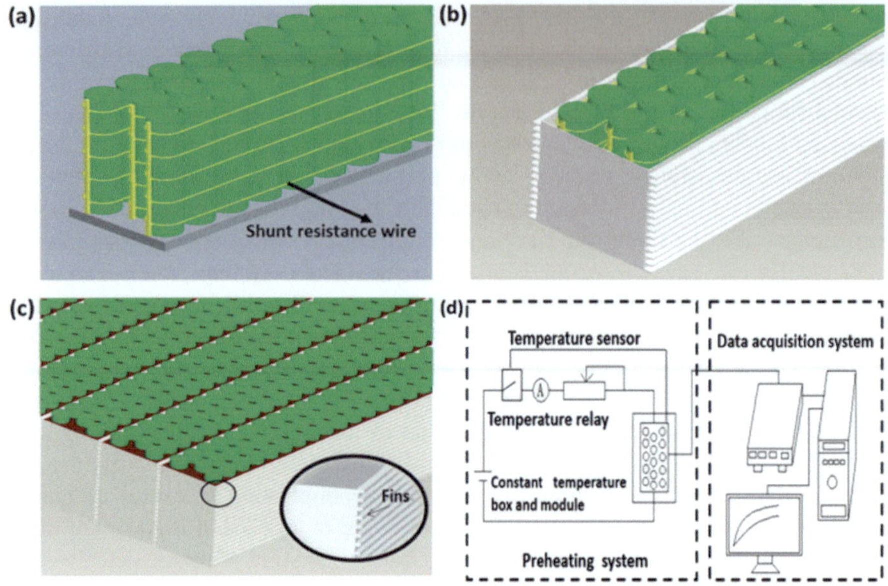

Fig. 5.3 Preheating system for batteries utilizing phase change materials (PCM). **a** Diagram of the layout with resistive wires wrapped around the batteries. **b** Arrangement of batteries within the battery module enclosure. **c** Comprehensive depiction of the battery pack made up of numerous battery modules. **d** Schematic of the experimental setup for preheating procedures [26]

efficient heat transfer and maintains a uniform temperature distribution within the battery pack.

PCM preheating systems typically achieve higher RTRs compared to air preheating, often in the range of 2–5 °C/min. For example, studies have shown that PCM preheating can achieve an RTR of up to 24 °C/min, significantly faster than air preheating. This higher RTR is attributed to the high latent heat of PCMs, which allows for rapid heat transfer to the battery cells. Energy consumption is another critical factor in PCM preheating systems. Unlike air and liquid preheating methods, PCM preheating does not require an external heat source during the phase change process, resulting in minimal energy consumption. The energy required to heat the PCM is typically provided by the battery's own heat generation during operation, making PCM preheating a highly energy-efficient solution. Temperature uniformity is a significant advantage of PCM preheating systems. PCMs have high latent heat, which allows for rapid and uniform heat transfer to the battery cells, reducing temperature gradients and improving battery performance and lifespan. PCM preheating does not require an external heat source during the phase change process, resulting in minimal energy consumption. PCM preheating systems are relatively simple and inexpensive to implement, making them attractive for cost-sensitive applications. PCM preheating systems can provide both heating and cooling functions, making them versatile for year-round thermal management. The low thermal conductivity of PCMs can limit the rate of temperature rise, resulting in longer heating times compared to liquid preheating systems. Repeated phase change cycles can lead to thermal degradation of the PCM, reducing its effectiveness over time. The PCM material and associated components add weight to the vehicle, potentially reducing its range and efficiency. Ensuring effective heat transfer between the PCM and the battery cells requires careful design and optimization of the thermal interface.

One approach is to enhance the thermal conductivity of PCMs by incorporating high thermal conductivity materials such as graphene, carbon nanotubes, and metal nanoparticles [18, 28, 29]. These materials can significantly improve the heat transfer efficiency of PCMs, reducing heating times and improving temperature uniformity. Another innovation is the development of composite PCMs, which combine PCMs with other materials to enhance their thermal properties and mechanical stability [30, 31]. For example, embedding PCMs in a graphite matrix or using expanded graphite as a reinforcement can improve the thermal conductivity and overall performance of the PCM. Furthermore, the development of smart control systems has enabled more precise management of the heating process. These systems can dynamically adjust the thermal interface and PCM configuration based on real-time monitoring of the battery's thermal state, optimizing energy consumption and temperature uniformity.

Al-Hallaj and Selman demonstrated that integrating PCM into the battery thermal management system could prevent the battery temperature from dropping below the PCM's melting point, thereby reducing capacity loss and extending battery life [27, 32]. Their simulation results showed that PCM preheating could maintain the battery temperature within an optimal range, significantly improving its performance in cold climates. Another study by Ling et al. investigated the warming-up effects of PCM on lithium-ion batteries operated at low temperatures. The results showed that PCM preheating could effectively reduce the capacity loss and extend the battery's cycle life by up to 76% at −10 °C [27, 33]. This study highlights the potential of PCM preheating in improving the overall performance and lifespan of LIBs in cold climates.

5.1.1.4 Peltier-Effect Heating

Peltier-effect heating is an innovative thermal management technique that leverages the thermoelectric properties of materials to generate heat or cooling directly within the battery system. This method is particularly advantageous for LIBs in EVs due to its ability to provide precise temperature control, high efficiency, and rapid response. By utilizing the Peltier effect, heat can be generated or absorbed at the junctions of two different conductive materials when an electric current is applied, allowing for both heating and cooling functions within the same system.

The Peltier effect is a thermoelectric phenomenon where heat is absorbed or released at the junction of two different conductors when an electric current flows through them. When a current flows from a material with a higher Seebeck coefficient to one with a lower Seebeck coefficient, heat is absorbed at the junction (cooling effect), and when the current direction is reversed, heat is released (heating effect).

This reversible process allows Peltier-effect devices to act as both heaters and coolers, making them highly versatile for battery thermal management. As shown in Fig. 5.4, a typical Peltier-effect heating system for LIBs consists of Peltier-effect elements (thermoelectric modules) integrated directly into the battery pack or placed in close proximity to the battery cells. These modules are composed of pairs of p-type and n-type semiconductors connected electrically in series and thermally in parallel. When an electric current is applied, one side of the module heats up while the other side cools down, allowing for precise temperature control of the battery cells.

Peltier-effect heating typically achieves moderate RTRs, often in the range of 1–2 °C/min, making it suitable for applications where rapid temperature changes are not critical. For example, studies have shown that a Peltier-effect system can heat a battery from −19.3 to −2.4 °C in approximately 48 min, achieving an RTR of approximately 0.35 °C/min. This moderate heating rate is sufficient for maintaining the battery temperature within an optimal range during cold starts or extended periods of low-temperature operation. The energy required to drive the Peltier-effect modules is directly related to the current and voltage applied. Typically, Peltier-effect heating consumes a small percentage of the battery's capacity, ranging from 2.5 to 5% depending on the initial SOC and the heating duration. The localized generation of heat or cooling allows for precise temperature control, reducing the risk of temperature gradients and uneven aging of the battery cells. Studies have shown that Peltier-effect heating can achieve a uniform temperature distribution within the battery, with temperature differences typically less than 3 °C.

Peltier-effect heating allows for precise temperature control, enabling both heating and cooling functions within the same system. Peltier-effect heating achieves moderate RTRs, making it suitable for applications where rapid temperature changes are not critical. The energy required to drive the Peltier-effect modules is relatively low, typically consuming a small percentage of the battery's capacity. Peltier-effect

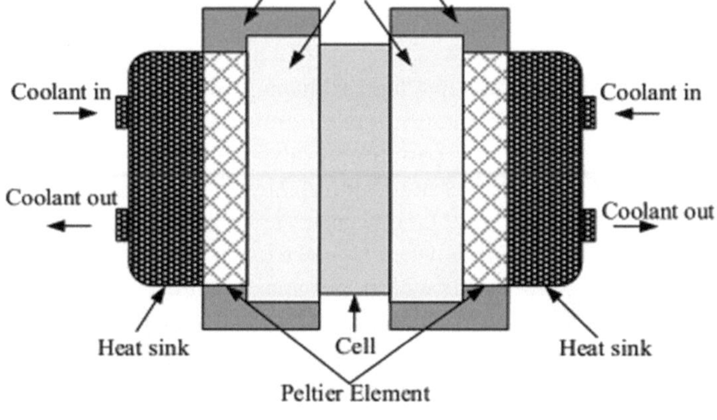

Fig. 5.4 Schematic of Peltier effect [17]

modules are compact and lightweight, making them suitable for integration into the limited space of EV battery packs. The need for Peltier-effect modules and control mechanisms to regulate the heating process can increase the complexity and cost of the system. The heating capacity of Peltier-effect modules is limited by their size and the current applied, which may not be sufficient for rapid temperature rise in large battery packs. The use of high current densities can pose safety risks, such as electrical shocks and thermal runaway, if not properly managed. Repeated thermal cycling can lead to degradation of the Peltier-effect modules, reducing their efficiency and lifespan.

One approach is to optimize the design and placement of the Peltier-effect modules to ensure efficient heat generation and uniform temperature distribution. For example, studies have explored the use of multiple Peltier-effect modules placed in strategic locations within the battery pack to enhance temperature control. Another innovation is the development of advanced control strategies to manage the heating process dynamically. These strategies involve monitoring the battery's thermal state in real-time and adjusting the current applied to the Peltier-effect modules accordingly to optimize energy consumption and temperature uniformity. For instance, Alaoui et al. [34] developed a control strategy that used a constant polarization voltage to achieve a good trade-off between short heating time and minimal damage to the battery life. Alaoui et al. demonstrated a Peltier-effect heating system that could heat a battery from 17 to 37 °C in the front box and from 17 to 29 °C in the rear box, achieving an RTR of approximately 1 °C/min. This moderate heating rate was sufficient for maintaining the battery temperature within an optimal range during cold starts or extended periods of low-temperature operation. Another study by Liu et al. [35] investigated the effects of different current magnitudes on the preheating efficiency of Peltier-effect elements. The results showed that the RTR of cells close to the Peltier-effect elements could reach 1 °C/min, while the RTR of cells farther away was approximately 0.6 °C/min. This study highlights the potential of optimizing the current applied to the Peltier-effect modules to enhance the performance and efficiency of the heating system.

5.1.2 Internal Heating

Internal heating methods generate heat directly inside the battery cells, resulting in faster and more uniform heating compared to external methods. These methods typically involve using the battery's own energy or an external current to generate heat through internal impedance. Internal heating methods encompass self-heating [36], DC preheating [6], and AC heating [17]. Nickel foil heating operates by integrating nickel foils into the interior of the battery to facilitate electrical current flow, which in turn generates thermal energy [36]. Although this method yields faster temperature increases and enhanced heating capabilities, it demands considerable changes to the battery cell's architecture. Conversely, discharge heating employs the current from discharging to elevate the battery's temperature. However, this approach is often associated with significant energy expenditure and less efficient heating performance.

5.1.2.1 Self-Heating

Self-heating techniques represent a significant advancement in the thermal management of LIBs for EVs, particularly in cold climates. Unlike external heating methods, which rely on heat transfer from an external source, self-heating leverages the battery's own energy to generate heat internally. This approach offers several advantages, including high heating efficiency, rapid temperature rise, and minimal energy consumption from external sources. However, it also presents unique challenges related to battery design, safety, and longevity. This section provides a comprehensive review of self-heating techniques, including their principles, system design, performance, advantages, disadvantages, and future trends.

As shown in Fig. 5.5, the fundamental principle of self-heating is to utilize the battery's internal resistance to generate heat through electrical current. When a current is applied to the battery, the internal resistance converts electrical energy into thermal energy, warming the battery cells from within. This internal heating process is highly efficient and can achieve rapid temperature rise, making it particularly suitable for cold climates where quick warming is crucial for battery performance [37]. A typical self-heating system involves modifying the battery structure to incorporate a heating element, such as a thin nickel foil or a resistive layer, which generates heat when an electrical current is applied. The heating element is usually embedded within the battery cell or placed in close proximity to the electrodes to ensure efficient heat transfer. The system also includes a switch or a control mechanism to activate the heating process when the battery temperature drops below a predefined threshold.

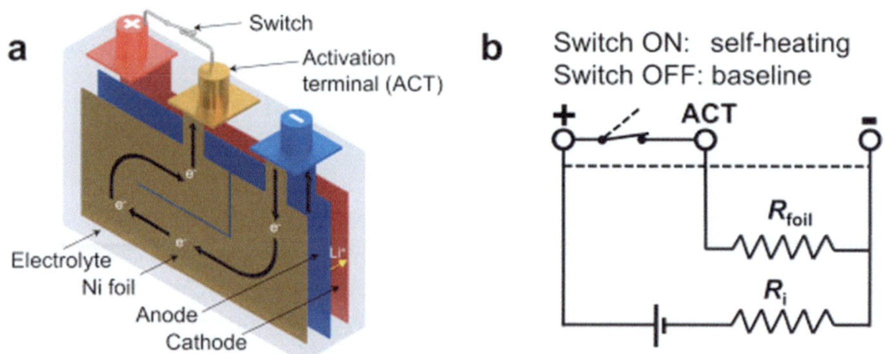

Fig. 5.5 Operating Mechanism of the Rapid Self-Heating Lithium-Ion Battery. **a** A schematic diagram illustrating the internal structure of the cell, featuring a strategically embedded nickel (Ni) foil and an electrical switch placed between the positive terminal and the activation terminal. **b** A detailed electrical circuit diagram showcasing the process by which the battery self-heats during operation [38]

Self-heating techniques typically achieve very high RTRs, often in the range of 10–60 °C/min, making them one of the fastest methods for preheating LIBs. For example, Wang et al. demonstrated that a self-heating lithium-ion battery (SHLB) could warm up from −30 to 0 °C in just 29.6 s, achieving an RTR of approximately 60 °C/min [37]. This rapid heating capability is crucial for ensuring that the battery reaches an optimal operating temperature quickly, thereby enhancing its performance and reducing the risk of lithium plating during charging. Since the heat is generated internally using the battery's own energy, the energy consumption is directly related to the amount of heat required to raise the battery temperature to the desired level. Typically, self-heating systems consume a small percentage of the battery's capacity, ranging from 3.8 to 5.5% of the cell capacity for a temperature rise from −30 to 0 °C. However, achieving uniform temperature distribution can be challenging, especially in large-format batteries, due to the potential for thermal gradients within the cell. To address this issue, researchers have explored various strategies, such as optimizing the placement and design of the heating element and using multiple heating elements to ensure uniform heat generation.

Self-heating does not require an external heat source, simplifying the system design and reducing the need for additional components. Self-heating requires modifications to the battery structure, such as embedding heating elements, which can increase the complexity and cost of battery manufacturing. The addition of heating elements and control mechanisms can increase the overall weight of the battery pack, potentially reducing the vehicle's range and efficiency.

One approach is to optimize the design and placement of the heating element to ensure efficient heat generation and uniform temperature distribution. For example, studies have explored the use of multiple heating elements, such as nickel foils, to reduce temperature gradients within the battery cell. This approach has been shown to significantly improve the heating efficiency and temperature uniformity of self-heating systems. Another innovation is the development of advanced control strategies to manage the heating process dynamically. Lei et al. [39] proposed an intermittent heating method, where the heating process is cycled on and off to reduce temperature gradients while maintaining a high RTR. Wang et al. [36] demonstrated a SHLB that could warm up from −30 to 0 °C in just 29.6 s, achieving an RTR of approximately 60 °C/min. This rapid heating capability significantly reduced the time required to reach an optimal operating temperature, enhancing the battery's performance in cold climates. Another study by Yang et al. [37] explored the use of multiple nickel foils in a self-heating battery design to improve temperature uniformity. The results showed that a two-sheet design could achieve an RTR of 96 °C/min from −20 to 0 °C, with only 2.9% battery capacity loss, compared to 4.1% for a single-sheet design. This study highlights the potential of optimizing the heating element design to enhance the performance of self-heating systems.

5.1.2.2 Direct Current (DC) Preheating

DC preheating is a method that involves applying a direct current to the battery to generate heat through its internal impedance. As shown in Fig. 5.6, this technique leverages the battery's own electrochemical reactions to produce heat, which can then be used to raise the temperature of the battery cells. DC preheating is particularly useful in cold climates where rapid warming of the battery is necessary to enhance performance and prevent issues such as lithium plating during charging.

The fundamental principle of DC preheating is to utilize the internal resistance of the battery to generate heat. When a direct current is applied to the battery, the internal resistance converts electrical energy into thermal energy, warming the battery cells from within. This internal heating process is highly efficient and can achieve rapid temperature rise, making it particularly suitable for cold climates where quick warming is crucial for battery performance [40, 41].

A typical DC preheating system involves applying a constant current to the battery through an external power source. The current is carefully controlled to ensure that the battery temperature rises to the desired level without causing over-discharge or other detrimental effects. The system also includes temperature sensors and control mechanisms to monitor and regulate the heating process, ensuring that the battery temperature remains within a safe and optimal range.

DC preheating typically achieves high RTRs, often in the range of 4–10 °C/min. For example, Qu et al. demonstrated that a DC preheating system could heat a battery from −10 to 10 °C in just 175 s, achieving an RTR of approximately 18.7 °C/min [42]. Typically, DC preheating consumes a small percentage of the battery's capacity,

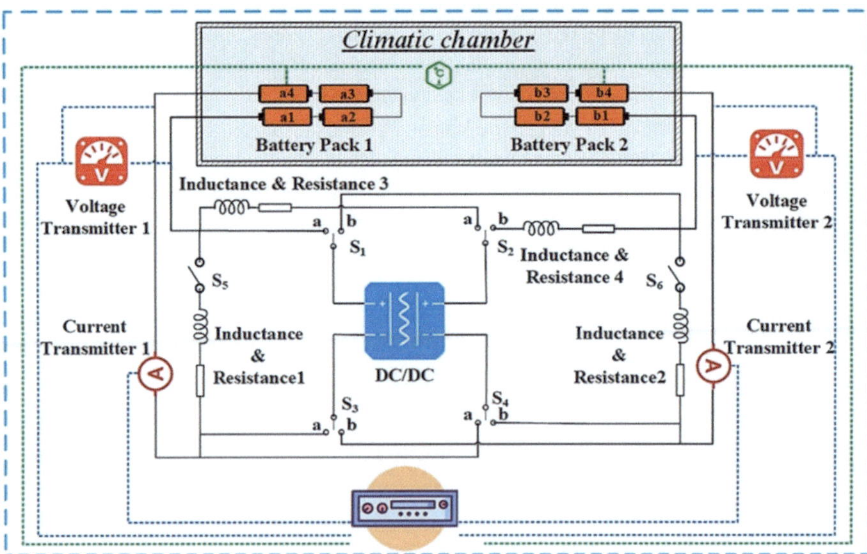

Fig. 5.6 DC preheating structure diagram [40]

ranging from 1.4% to 30% depending on the initial SOC and the heating duration [42]. Researchers have explored various strategies, such as optimizing the current profile and using multiple heating elements to ensure uniform heat generation.

The internal generation of heat using the battery's own energy results in minimal energy consumption, typically consuming a small percentage of the battery's capacity [40–42]. The high current densities required for DC preheating can potentially accelerate battery degradation and reduce its lifespan. The need for precise control of the heating current and temperature monitoring can increase the complexity and cost of the system. DC preheating is most effective for batteries with high initial SOC. For batteries with low SOC, the energy consumption can be significant, potentially reducing the overall efficiency of the system.

One approach is to optimize the current profile to balance the heating rate and energy consumption. For example, Ruan et al. proposed an optimal DC preheating technique that used a multi-objective genetic algorithm to balance the heating time and capacity loss [42]. This approach achieved a rapid temperature rise while minimizing the impact on battery lifespan. Dai et al. developed a control strategy that used a constant polarization voltage to achieve a good trade-off between short heating time and minimal damage to the battery life[43].

DC preheating techniques have been widely studied and tested in various laboratory settings due to their potential for improving battery performance and lifespan. For example, Qu et al. demonstrated a DC preheating system that could heat a battery from −10 to 10 °C in just 175 s, achieving an RTR of approximately 18.7 °C/min [42]. Another study by Ruan et al. explored the use of a multi-objective genetic algorithm to optimize the DC preheating process. The results showed that the optimized technique could heat the battery from −30 to 2.1 °C in just 103 s, with a capacity loss of only 1.4% after 500 cycles. This study highlights the potential of optimizing the heating current to enhance the performance and longevity of DC preheating systems.

5.1.2.3 Alternating Current (AC) Preheating

AC preheating is an innovative method for warming LIBs in EVs, particularly in cold climates. This technique involves applying an alternating current to the battery terminals to generate heat through the internal impedance of the battery. Unlike DC preheating, which can cause lithium plating and other degradation issues, AC preheating avoids these problems by rapidly oscillating the current direction, preventing lithium deposition. This method offers several advantages, including high heating efficiency, uniform temperature distribution, and minimal impact on battery health.

The fundamental principle of AC preheating is to utilize the internal impedance of the battery to generate heat. As shown in Fig. 5.7, when an alternating current is applied to the battery, the rapid oscillation of the current direction causes lithium ions to diffuse in and out of the electrodes repeatedly. This process generates heat through the internal resistance of the battery, warming the cells from within. The heat generation is proportional to the square of the current amplitude and the frequency

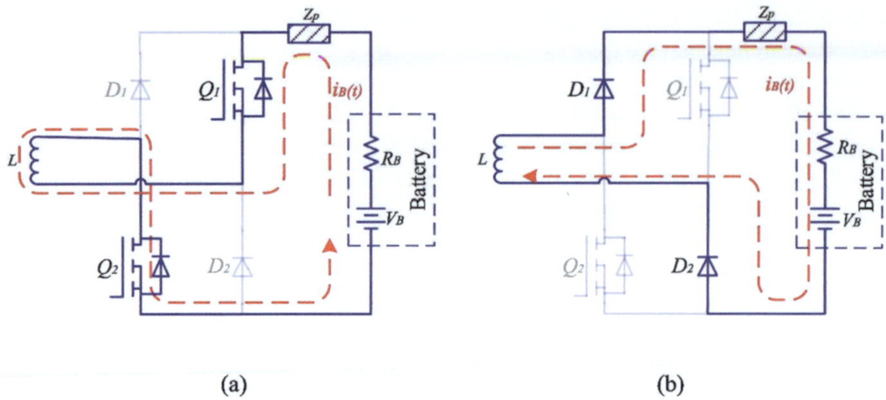

Fig. 5.7 High frequency AC preheating schematic diagram [46]

of the AC signal, making it highly controllable and efficient [44, 45]. A typical AC preheating system involves applying an AC signal to the battery terminals using an external power source. The system includes a frequency generator to produce the AC signal, a power amplifier to boost the signal to the required amplitude, and control mechanisms to regulate the heating process. The AC signal is carefully controlled to ensure that the battery temperature rises to the desired level without causing over-heating or other detrimental effects.

AC preheating typically achieves high RTRs, often in the range of 2–5 °C/min. For example, studies have shown that an AC preheating system can heat a battery from − 20 to 5 °C in approximately 800 s, achieving an RTR of approximately 2.21 °C/min [47]. Typically, AC preheating consumes a small percentage of the battery's capacity, ranging from 0.5 to 2% depending on the initial SOC and the heating duration.

AC preheating achieves very high RTRs, often in the range of 2–5 °C/min, making it one of the most efficient methods for preheating LIBs. AC preheating avoids lithium plating and other degradation issues associated with DC preheating, making it a safer and more battery-friendly solution. AC preheating requires access to an external power source, which may not always be available in all EV applications.

Ge et al. proposed a temperature-adaptive AC preheating method that adjusted the current amplitude based on the battery temperature to achieve a good trade-off between short heating time and minimal damage to the battery life. This approach achieved a rapid temperature rise while minimizing the impact on battery lifespan. Ruan et al. developed a control strategy that used a constant polarization voltage to achieve a good trade-off between short heating time and minimal damage to the battery life [47].

AC heating techniques offer a diverse range of waveform options [41, 48–50], with SC and BPC being the most widely employed methods, as illustrated in Fig. 5.8. Initially, researchers focused on SC heating, with Zhang et al. [51] and Zhu et al. [52] experimenting with different frequencies and amplitudes. Their studies revealed that

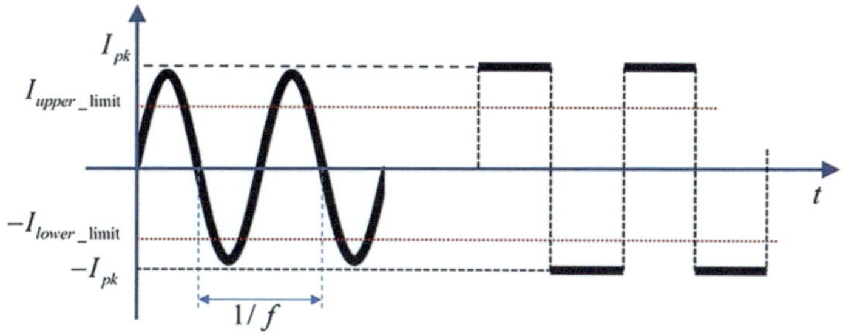

Fig. 5.8 Two types of AC heating methods: SC and BPC waveforms [52]

SC heating could effectively increase the battery temperature from sub-zero conditions to an optimal range within a short duration. Building upon this, Zhang et al. [53] and Guo et al. [54] concentrated on improving the magnitude of SC heating, whereas Ruan et al. [55] aimed at fine-tuning the frequency to boost the overall effectiveness of the heating process. In comparison, Zhu et al. [52] and Zhang et al. [56] assessed the performance of SC heating against BPC heating and determined that BPC heating delivers better thermal efficiency. Li et al. [57] performed parameter tuning for the equivalent circuit model of the battery by employing the HPPC test. They also developed a heating approach for the BPC that operates at frequencies lower than 1 Hz. On the other hand, Zhang et al. [56] and Wu et al. [58] showed that applying higher-frequency currents with increased amplitude enhanced the heating effectiveness. They introduced a high-frequency BPC approach that adjusted the amplitude based on temperature, facilitating quick temperature rise to the target value. Moreover, Qin et al. [50, 59–61] extensively studied BPC heating, demonstrating its high heating efficiency and minimal battery damage when designed appropriately. They further confirmed its practicality across both cell and module scales, contributing to the development of BPC heating technology. While symmetrical BPC heating has garnered significant attention, studies on asymmetrical BPC heating are still limited, hindering the complete optimization of heating methods.

5.1.2.4 Pulse Preheating

Pulse preheating is an innovative method for warming LIBs in EVs, particularly in cold climates. This technique involves applying a pulsed current to the battery to generate heat through its internal impedance. Unlike continuous DC or AC preheating methods, pulse preheating uses intermittent current pulses, which can mitigate some of the issues associated with continuous heating, such as battery degradation and energy consumption. This section provides a comprehensive review of pulse preheating, including its principles, system design, performance, advantages, disadvantages, and future trends.

The fundamental principle of pulse preheating is to utilize the internal resistance of the battery to generate heat through intermittent current pulses. When a pulsed current is applied to the battery, the internal resistance converts electrical energy into thermal energy, warming the battery cells from within. The intermittent nature of the current pulses allows for controlled heat generation, reducing the risk of over-heating and battery degradation [52].

As shown in Fig. 5.9, a typical pulse preheating system involves applying a pulsed current to the battery through an external power source. The system includes a pulse generator to produce the current pulses, a power amplifier to boost the signal to the required amplitude, and control mechanisms to regulate the heating process. The pulse parameters, such as amplitude, frequency, and duty cycle, are carefully controlled to ensure that the battery temperature rises to the desired level without causing over-discharge or other detrimental effects.

Pulse preheating typically achieves high RTRs, often in the range of 2–5 °C/min, making it one of the most efficient methods for preheating LIBs. For example, studies have shown that a pulse preheating system can heat a battery from −10 to 10 °C in approximately 175 s, achieving an RTR of approximately 4.87 °C/min. Typically, pulse preheating consumes a small percentage of the battery's capacity, ranging from 0.5 to 2% depending on the initial SOC and the heating duration. This low energy consumption makes pulse preheating a highly efficient solution for preheating LIBs. Studies have shown that pulse preheating can achieve a uniform temperature distribution within the battery, with temperature differences typically less than 3 °C [52].

Pulse preheating achieves very high RTRs, often in the range of 2–5 °C/min. Pulse preheating avoids lithium plating and other degradation issues associated with continuous DC preheating, making it a safer and more battery-friendly solution. The need for a pulse generator and control mechanisms to regulate the heating process can increase the complexity and cost of the system. The addition of components such as pulse generators and power amplifiers can increase the overall weight of the system.

One approach is to optimize the pulse parameters, such as amplitude, frequency, and duty cycle, to balance the heating rate and energy consumption. Wu et al. proposed a capacity degradation minimization-oriented optimization for pulse preheating, which achieved a rapid temperature rise while minimizing the impact on battery lifespan [63]. This approach demonstrated that optimizing the pulse parameters could significantly improve the efficiency and longevity of pulse preheating systems. Zhu et al. [52] demonstrated a pulse preheating system that could heat a battery from −24 to 25.6 °C in just 600 s using a rectangular pulse waveform

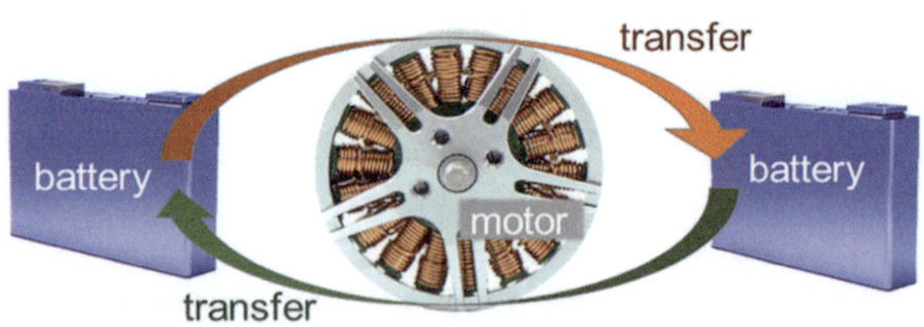

Fig. 5.9 Schematic diagram of pulse preheating system [62]

with a frequency of 30 Hz and an amplitude of 10 A. Another study by Wu et al. [58] proposed a capacity degradation minimization-oriented optimization for pulse preheating, which achieved a rapid temperature rise while minimizing the impact on battery lifespan. The results showed that the optimized technique could heat the battery from −20 to 5 °C in just 308 s, with a capacity loss of only 0.035% after 30 cycles. This study highlights the potential of optimizing the pulse parameters to enhance the performance and longevity of pulse preheating systems.

5.2 Low Temperature Heating Constraints

When designing a low-temperature AC heating strategy for LIBs, the heating constraint plays a crucial role in ensuring a non-destructive heating. Currently, the failure modes associated with low-temperature AC heating for LIBs primarily include lithium plating, overcharge, and over-discharge [52, 59, 64]. These three failure modes will not only reduce battery life but also pose safety risks.

5.2.1 Lithium Plating Criterion

Lithium plating can occur in LIBs when charged at low temperatures. This plating process can have two detrimental effects: first, it consumes active lithium ions, reducing battery capacity; second, it leads to the formation of lithium dendrites, which can pierce the separator and cause internal short circuits. Although overcharge and over-discharge can be effectively managed by regulating terminal voltage, the precise approach to preventing lithium plating is still not well-established.

Ge et al. [65] explored the factors influencing lithium deposition during the charging phase of LIBs. They discovered that maintaining the potential difference between the anode's solid and liquid phases above the equilibrium value for lithium plating effectively prevents the formation of lithium deposits. Building on this work, Ge et al. [66], Guo et al. [44], and Zhang et al. [56] further developed the concept by using the lithium deposition threshold to establish the optimal AC heating intensity, effectively mitigating the risk of lithium plating. In an alternative method, Li et al. [57] proposed an integrated electro-thermal model for a three-electrode system, utilizing an anode potential drop below 0 V as a critical indicator for forecasting lithium plating. However, further investigations have shown that plating may still happen even if the graphite anode potential remains above 0 V [67, 68]. On the other hand, studies by Gao et al. [69] and Bhandari et al. [70] indicate that lithium deposition occurs only when the potential of the graphite anode falls beneath a critical limit, even if it starts at a potential lower than 0 V. The ongoing discussion regarding the exact potential threshold for lithium plating underscores the challenge of establishing a consistent and dependable standard to avoid battery damage, especially during the BPC heating phase.

Lithium deposition takes place on the graphite anode, highlighting the importance of concentrating on the anode in relevant research. The diagram in Fig. 5.10a illustrates the setup for measuring the potential of the graphite anode using a three-electrode configuration. At the graphite anode, the interactions of lithium ions mainly encompass the formation of the SEI layer and the charge transfer occurring during BPC heating. Additionally, lithium tends to deposit at the boundary between the graphite surface and the SEI film, where it competes with the ongoing charge transfer.

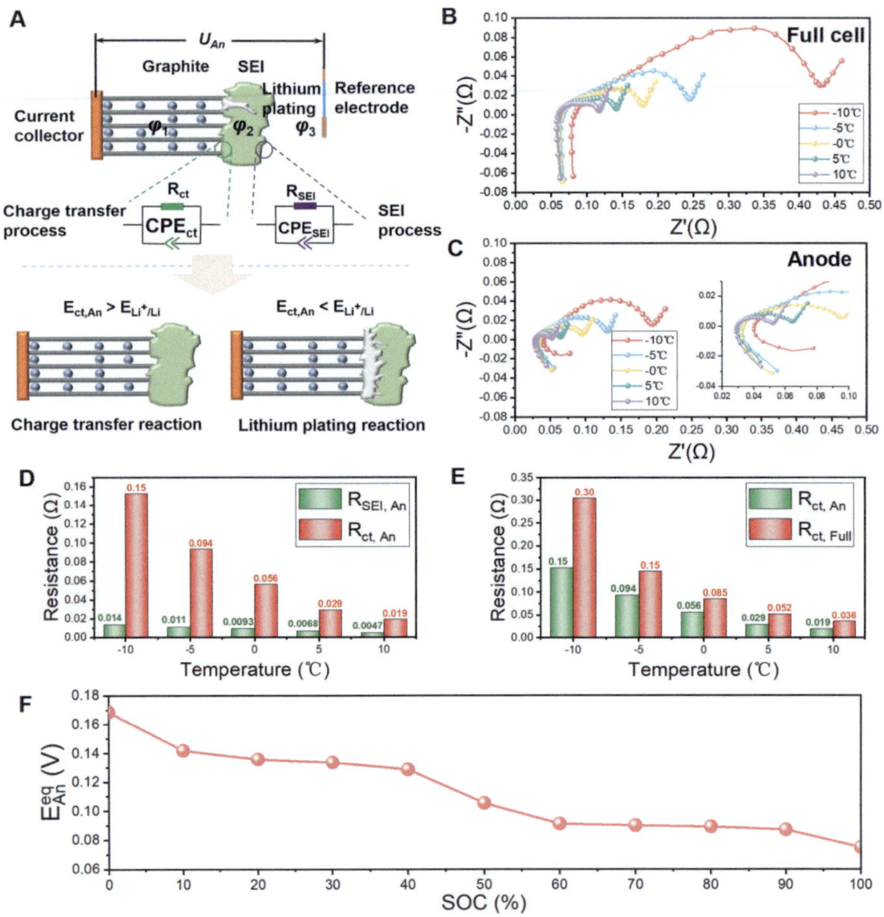

Fig. 5.10 Illustration and calculations related to lithium deposition and the three-electrode setup. **a** Diagram depicting a three-electrode battery used to assess the anode potential, emphasizing key reactions at the graphite anode. **b** EIS measurements on the entire cell at 10% SOC under varying temperature conditions. **c** EIS measurements focused on the anode at 10% SOC across different temperatures. **d** Variation of the value of K_1 with temperature. **e** Variation of the value of K_2 with temperature. **f** Graphite anode equilibrium potential at various SOC levels [71]

Please resupply this artwork (Figs. 5.10, 5.11, 5.12, 5.14, 5.15, 5.16 and 5.17) in a format suitable for printing [Figures text size are too small].Thank you very much for your suggestion. We have revised the text within the images and made every effort to ensure high-resolution outputs. However, some of the original images were difficult to obtain, so we had to base our modifications on figures published in online journals. If the clarity of the original images was limited, this may have affected the quality of the modified versions. Nevertheless, we have done our best to enhance the image quality to ensure that it does not hinder the reader's understanding. Please refer to the uploaded file (attachments) for the modified image.

The lithium plating reaction occurs when the potential of the lithium plating reaction $E_{Li^+/Li}$ exceeds that of the charge transfer reaction $E_{ct,An}$. Conversely, when the potential of the lithium plating reaction $E_{Li^+/Li}$ is less than the potential of the charge reaction $E_{ct,An}$, the charge transfer reaction occurs. According to the thermodynamic and kinetic definitions, the condition under which the lithium plating reaction occurs is

$$E_{ct,An}^{eq} + \eta_{ct,An} \leq E_{Li^+/Li}^{eq} + \eta_{Li^+/Li} \tag{5.1}$$

The equilibrium potential and overpotential for the charge transfer at the graphite anode are denoted as $E_{ct,An}^{eq}$ and $\eta_{ct,An}$, respectively. For the lithium plating reaction, $E_{Li^+/Li}^{eq}$ and $\eta_{Li^+/Li}$ represent its equilibrium potential and overpotential.

During the charging process, both $\eta_{ct,An}$ and $\eta_{Li^+/Li}$ are negative. In a three-electrode setup, all potentials are referenced against lithium, making the equilibrium potential of the lithium plating reaction, $E_{Li^+/Li}^{eq}$, equal to 0. Then, Eq. (5.1) can be written as

$$\left|\eta_{ct,An}\right| \geq E_{ct,An}^{eq} + \left|\eta_{Li^+/Li}\right| \tag{5.2}$$

Lithium plating takes place when the thermodynamic condition $\left|\eta_{Li^+/Li}\right| > 0$ is met [72]. Consequently, the absence of lithium plating occurs when this condition is not fulfilled

$$\left|\eta_{ct,An}\right| < E_{ct,An}^{eq} \tag{5.3}$$

In a three-electrode battery system, based on the graphite anode model presented earlier, the anode's overpotential arises from factors such as the internal resistance, SEI layer dynamics, and charge transfer mechanisms. As a result, the estimation of $\eta_{ct,An}$ in this study can be

$$\left|\eta_{ct,An}\right| \approx \frac{\left(V_{p,An} - IR_{0,An}\right)R_{2,An}}{R_{1,An} + R_{2,An}} \tag{5.4}$$

In Fig. 5.10a, assuming sufficient time has passed for the battery to settle, it is presumed that the potential within each phase becomes uniform. Consequently, the equilibrium potential for each phase can be described as [73]

$$\begin{cases} E_{An}^{eq} = \varphi_1^{eq} - \varphi_3^{eq} \\ E_{ct,An}^{eq} = \varphi_1^{eq} - \varphi_2^{eq} \\ E_{SEI,An}^{eq} = \varphi_2^{eq} - \varphi_3^{eq} \end{cases} \tag{5.5}$$

where φ_1^{eq}, φ_2^{eq}, and φ_3^{eq} represent the equilibrium potentials of graphite, the SEI layer, and the electrolyte in proximity to the reference electrode, respectively. Meanwhile, E_{An}^{eq} denotes the equilibrium potential of the anode. The connection between E_{An}^{eq}, $E_{ct,An}^{eq}$, and $E_{SEI,An}^{eq}$ can be formulated as [73]

$$\begin{cases} E_{An}^{eq} = E_{SEI,An}^{eq} + E_{ct,An}^{eq} \\ \frac{E_{ct,An}^{eq}}{E_{SEI,An}^{eq}} \approx \frac{R_{ct,An}^{eq}}{R_{SEI,An}^{eq}} \end{cases} \tag{5.6}$$

where $R_{SEI,An}^{eq}$ and $R_{ct,An}^{eq}$ represent the resistances associated with the SEI formation and the charge transfer processes when the system is at equilibrium. By integrating Eqs. (5.3), (5.4), and (5.6), the condition to prevent lithium plating can be written as

$$\frac{\left(V_{p,An} - IR_{0,An}\right)R_{2,An}}{R_{1,An} + R_{2,An}} < \frac{E_{An}^{eq} R_{2,An}^{eq}}{R_{1,An}^{eq} + R_{2,An}^{eq}} \tag{5.7}$$

To bridge the gap between theoretical insights from the three-electrode model and the practical limitations of commercial two-electrode batteries, it is necessary to develop a lithium plating criterion applicable to full cells. While the three-electrode model provides detailed anode behavior and allows precise monitoring of lithium plating, most commercial batteries are two-electrode systems. These systems lack a direct reference electrode, making accurate anode potential measurements challenging and limiting the direct application of the three-electrode criterion. Consequently, a practical approach is required to adapt the lithium plating criterion from the three-electrode model to full cells. This adaptation should consider both electrochemical reactions and charge transfer processes in both systems, accounting for differences in their electrical and thermal behaviors. By establishing a robust and transferable criterion, we can apply the insights from the three-electrode model to the design and optimization of heating strategies for two-electrode batteries, thereby improving their performance and lifespan in low-temperature conditions. In a full battery, Eq. (5.3) can be expressed as:

$$\left|\eta_{ct,An}\right| < \frac{E_{An}^{eq} R_{ct,An}^{eq}}{R_{SEI,An}^{eq} + R_{ct,An}^{eq}} = E_{An}^{eq} K_1 \tag{5.8}$$

In a complete electrochemical cell, the charge transfer process encompasses both the cathode and anode components. The overall resistance to charge transfer in the full cell is greater than that observed at the anode by itself. As depicted in Fig. 5.10e, the ratio of the anode's charge transfer resistance to the total cell resistance, referred to as K_2, is determined to evaluate the accuracy of the resistance analysis. As a result, at an identical current, the voltage drop due to charge transfer at the graphite anode is smaller compared to that of the entire cell. Therefore, Eq. (5.8) can be adjusted as follows

$$\left|\eta_{ct,Full}\right| < E_{An}^{eq} K_1 \tag{5.9}$$

where $\eta_{ct,Full}$ is the voltage of the charge transfer process in the full cell [71]. The validity of Eq. (5.9) establishes a sufficient condition for Eq. (5.1) to be true. Therefore, Eq. (5.9) can serve as a lithium plating criterion oriented to BPC heating for the full cell. The next task is to identify the parameter values for E_{An}^{eq} and K_1. According to the study in [74], as the state of charge (SOC) of the graphite anode decreases, the contribution of the SEI formation process relative to the charge transfer resistance becomes progressively smaller. As a result, the lowest K_1 value is determined when the battery's SOC is at a low level. To achieve this, a three-electrode cell with a 10% SOC is selected to pinpoint the minimum K_1 value. The detailed outcomes are presented in Fig. 5.10b–e. The values of K_1 values vary with temperature, with the lowest recorded at 0.8. To minimize potential measurement inaccuracies and prevent the occurrence of lithium plating, the K_1 value is cautiously adjusted to 0.6. Additionally, E_{An}^{eq} can be determined through the construction of coin cells, following the method outlined in [75]. In this case, E_{An}^{eq} is directly measured using a three-electrode setup, with the results displayed in Fig. 5.10f.

5.2.2 *Experimental Verifications*

Earlier, we established a criterion for lithium plating during AC heating, utilizing both three-electrode and two-electrode battery models. This criterion provides a theoretical foundation for understanding and preventing lithium plating during low-temperature heating processes. Since the theoretical analysis was derived using the three-electrode system, it is essential to validate its accuracy and practical applicability through experimental methods. Therefore, in this section, we present experimental verifications using three-electrode batteries. Details about the test samples and protocols can be found in Sect. 4.3 and will not be repeated here. By utilizing the three-electrode configuration, we can directly monitor the anode potential, enabling precise detection of lithium plating under various heating conditions. This experimental approach not only validates the theoretical criterion established in Sect. 5.2.1 but also provides empirical evidence to ensure that the heating strategy is both non-destructive and effective for practical applications. The experimental results will further refine our understanding of lithium plating and help optimize heating strategies for lithium-ion batteries. It should be emphasized that Table 5.1 presents the amplitudes of different constraints for the symmetrical BPC at a frequency of 1 Hz and a temperature of 10 °C. From the table, we observe that the minimum current amplitude corresponding to the terminal voltage constraint is 1.91 A. The current amplitudes for lithium plating and for the condition where $V_{An} < 0$ V are 1.37 and

0.98 A. To validate the lithium plating criterion and prevent issues related to overcharging or excessive discharge, current amplitudes of 1.4 and 1.6 A are selected for the experimental tests.

As depicted in Fig. 5.11a, when a current amplitude of 1.4 A is applied, the battery's capacity remains unchanged. In contrast, at a current amplitude of 1.6 A, the battery's capacity progressively diminishes as the excitation duration increases. Following 500 h of operation, the battery's capacity reduces to roughly 0.9672 Ah. In order to explore the root cause of the observed capacity loss, both the experimental batteries and a fresh one were dismantled for analysis. The findings reveal that the graphite anode of the new battery, as well as the anode exposed to a 1.4 A current amplitude, exhibited no notable differences. In contrast, the graphite anode subjected to a current amplitude of 1.6 A, which is identified as lithium plating, as reported by Petzl et al. [76]. SEM observations showed that the material takes on a branching structure, resembling the lithium dendrites described by Zhang et al. [77]. This metallic substance is recognized as lithium and its associated compounds. As displayed in Table 5.1, the current amplitude for lithium deposition in the symmetrical BPC, when subjected to a 1 Hz frequency at 10 °C, is 1.37 A. At a current amplitude of 1.4 A, the battery exhibits minimal capacity loss. This behavior can be attributed to the development of lithium deposition at varying current levels, as shown in Fig. 5.11b. At lower current levels, the graphite anode participates in a charge exchange process while charging. As the current intensity rises, lithium starts to deposit onto the anode. Nevertheless, as a result of the stripping mechanism [78], the deposited lithium undergoes oxidation during discharge, especially when its quantity remains minimal. the lithium that has deposited can fully revert. However, as the current intensity escalates, the deposited lithium consists of both reversible and non-functional ("dead") lithium. As a result, during discharge, the deposited lithium fails to fully oxidize. Over time, the buildup of inactive lithium contributes to a decrease in the overall capacity of the battery. Therefore, the current amplitude that induces capacity degradation during BPC heating is larger than the threshold for lithium plating. Nonetheless, this experiment further validates the reliability of the lithium plating criterion.

Table 5.1 The magnitudes of various limitations for the symmetrical BPC at a 1 Hz frequency and a temperature of −10 °C [73]

Lithium plating	$V_{full} = V_{min}$	$V_{full} = V_{max}$	$V_{An} = 0$ V
1.37 A	1.91 A	2.17 A	0.98 A

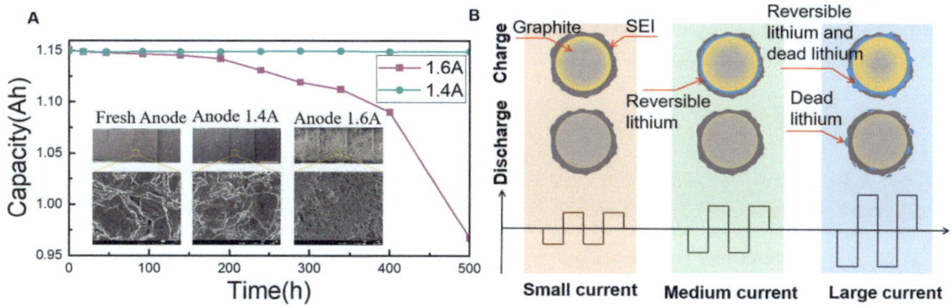

Fig. 5.11 At -10 °C, lithium deposition was confirmed under varying current amplitudes in a 1 Hz symmetrical BPC. **a** The battery capacity was observed to change with different current amplitudes, accompanied by insights from disassembling the battery; **b** The pattern of lithium deposition was tracked across a range of current levels. (Graphical data for the graphite shown in this figure is derived from Ren et al. [78])

5.3 Bidirectional Pulsed Current Heating

5.3.1 Heating Strategy Design

When implementing the low-temperature AC heating method, two key issues arise: voltage constraints at the terminals and the potential for lithium plating. This voltage limits are managed by establishing defined upper and lower thresholds, ensuring safe operation within these boundaries:

$$U_{\min} \leq U_t \leq U_{\max} \tag{5.10}$$

where the charge and discharge cutoff voltages are set at 4.2 and 3 V, denoted as U_{max} and U_{min}. While the terminal voltage limitation is straightforward, establishing the conditions for lithium plating involves more complexity.

Figure 5.12 demonstrates the process of the BPC heating technique explored in this study, outlining the primary procedure and the generation of the BPC map. This process is divided into two stages: "online look-up table" and "offline calculation." The second stage involves computing the optimal heating strategy for different SOCs and temperatures beforehand, with the results stored in a system memory table for future reference. In practical use, the system accesses an first to obtain the appropriate strategy based on the battery's present temperature and SOC. This approach offers better resource efficiency compared to real-time optimization methods and is easier to deploy. The process starts by determining the battery's initial SOC SOC_{start}, temperature T_{start}, and the objective temperature T_{obj} are initially determined. The system then checks if the battery's temperature T has reached T_{obj}. If it has, the heating process ends. Otherwise, the system assesses whether the temperature range, or "interval," has changed. These intervals correspond to discrete temperature ranges used in the BPC map, such as [−10 °C, −5 °C), [−5 °C, 0 °C), [0 °C, 5 °C), [5 °C,

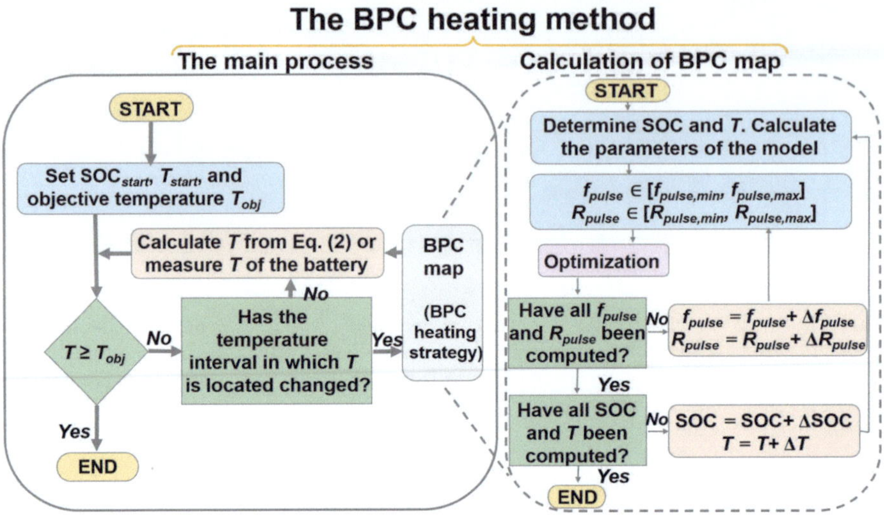

Fig. 5.12 The BPC heating method [71]

10 °C) and [10 °C, 15 °C). The BPC heating strategy for each temperature range is set according to the BPC map at the lowest temperature of that range. If the temperature range stays the same, the battery follows the initial BPC heating plan. All calculations in this chapter rely on the electro-thermal coupling models presented in Chap. 4.

5.3.2 Experimental Verifications

In this part, we delve into the experimental validation and optimization of the BPC heating strategy, building upon the theoretical framework established in the preceding sections [73]. We examine how factors like lithium plating and voltage limitations affect heating performance and battery longevity. The results not only confirm the effectiveness of the optimized BPC strategy in enhancing heating rates but also demonstrate its minimal impact on battery capacity and lifespan. This section ultimately underscores the potential of the optimized BPC heating strategy as a robust solution for improving the low-temperature performance of lithium-ion batteries, paving the way for broader applications in electric vehicles and energy storage systems. Similarly, the test samples and protocols used in this section are described in detail in the previous chapter.

5.3.2.1 The Effect of Various Limitations on Battery Current Flow and Heat Production

Figure 5.13a demonstrates the application of symmetrical BPC in the development of heating strategies [56, 58, 59]. To investigate the relationship between the maximum BPC heating magnitude and frequency under diverse conditions, a pulse ratio R_{pulse} of 0.5 is specified. Figure 5.13b–e depict the variations in current and heat production as a function of frequency at different temperatures, namely −15 −10, 5, 0, 5, and 10 °C, while considering various constraints. In addition to the lithium plating threshold, the figures incorporate constraints related to the maximum charge voltage (V_{max}), the minimum discharge voltage (V_{min}), and the condition that V_{An} must be less than zero. Referring to the data presented in Fig. 5.13b, it becomes evident that the current levels tied to each constraint experience an upward trend when subjected to increased frequency and temperature. This correlation extends to the heat generation associated with these constraints, which is illustrated in Fig. 5.13d. When $V_{An} < 0$, the current amplitude is found to be at its lowest under the given temperature and frequency conditions. This observation implies that although $V_{An} < 0$ is commonly used to signal lithium plating during charging [57], it is not a suitable criterion for identifying lithium plating when BPC heating is applied. Furthermore, as depicted in Fig. 5.13e, the condition of $V_{An} < 0$ corresponds to the minimal level of heat generation. meaning that using this constraint in BPC heating design would result in very low heating efficiency. In Fig. 5.13c, when frequency exceeds approximately 70 Hz, the current associated with the lithium plating constraint surpasses that of the terminal voltage constraint at different temperatures. This shows that, at frequencies above 70 Hz, over-discharge becomes the primary failure mode rather than lithium plating. Conversely, at frequencies below 70 Hz, as seen in Fig. 5.13c, the intersection points of the current related to lithium plating limitations coincide with those of the currents associated with voltage constraints at temperatures of −15, −10, −5, and 0 °C. Nevertheless, when the temperature surpasses 0 °C, the intersection point disappears, indicating that over-discharge becomes the dominant failure mode above this temperature. In the context of BPC heating, the low SOC level causes the system to hit the discharge termination voltage before other thresholds. Consequently, the convergence point of the current trajectories associated with lithium deposition and discharge termination voltage serves as a key indicator of the onset of failure. Below the intersection frequency, lithium plating occurs, while above it, over-discharge is the dominant failure mode.

To explore how R_{pulse} impacts thermal generation, the f_{pulse} is maintained at a constant 100 Hz. Figure 5.14b and d illustrate the fluctuations in current and thermal output corresponding to various R_{pulse} values, all within the confines of a terminal voltage limit. As the thermal level ascends, both the electrical flow and thermal output escalate correspondingly. Meanwhile, when R_{pulse} is augmented, the current during the charging phase progressively climbs, while the current during the discharging phase diminishes. The reason for this lies in the impact of R_{pulse} on the discharge process.

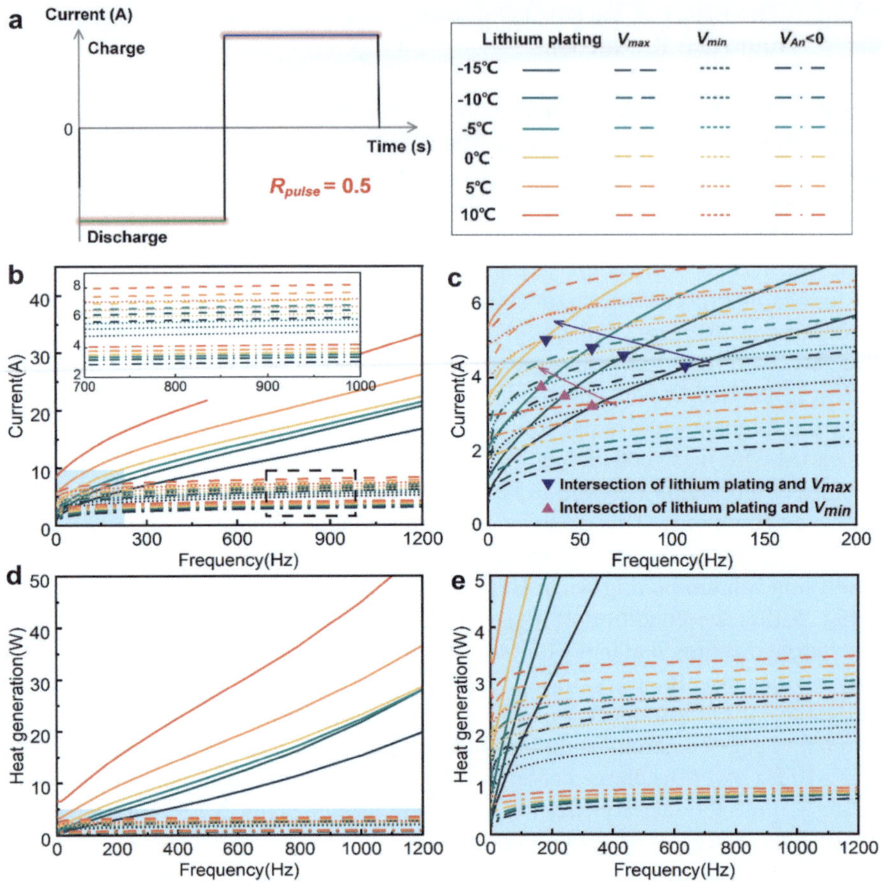

Fig. 5.13 Current and heat generation for each constraint at different frequencies when R_{pulse} is 0.5. **a** BPC waveform with R_{pulse} of 0.5; **b**, **c** Current for each constraint at varying frequencies; **d**, **e** Heat generation for each constraint at varying frequencies [73]

A smaller R_{pulse} value causes the discharge period to be briefer, which in turn drives up the discharge current when the specified cut-off voltage is attained. This phenomenon occurs because a smaller resistance allows for a more rapid flow of current during the discharge phase. As a result, the system reaches the specified cut-off voltage more quickly. The cut-off voltage is a critical threshold that signifies the end of the discharge process to prevent over-discharge and potential damage to the system. When this cut-off voltage is attained more rapidly due to a smaller R_{pulse}, the discharge current experiences a corresponding increase. On the other hand, the charge current exhibits the opposite trend. When R_{pulse}, a shift occurs in the terminal voltage constraint, with R_{pulse} values below 0.55 being dominated by the discharge cut-off voltage and those above 0.55 shifting the dominance to the charge cut-off voltage. As depicted in Fig. 5.14d, a specific R_{pulse} value is associated with the peak level

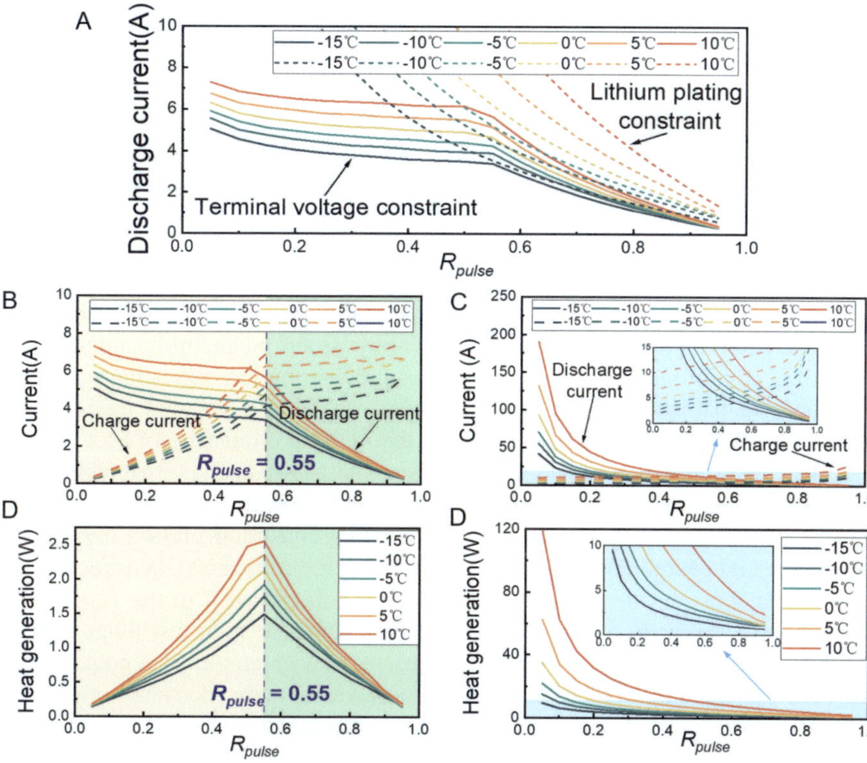

Fig. 5.14 An analysis of discharge current behavior under lithium plating and terminal voltage limits at varying R_{pulse} values, with f_{pulse} fixed at 100 Hz. **a** A comparison of discharge currents dictated by lithium plating and terminal voltage constraints across different R_{pulse} settings; The electrical current and thermal generation under the terminal voltage restriction are illustrated in sections (**b**) and (**d**), respectively. The current flow and heat output under the lithium plating constraint are presented in sections (**c**) and (**e**) [73]

of heat production. This particular value remains constant regardless of temperature fluctuations. This observation implies that symmetrical BPC does not necessarily lead to the maximum generation of heat. In Fig. 5.14c and d, the fluctuations in current and heat production are examined under the lithium plating limitation.

Analogous to the scenario with the terminal voltage restriction, an elevation in R_{pulse} correlates with an increase in the charge current and a concurrent decrease in the discharge current. In comparison to the terminal voltage constraint, heat generation actually rises when R_{pulse} is reduced under the lithium plating condition. In order to better analyze the failure modes under varying R_{pulse} conditions, Fig. 5.14a illustrates the discharge current trajectories corresponding to the constraints of lithium deposition and terminal voltage. Under identical temperature conditions, the discharge current governed by lithium deposition limitations is observed to be significantly greater than that dictated by terminal voltage boundaries. From the analysis, it can be seen that at an f_{pulse} setting of 100 Hz, over-discharge is the key failure mechanism when R_{pulse} values less than 0.55. In contrast, overcharge becomes the leading failure mode once R_{pulse} rises above 0.55.

To broaden the scope of the research and provide a more comprehensive understanding of how different constraints affect battery current and heat generation, it is crucial to extend the analysis from a fixed SOC to a wider range of SOC levels. While Figs. 5.13 and 5.14 offer in-depth insights into the behavior of battery current and heat generation under various constraints at a specific SOC, Fig. 5.15 builds on this by demonstrating how these constraints impact battery performance across different SOC levels. This shift from a fixed to a variable SOC analysis allows for a more thorough evaluation of the heating strategy's effectiveness and adaptability under diverse operational conditions. Figure 5.15 illustrates the distribution of heating power for different BPC parameters under the constraints of the BPC heating method. The visualizations utilize diverse color schemes to signify the intensity of heat generation. The backdrop is partitioned into three specific zones, each corresponding to a different constraint: U_{max} constraint, U_{min} constraint, and lithium plating constraint. In the lower-frequency domain, the lithium plating constraint plays a major role, while in the higher-frequency area, the terminal voltage constraint takes precedence. Notably, the R_{pulse} in the U_{max} dominant region is larger than in the U_{min} dominant region. A larger R_{pulse} value causes the discharge pulse to last longer within each BPC period. This extended duration results in a greater charge pulse amplitude than the discharge pulse, which in turn makes it more likely to activate the U_{min} constraint. When R_{pulse} is smaller, the U_{min} constraint is more likely to be triggered. At a steady SOC, decreasing the temperature enlarges the area where the lithium plating constraint prevails, particularly in the direction of lower R_{pulse} and higher frequencies. Meanwhile, the area influenced by the terminal voltage constraint becomes smaller. However, the R_{pulse} at the boundary between the U_{max} and U_{min} constraint-dominated regions remains largely unchanged. When the SOC is reduced, the lithium plating constraint region first expands and then contracts, while the U_{max} constraint region decreases and the U_{min} constraint region grows. At 100% SOC, the lithium plating constraint does not dominate any region, and only the U_{max} constraint region exists. This occurs because, at 100% SOC, the terminal voltage aligns closely with U_{max}, rendering the U_{max} constraint the primary determinant.

Regarding heating power, it increases with frequency under the same conditions. As the frequency rises, Z' decreases, leading to larger pulse amplitudes for each constraint. This highlights the dominant role of current in heating power calculation. With an escalation in R_{pulse}, the thermal energy output initially surges before eventually tapering off. The pinnacle of thermal energy generation coincides with the transition point of the primary limiting factor. The phenomenon is attributed to the interplay between variations in pulse amplitude and the alterations in R_{pulse}. When the system reaches 100% SOC, the highest level of heating power is attained when R_{pulse} is set at 0.05. This is because the U_{max} constraint dominates, with terminal voltage nearly equal to U_{max}, which makes the A_{cha} almost 0 A. As R_{pulse} decreases, A_{dis} is increases, thereby raising the calculated heating power. When the temperature remains steady, the optimal R_{pulse} for achieving maximum heating power tends to

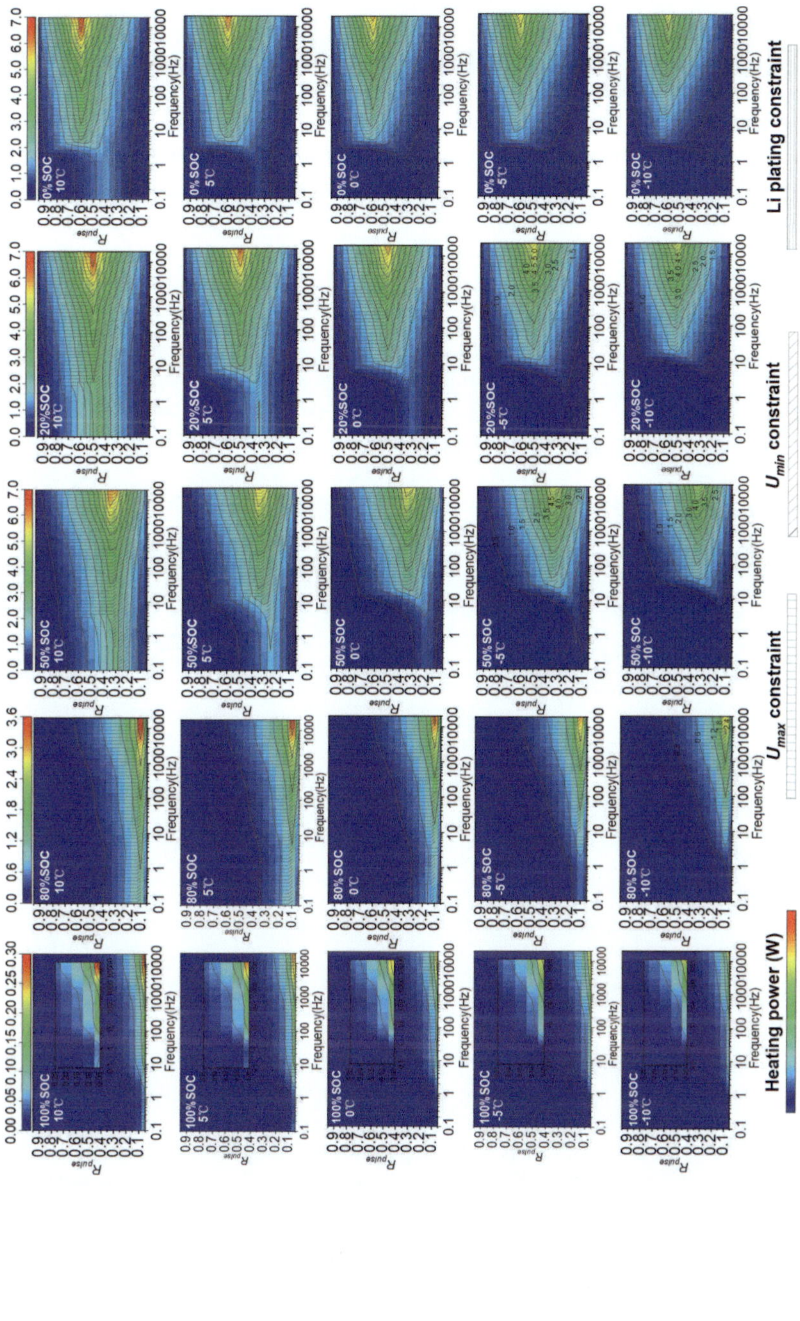

Fig. 5.15 Heat production in batteries under the influence of diverse BPC settings, constrained by the thermal effects of BPC [71]

grow higher as the SOC level falls. This correlation highlights the transition in the key influencing factor area. Furthermore, heating power increases with temperature at the same SOC, as Z′ decreases and pulse amplitude rises with higher temperatures under constant constraints. Since heating power depends on the square of current, the battery's heating power also increases.

To further analyze the variations in BPC parameters, Fig. 5.16 shows the changes in maximum heating power and its corresponding BPC parameters with frequency at different temperatures. As seen in Fig. 5.16a, d and g, maximum heating power increases with both frequency and temperature, consistent with previous observations. In the Fig. 5.16b, e and h, the R_{pulse} measurements climb steadily as the frequency rises, regardless of the temperature. However, these values tend to coalesce toward a uniform figure, with the notable exception being the case at full capacity

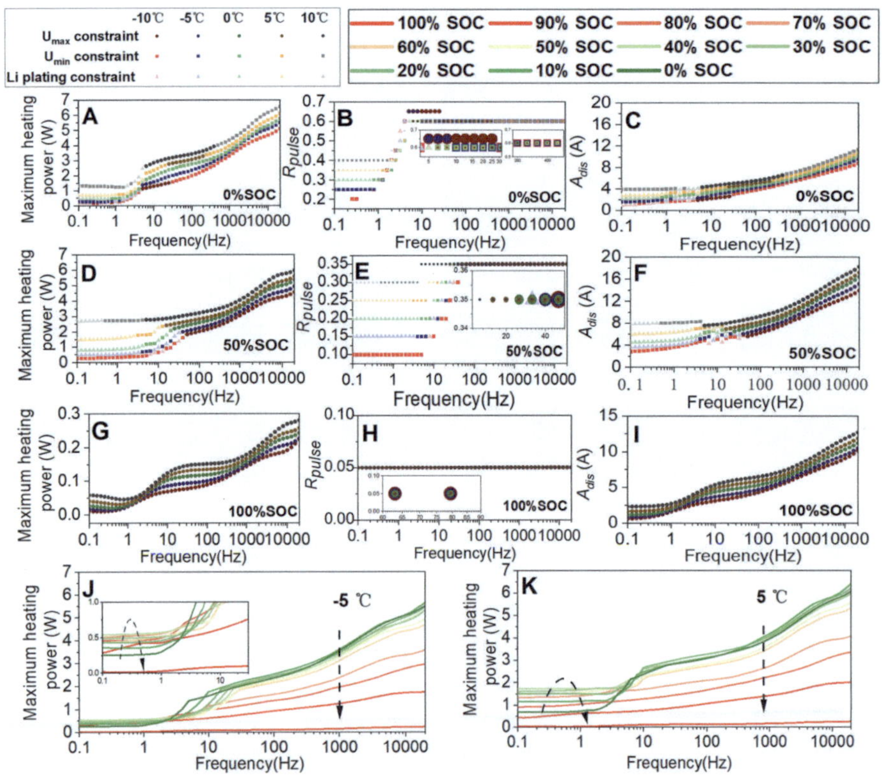

Fig. 5.16 Variation of maximum heating power and its corresponding BPC parameters with frequency at different temperatures. Panels **a**, **d**, and **g** show maximum heating power at 0% SOC, 50% SOC, and 100% SOC, respectively. Diagrams **b**, **e** and **h** correspond to the R_{pulse} measurements for peak heating power at 0%, 50%, and 100% charge levels, respectively. Additionally, diagrams **c**, **f** and **i** highlight the A_{dis} values linked to maximum heating power at 0%, 50%, and 100% charge levels. Diagram **j** provides an analysis of how maximum heating power fluctuates with frequency across various charge states at −5 and 5 °C [71]

(100% SOC). In the illustrations designated as Fig. 5.16c, f and i, it is observed that the behavior of A_{dis} closely mirrors that of the peak heating power. Specifically, at a 50% SOC, A_{dis} diminishes as the primary constraint transitions to the U_{max} constraint, driven by an increase in frequency. Nevertheless, this phenomenon does not alter the general upward trajectory of the heating power, emphasizing that both current and R_{pulse} influence the heating power calculation. In the lower frequency range, specifically under around 50 Hz, R_{pulse} undergoes intricate alterations in primary restrictive elements across diverse temperature levels. These changes pose significant challenges for the formulation of effective approaches in low-frequency BPC heating processes. In conclusion, as depicted in Fig. 5.16j and k, the variation of peak heating power with frequency is examined at −5 and 5 °C for multiple SOCs. When the frequency is beneath 3 Hz, the peak heating power initially surges before tapering off as the SOC escalates. Conversely, at frequencies exceeding 3 Hz, the heating power exhibits minimal fluctuation when SOC is confined between 0 and 40%, but undergoes a pronounced reduction as SOC keeps ascending.

5.3.2.2 An Examination of the Differences in Heating Power Between BPC Heating that Accounts for the Impact of R_{pulse} and Conventional BPC Heating

Figure 5.17 presents a comparison of heating power between conventional BPC heating and optimized BPC heating that factors in R_{pulse}. In standard BPC heating, R_{pulse} is generally set at 0.5. To assess the difference, we have determined the ratio of the peak heating power (P_{max}) with R_{pulse} effects to the heating power of traditional BPC heating ($P_{0.5}$) across different frequencies at −5 and 5 °C. The data demonstrates that Pmax is substantially greater than $P_{0.5}$ across all tested frequencies. Particularly when SOC exceeds 80%, P_{max} is more than six times $P_{0.5}$. For charge levels below 80%, the relationship between P_{max} and $P_{0.5}$ is more significant at lower frequencies. At −5 °C, the value of P_{max} relative to $P_{0.5}$ is over 2, while at 5 °C, it nears 2. With increasing frequency, the ratio of P_{max} to $P_{0.5}$ becomes more stable, falling between 1 and 2. These findings suggest that optimizing for R_{pulse} enhances BPC heating power, particularly at higher SOCs and lower heating frequencies.

5.3.2.3 Validation of Optimized Heating Strategies in Different SOCs

The best methods for increasing battery temperature from −10 to 10 °C across different levels of charge are detailed in Table 5.2. To keep things straightforward, the table focuses solely on the most efficient heating techniques for batteries at 0, 50, and 90% charge levels. The 0% charge level represents a completely depleted state, where the battery may require special handling to ensure safe and efficient heating without causing damage. In particular, the frequency settings for these heating methods match the highest values within the refined frequency spectrum. Furthermore, the pulse resistance (R_{pulse}) associated with the ideal strategy remains

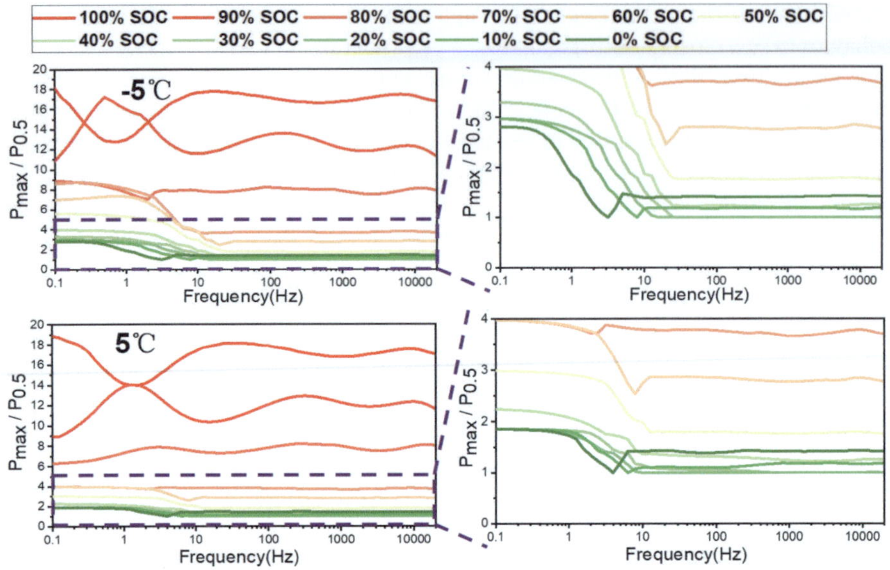

Fig. 5.17 The ratio of the maximum heating power P_{max} considering the effect of R_{pulse} to the traditional BPC heating power $P_{0.5}$ at −5 °C (**a**) and 5 °C (**b**) [71]

unchanged throughout different temperature ranges, while only the magnitude of the charging and discharging pulses fluctuates. To confirm the reliability of the temperature increase estimates, Table 5.3 outlines the ideal heating method at 50% charge level, operating at a frequency of 100 Hz. This specific frequency is used for validation purposes, as the power amplifier's maximum output current is capped at 10 A. This constraint means that the heating strategies detailed in Table 5.2 cannot be directly tested. Additionally, the decision to keep the pulse resistance (R_{pulse}) constant while varying the magnitude of the charging and discharging pulses is a strategic choice. By maintaining a consistent R_{pulse}, the study can isolate the effects of pulse magnitude on temperature increase. This approach allows for a more accurate assessment of how different pulse magnitudes influence the heating efficiency without the complicating factor of varying resistance levels. This methodological consistency is essential for generating reliable and comparable data across different charge levels and temperature ranges.

Figure 5.18a illustrates a comparison between the observed and modeled temperatures for the best heating technique at 100 Hz on a battery with 50% charge. The temperature rises from −10 to 10 °C at an average rate of 4.23 °C per minute, with the experimental and theoretical temperature curves closely aligning. The accuracy of the electro-thermal interaction model in simulating the thermal behavior of the battery during heat application has been verified. Illustrations Fig. 5.18b and c showcase the computed temperature increments for distinct charge levels, covering a battery temperature spectrum from −10 °C to 10 °C. The analysis of heating power previously conducted reveals that for charge states from 0 to 40%, the temperature

Table 5.2 An overview of the most effective BPC heating methods is provided for batteries at 0% charge, 50% charge, and 90% charge levels [71]

SOC	Temperature range	f_{pulse}	R_{pulse}	A_{dis} and A_{cha}
0%	−10~−5 °C −5~0 °C 0~5 °C 5~10 °C	20 kHz	0.60	8.77 A, 13.16 A 9.56 A, 14.33 A 10.11 A, 15.16 A 10.68 A, 16.02 A
50%	−10~−5 °C −5~0 °C 0~5 °C 5~10 °C	20 kHz	0.35	13.93 A, 7.50 A 15.14 A, 8.15 A 16.39 A, 8.82 A 17.17 A, 9.24 A
90%	−10~−5 °C −5~ 0 °C 0~5 °C 5~10 °C	20 kHz	0.05	26.98 A, 1.42 A 29.10 A, 1.53 A 31.36 A, 1.65 A 33.05 A, 1.74 A

Table 5.3 A comprehensive breakdown of the ideal BPC heating approach is provided for a 50% charged battery operating at a frequency of 100 Hz [71]

SOC	Temperature range	f_{pulse}	R_{pulse}	A_{dis} and A_{cha}
50%	−10~−5 °C −5~0 °C 0~ 5 °C 5~ 10 °C	100 Hz	0.35	6.22 A, 3.35 A 6.84 A, 3.68 A 7.71 A, 4.15 A 8.27 A, 4.45 A

escalation is nearly constant, averaging at 11.28 °C per minute. However, as the battery's charge level surpasses this range, the rate of temperature rise begins to decline. At a charge level of 90%, the rate plummets to 2.88 °C per minute. Furthermore, attaining a 10 °C temperature increase using the optimal BPC heating method becomes impractical at 100% SOC, primarily due to the restrictions of the maximum permissible charging voltage. As a result, heating at 100% SOC is not exhibited in this study.

5.3.2.4 Effect of Low Temperature Heating Strategy on Battery Degradation Performance

Battery performance degradation validation for BPC heating strategy. In the context of 60 BPC thermal cycles executed based on the methodology, Fig. 5.19 provides insights into the battery's performance metrics. Specifically, it highlights the consistent maintenance of battery capacity, with negligible fluctuations observed across the entire cycle duration. Moreover, the EIS analysis indicates that the battery's impedance arc remains relatively stable, showing no substantial changes even as the cycle count rises. This finding implies that the implemented BPC heating method does not induce any discernible decline in the battery's operational efficiency.

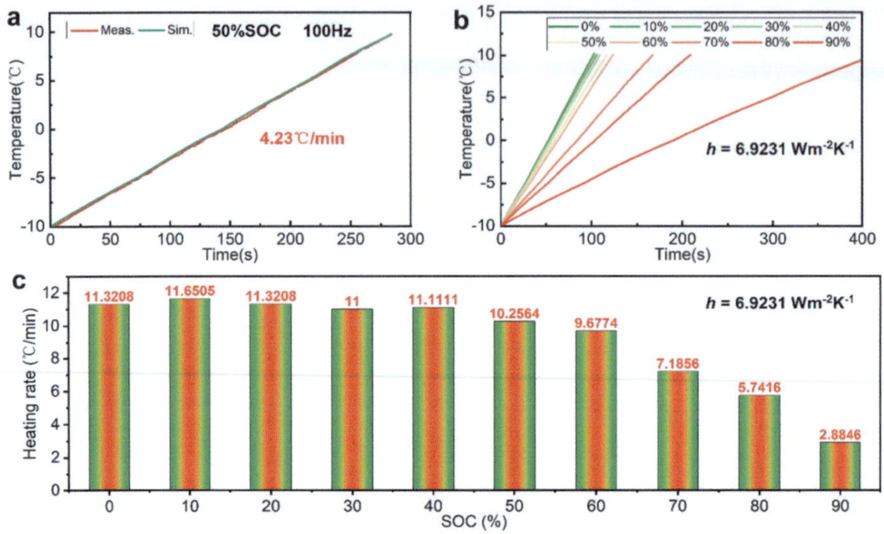

Fig. 5.18 Investigation into the ideal BPC heating tactics at diverse battery charge levels. **a** Observed and modeled temperature profiles for the top-performing heating technique at 100 Hz, with the battery at a 50% charge capacity. **b** Temperature gains resulting from the best heating strategies across various charge states. **c** The efficiency of heating, in terms of speed, for the optimal heating protocols at different charge levels [71]

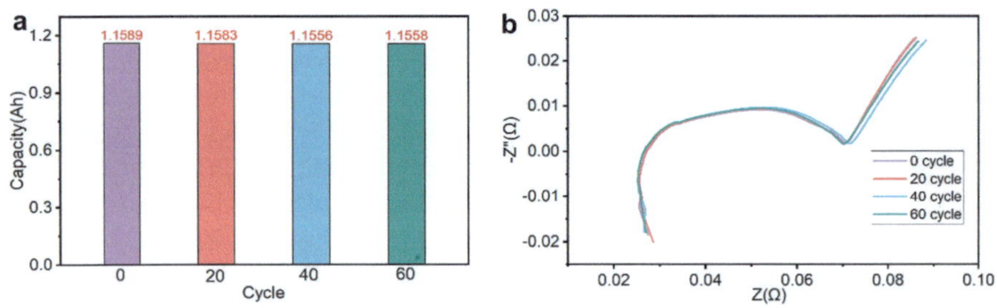

Fig. 5.19 The outcomes of the battery capacity assessment (labeled as **a**) and the EIS evaluation (labeled as **b**) throughout 60 thermal cycles [71]

In this research, as visualized in Fig. 5.20, a non-intrusive BPC thermal framework is presented. It encompasses four key elements: an integrated electro-thermal model, determination of constraints, the creation of a BPC heating strategy, and the practical applications of BPC heating. At the heart of this framework lies the electro-thermal coupled model, which is instrumental in calculating battery heat production, terminal voltage levels, and tracking lithium plating. This model is composed of a fractional-order equivalent circuit model that is integrated with a lumped thermal model to form its architecture. Additionally, it involves parameter tuning based on EIS assessments and experiments that measure temperature increases. Incorporating graphite

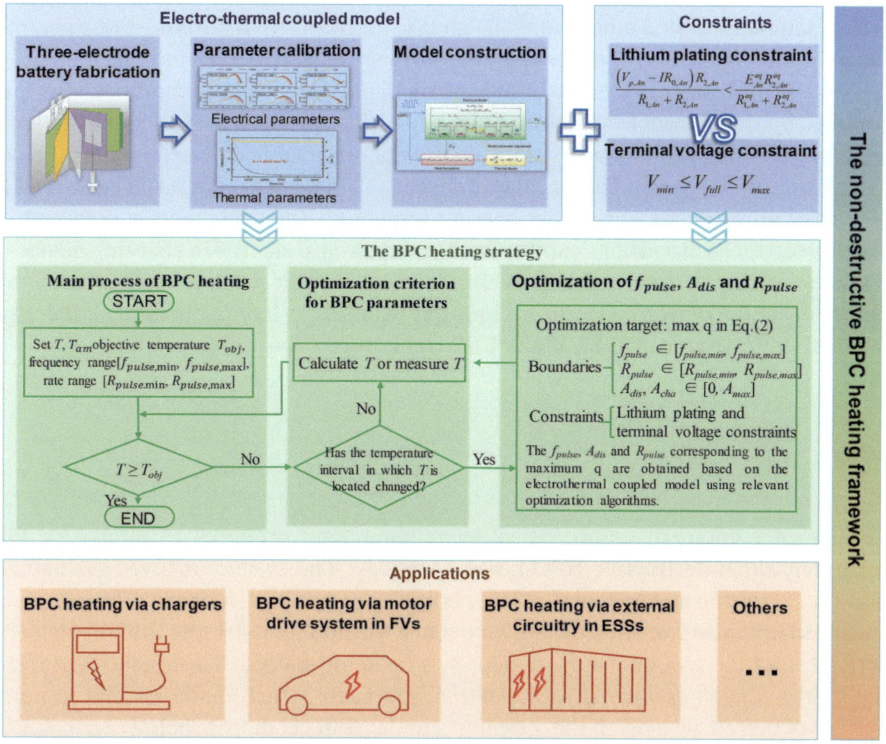

Fig. 5.20 The BPC heating system, which is non-damaging [73]

anode parameters, the model is designed with a three-electrode battery system, effectively merging electrical and thermal factors. These elements are essential for accurately simulating lithium plating behavior. Meanwhile, the framework's constraints, which focus on lithium plating and terminal voltage, serve as critical safeguards to prevent adverse effects on the battery during the heating procedure. Upon defining the electro-thermal framework and associated limitations, the BPC thermal management approach is devised. It integrates several critical components: the principal heating operation, the evaluation standards for refining BPC settings, and the fine-tuning of variables like pulse frequency, discharge area, and pulse resistance. The main heating process involves determining initial values and acceptable ranges for f_{pulse} and $R_{pulse,}$ which are based on the specific BPC generator. The ranges of f_{pulse} and R_{pulse} need to be determined based on the specific BPC generator. Heating continues until the target temperature is reached, with temperature monitored throughout. The optimization criteria assess whether the BPC parameters need adjustment, taking into account the temperature-dependent variations in battery characteristics. For example, During the thermal elevation phase from −15 to 10 °C, the BPC settings are recalibrated at each 5 °C increment. Additionally, the refinement protocol entails assessing the prevailing battery temperature and pinpointing the precise moments for modifying these parameters. In the enhancement procedure, the primary objective is to modify the f_{pulse}, A_{dis} and R_{pulse} to amplify the heat-producing capability q. These modifications are confined by the allowable spectrum of f_{pulse} and R_{pulse} values, as well as restrictions aimed at avoiding lithium deposition and maintaining stable terminal voltage. Upon completion of this optimization, the refined BPC parameters are utilized to make necessary adjustments during the heating sequence. The framework's potential is highlighted by its broad applicability in diverse settings, such as in charging devices, EVs, and energy storage solutions. This non-intrusive BPC heating method substantially boosts the performance of lithium-ion batteries at low temperatures, thereby facilitating the expansion of EV and ESS technologies into colder regions.

5.4 Conclusions

This chapter compares various heating methods and proposes a framework for designing a non-destructive BPC heating strategy. The system involves evaluating lithium deposition and endpoint voltage boundaries, together with the creation of an enhanced optimization strategy. An innovative method for detecting lithium plating, derived from the graphite anode model, has been introduced. This method exhibits superior precision compared to the earlier approach, which indicated that lithium plating occurred whenever the graphite anode voltage dipped below zero volts. Battery aging experiments validate the proposed criterion. Employing a holistic, full-cell assessment technique for detecting lithium plating, combined with restrictions based on charge and discharge threshold voltages, a thermal regulation method has been devised. This method takes into consideration the temperature-induced alterations in BPC characteristics. The lithium plating constraint is critical for designing low-frequency AC heating, while terminal voltage constraints are more relevant for high-frequency AC heating. A more detailed investigation into BPC settings and their correlation with battery thermal output reveals that the thermal energy produced escalates in tandem with the frequency of BPC heating cycles. Moreover, it has been concluded that employing a balanced BPC heating method does not yield the best results. When considering a battery at a 50% state of charge, the most efficient strategy involves an R_{pulse} value of 0.35, paired with a BPC frequency set at 100 Hz. The BPC heating framework achieves an average heating rate of 4.23 °C/min for battery temperatures ranging from −10 to 10 °C. Additionally, after 60 heating cycles, this strategy has no negative impact on battery capacity. This method's thermal output significantly outperforms strategies that ignore the pulse duration ratio during charging and discharging. In-depth theoretical examination indicates that once the battery's charge level exceeds 80%, the thermal output is amplified by over six times compared to standard conditions. Conversely, when the charge level falls below 40%, the average warming speed from −10 to 10 °C soars to 11.28 °C per minute. Even at a charge level of 90%, the warming rate stabilizes at 2.88 °C per minute. This work successfully addresses the challenges posed by low temperatures on the performance of lithium-ion batteries, providing a viable solution for low-temperature charging in EVs and ESSs. These advancements support the broader goal of promoting a green, decarbonized society.

References

1. Hu X, Zheng Y, Howey D A, et al. Battery warm-up methodologies at subzero temperatures for automotive applications: Recent advances and perspectives [J]. Progress in Energy and Combustion Science, 2020, 77: 100806–100834.
2. Zhu T, Min H, Yu Y, et al. An Optimized Energy Management Strategy for Preheating Vehicle-Mounted Li-Ion Batteries at Subzero Temperatures [J]. Energies 2017, 10(2): 243–266.
3. Wang Y, Liu P, Jin X, et al. Design and experiment of a novel stepwise preheating system for battery packs coupled with non-dissipative balancing function based on supercapacitors [J]. Journal of Energy Storage, 2023, 66(30): 107444–107461.
4. Lin C, Kong W, Tian Y, et al. Sandwich self-heating structure-based lithium-ion battery system and its application in the fuel cell bus for Beijing Winter Olympic Games [J]. Energy Convers Manage, 2023, 284: 116977–116991.
5. Ling Z, Wen X, Zhang Z, et al. Warming-Up Effects of Phase Change Materials on Lithium-Ion Batteries Operated at Low Temperatures [J]. Energy Technology, 2016, 4(9): 1071–1076.
6. Ji Y, Wang C Y. Heating strategies for Li-ion batteries operated from subzero temperatures [J]. Electrochimica Acta, 2013, 107: 664–674.
7. Lin C, Kong W, TIAN Y, et al. Heating Lithium-Ion Batteries at Low Temperatures for Onboard Applications: Recent Progress, Challenges and Prospects [J]. Automotive Innovation, 2022, 5: 3–17.
8. Wu S, Xiong R, Li H, et al. The state of the art on preheating lithium-ion batteries in cold weather [J]. Journal of Energy Storage, 2020, 27: 101059–101072.
9. Wang Y, Zhang X, Chen Z. Low temperature preheating techniques for Lithium-ion batteries: Recent advances and future challenges [J]. Applied Energy, 2022, 313: 118832–118855.
10. Piao N, Gao X, Yang H, et al. Challenges and development of lithium-ion batteries for low temperature environments [J]. eTransportation, 2022, 11: 100145–100163.
11. Pesaran A A. Battery thermal models for hybrid vehicle simulations [J]. Journal of Power Sources, 2002, 110(2): 377–382.
12. Hu X S, Zheng Y S, Howey D A, et al. Battery warm-up methodologies at subzero temperatures for automotive applications: Recent advances and perspectives [J]. Progress in Energy and Combustion Science, 2020, 77: 100806–100834.
13. Waag W, KÄBITZ S, Sauer D U. Experimental investigation of the lithium-ion battery impedance characteristic at various conditions and aging states and its influence on the application [J]. Applied Energy, 2013, 102: 885–897.
14. Jin B A, Liu F F, An Z W, et al. Thermal Simulation of Power Lithium-ion Battery Under Low Temperature and Preheating Condition [C]. Proceedings of 2020 5th International Conference on Smart Grid and Electrical Automation (ICSGEA). Zhangjiajie, China: IEEE, 2020: 51–54.
15. Abdelkareem M A, Maghrabie H M, Abo-Khalil A G, et al. Thermal management systems based on heat pipes for batteries in EVs/HEVs [J]. Journal of Energy Storage, 2022, 51: 104384–104416.
16. Wang Y B, Rao Z, Liu S C, et al. Evaluating the performance of liquid immersing preheating system for Lithium-ion battery pack [J]. Applied Thermal Engineering, 2021, 190: 116811–116821.
17. Wu S J, Xiong R, Li H L, et al. The state of the art on preheating lithium-ion batteries in cold weather [J]. Journal of Energy Storage, 2020, 27: 101059–101072.

18. Pirasaci T, Sunol A. Potential of phase change materials (PCM) for building thermal performance enhancement: PCM-composite aggregate application throughout Turkey [J]. Energy, 2024, 292: 130589–130605.
19. Guo J, Jiang F M. A novel electric vehicle thermal management system based on cooling and heating of batteries by refrigerant [J]. Energy Conversion and Management, 2021, 237: 114145–114158.
20. Li G, Huang X D, Fu X F, Design research on battery heating and preservation system based on liquid cooling mode [J]. Journal of Hunan University. Natural Sciences, 2017, 44: 26–33.
21. Wang T Z, Wu X G, Xu S B, et al. Performance of plug-in hybrid electric vehicle under low temperature condition and economy analysis of battery pre-heating [J]. Journal of Power Sources, 2018, 401: 245–254.
22. Sheng L, Zhang H Y, Su L, et al. Effect analysis on thermal profile management of a cylindrical lithium-ion battery utilizing a cellular liquid cooling jacket [J]. Energy, 2021, 220: 119725–119737.
23. Liu H Q, Chika E, Zhao J Y. Investigation into the effectiveness of nanofluids on the mini-channel thermal management for high power lithium ion battery [J]. Applied Thermal Engineering, 2018, 142: 511–523.
24. Huo Y T, Rao Z H, Liu X J, et al. Investigation of power battery thermal management by using mini-channel cold plate [J]. Energy Conversion and Management, 2015, 89: 387–395.
25. Zhao J T, Rao Z H, Li Y M. Thermal performance of mini-channel liquid cooled cylinder based battery thermal management for cylindrical lithium-ion power battery [J]. Energy Conversion and Management, 2015, 103: 157–165.
26. Zhong G J, Zhang G Q, Yang X Q, et al. Researches of composite phase change material cooling/resistance wire preheating coupling system of a designed 18650-type battery module [J]. Applied Thermal Engineering, 2017, 127: 176–183.
27. He F Q, Li X X, Zhang G Q, et al. Experimental investigation of thermal management system for lithium ion batteries module with coupling effect by heat sheets and phase change materials [J]. International Journal of Energy Research, 2018, 42(10): 3279–3288.
28. Sefidan A M, Sojoudi A, Saha S C. Nanofluid-based cooling of cylindrical lithium-ion battery packs employing forced air flow [J]. International Journal of Thermal Sciences, 2017, 117: 44–58.
29. Goli P, Legedza S, Dhar A, et al. Graphene-enhanced hybrid phase change materials for thermal management of Li-ion batteries [J]. Journal of Power Sources, 2014, 248: 37–43.
30. Li X, Xia J F, Li X, et al. Enhanced Comprehensive Performance of Polyethylene Glycol Based Phase Change Material with Hybrid Graphene Synthesized by Ball Milling [J]. Chemistry Letters, 2018, 47(6): 797–799.
31. Mills A, Farid M, Selman J R, et al. Thermal conductivity enhancement of phase change materials using a graphite matrix [J]. Applied Thermal Engineering, 2006, 26(14–15): 1652–1661.
32. Jaguemont J, Boulon L, Dubé Y. A comprehensive review of lithium-ion batteries used in hybrid and electric vehicles at cold temperatures [J]. Applied Energy, 2016, 164: 99–114.
33. Huo Y T, Rao Z H. The numerical investigation of nanofluid based cylinder battery thermal management using lattice Boltzmann method [J]. International Journal of Heat and Mass Transfer, 2015, 91: 374–384.
34. ALAOUI C. Solid state heater cooler: Design and evaluation[D]. Lowell: University of Massachusetts Lowell, 1998. ProQuest Dissertations & Theses, 1390918.
35. Liu X H, Ai W L, Marlow M N, et al. The effect of cell-to-cell variations and thermal gradients on the performance and degradation of lithium-ion battery packs [J]. Applied Energy, 2019, 248: 489–499.
36. Wang C Y, Zhang G S, Ge S H, et al. Lithium-ion battery structure that self-heats at low temperatures [J]. Nature, 2016, 529(7587): 515–518.
37. Yang X G, Zhang G S, Wang C Y. Computational design and refinement of self-heating lithium ion batteries [J]. Journal of Power Sources, 2016, 328: 203–211.

38. Zhang G S, Ge S H, Xu T, et al. Rapid self-heating and internal temperature sensing of lithium-ion batteries at low temperatures [J]. Electrochimica Acta, 2016, 218: 149–155.
39. Lei Z G, Zhang Y W, Lei X G. Improving temperature uniformity of a lithium-ion battery by intermittent heating method in cold climate [J]. International Journal of Heat and Mass Transfer, 2018, 121: 275–281.
40. Cai F Y, Chang H W, Yang Z B, et al. Experimental study on self-heating strategy of lithium-ion battery at low temperatures based on bidirectional pulse current [J]. Applied Energy, 2024, 354: 122232–122255.
41. Qu Z G, Jiang Z Y, Wang Q. Experimental study on pulse self-heating of lithium-ion battery at low temperature [J]. International Journal of Heat and Mass Transfer, 2019, 135: 696–705.
42. Wu Z P, Xu Q F, Wang J N, et al. Preparation of Large Area Double-walled Carbon Nanotube Macro-films with Self-cleaning Properties [J]. Journal of Materials Science & Technology, 2010, 26(1): 20–26.
43. Kim T, Kim Y W, Lee H S, et al. Uniformly Interconnected Silver-Nanowire Networks for Transparent Film Heaters [J]. Advanced Functional Materials, 2013, 23(10): 1250–1255.
44. Guo S S, Xiong R, Wang K, et al. A novel echelon internal heating strategy of cold batteries for all-climate electric vehicles application [J]. Applied Energy, 2018, 219: 256–263.
45. Guo S S, Yang R X, Shen W X, et al. DC-AC hybrid rapid heating method for lithium-ion batteries at high state of charge operated from low temperatures [J]. Energy, 2022, 238: 121809–121819.
46. Hu Z X B, Liu F R, Chen P, et al. Experimental study on the mechanism of frequency-dependent heat in AC preheating of lithium-ion battery at low temperature [J]. Applied Thermal Engineering, 2022, 214: 121809–121821.
47. Celle C, Mayousse C, Moreau E, et al. Highly flexible transparent film heaters based on random networks of silver nanowires [J]. Nano Research, 2012, 5(6): 427–433.
48. Hu Z X, Li Y, Liu F R, et al. Thermal Characteristics Investigation of Lithium-Ion Battery Under High-Frequency AC Excitation in Low-Temperature Environment [J]. Ieee Transactions on Transportation Electrification, 2022, 8(1): 407-419.
49. Zhang Y X, Yang Y N, Shang Y L, et al. A high frequency AC heater based on switched capacitors for lithium-ion batteries at low temperature [J]. Journal of Energy Storage, 2021, 42: 102977–102987.
50. Qin Y D, Chen X R, Tomaszewska A, et al. Lithium-ion batteries under pulsed current operation to stabilize future grids [J]. Cell Reports Physical Science, 2022, 3(1): 100708–100733.
51. Zhang J B, Ge H, Li Z, et al. Internal heating of lithium-ion batteries using alternating current based on the heat generation model in frequency domain [J]. Journal of Power Sources, 2015, 273: 1030–1037.
52. Zhu J G, Sun Z C, Wei X Z, et al. An alternating current heating method for lithium-ion batteries from subzero temperatures [J]. International Journal of Energy Research, 2016, 40(13): 1869–1883.
53. Zhang L, Fan W T, Wang Z P, et al. Battery heating for lithium-ion batteries based on multi-stage alternative currents [J]. Journal of Energy Storage, 2020, 32: 101885–101894.
54. Guo S S, Xiong R, Shen W X, et al. Aging investigation of an echelon internal heating method on a three-electrode lithium ion cell at low temperatures [J]. Journal of Energy Storage, 2019, 25: 100878–100885.
55. Ruan H J, Jiang J C, Sun B X, et al. A rapid low-temperature internal heating strategy with optimal frequency based on constant polarization voltage for lithium-ion batteries [J]. Applied Energy, 2016, 177: 771–782.
56. Zhang Z P, Yang X L, Jian J T, et al. Rapid and deposition-free preheating of lithium-ion cell with square wave current [J]. International Journal of Energy Research, 2022, 46(10): 13986–14004.
57. Li W, Xie Y, Hu X S, et al. An Internal Heating Strategy for Lithium-Ion Batteries Without Lithium Plating Based on Self-Adaptive Alternating Current Pulse [J]. Ieee Transactions on Vehicular Technology, 2023, 72(5): 5809–5823.

58. Wu X G, Cui Z H, Chen E S, et al. Capacity degradation minimization oriented optimization for the pulse preheating of lithium-ion batteries under low temperature [J]. Journal of Energy Storage, 2020, 31: 101746–101758.
59. Qin Y D, Du J Y, Lu L G, et al. A rapid lithium-ion battery heating method based on bidirectional pulsed current: Heating effect and impact on battery life [J]. Applied Energy, 2020, 280: 115957–115971.
60. Qin Y D, Xu Z C, Xiao S R, et al. Temperature consistency-oriented rapid heating strategy combining pulsed operation and external thermal management for lithium-ion batteries [J]. Applied Energy, 2023, 335: 120659–120676.
61. Qin Y D, Xu Z C, Wu Y Q, et al. Temperature distribution of lithium ion battery module with inconsistent cells under pulsed heating method [J]. Applied Thermal Engineering, 2022, 212: 118529–118548.
62. Li Y L, Gao X L, Qin Y D, et al. Drive circuitry of an electric vehicle enabling rapid heating of the battery pack at low temperatures [J]. Iscience, 2021, 24(1): 101921–101944.
63. Wang Q, Jiang B, xue Q F, et al. Experimental investigation on EV battery cooling and heating by heat pipes [J]. Applied Thermal Engineering, 2015, 88: 54–60.
64. Waldmann T, Hogg B I, Wohlfahrt-Mehrens M. Li plating as unwanted side reaction in commercial Li-ion cells - A review [J]. Journal of Power Sources, 2018, 384: 107–124.
65. Ge H, Aoki T, Ikeda N, et al. Investigating Lithium Plating in Lithium-Ion Batteries at Low Temperatures Using Electrochemical Model with NMR Assisted Parameterization [J]. Journal of the Electrochemical Society, 2017, 164(6): A1050–A1060.
66. Ge H, Huang J, Zhang J B, et al. Temperature-Adaptive Alternating Current Preheating of Lithium-Ion Batteries with Lithium Deposition Prevention [J]. Journal of the Electrochemical Society, 2016, 163(2): A290–A A299.
67. Wang B, Le Fevre L W, Brookfield A, et al. Resolution of Lithium Deposition versus Intercalation of Graphite Anodes in Lithium Ion Batteries: An In Situ Electron Paramagnetic Resonance Study [J]. Angewandte Chemie-International Edition, 2021, 60(40): 21860–21867.
68. Wang H S, Zhu Y Y, Kim S C, et al. Underpotential lithium plating on graphite anodes caused by temperature heterogeneity [J]. Proceedings of the National Academy of Sciences of the United States of America, 2020, 117(47): 29453–29461.
69. Gao T, Han Y, Fraggedakis D, et al. Interplay of Lithium Intercalation and Plating on a Single Graphite Particle [J]. Joule, 2021, 5(2): 393-414.
70. Bhandari A, Peng C, Dziedzic J, et al. Li nucleation on the graphite anode under potential control in Li-ion batteries [J]. Journal of Materials Chemistry A, 2022, 10(21): 11426–11436.
71. Huang R J, Wei G, Wang X Y, et al. A non-destructive heating method for lithium-ion batteries at low temperatures [J]. Renewable & Sustainable Energy Reviews, 2024, 205: 114868–114886.
72. Legrand N, Knosp B, Desprez P, et al. Physical characterization of the charging process of a Li-ion battery and prediction of Li plating by electrochemical modelling [J]. Journal of Power Sources, 2014, 245: 208–216.
73. Huang R J, Wei G, Zhou X Y, et al. Targeting the low-temperature performance degradation of lithium-ion batteries: A non-destructive bidirectional pulse current heating framework [J]. Energy Storage Materials, 2024, 65: 103173–103191.
74. Manikandan B, Ramar V, Yap C, et al. Investigation of physico-chemical processes in lithium-ion batteries by deconvolution of electrochemical impedance spectra [J]. Journal of Power Sources, 2017, 361: 300–309.
75. Huang R J, Wang X Y, Jiang B, et al. Revealing the electrochemical impedance characteristics of lithium-ion battery (nickel-cobalt-aluminum vs. graphite) under various alternating current amplitudes [J]. Journal of Power Sources, 2023, 566: 232929–232943.
76. Petzl M, Danzer M A. Nondestructive detection, characterization, and quantification of lithium plating in commercial lithium-ion batteries [J]. Journal of Power Sources, 2014, 254: 80–87.
77. Zhang G X, Wei X Z, Chen S Q, et al. Investigation the Degradation Mechanisms of Lithium-Ion Batteries under Low-Temperature High-Rate Cycling [J]. Acs Applied Energy Materials, 2022, 5(5): 6462–71.

78. Ren D S, Smith K, Guo D X, et al. Investigation of Lithium Plating-Stripping Process in Li-Ion Batteries at Low Temperature Using an Electrochemical Model [J]. Journal of the Electrochemical Society, 2018, 165(10): A2167–A2178.
79. Xie Y, Li W, Song Z Y, et al. A health-aware AC heating strategy with lithium plating criterion for batteries at low temperatures [J]. Ieee Transactions on Industrial Informatics, 2024, 20(2): 2295–2306.

Chapter 6
AC Heating System Implementation

Abstract As energy demands and environmental concerns continue to rise, new energy vehicles powered by lithium-ion batteries (LIBs) have emerged as a major trend in the automotive industry. With their high operating voltage, large energy density, long cycle life, and fast charging capabilities, LIBs have become a crucial energy source for electric vehicles (EVs). Beyond transportation, these batteries also hold significant potential for applications in portable electronics, military equipment, and aerospace technologies. Despite their numerous advantages, LIBs face considerable challenges in low-temperature environments. Issues such as reduced charging efficiency, increased internal resistance, and the potential formation of lithium dendrites can compromise both performance and safety, ultimately limiting their applicability. To address these challenges, effective thermal management is essential for maintaining battery longevity, ensuring operational safety, and optimizing overall energy efficiency. Among various thermal management strategies, AC heating technology has emerged as a promising solution due to its high efficiency and effectiveness. Building upon the theoretical foundations established in Chaps. 4 and 5, which explored the electro-thermal coupling model of batteries, existing heating methods, and the limitations of low-temperature heating, this chapter shifts the focus toward the practical implementation of AC heating systems. A comprehensive review of current AC heating solutions is presented, categorizing them into energy injection-based and non-energy injection-based approaches. Using an integrated charger design as a case study, a detailed examination of system topology, control strategies, and simulation processes is provided, with a particular emphasis on battery-integrated AC heating technology. This discussion offers valuable insights into the thermal management performance of LIBs in EVs and other energy storage applications. Additionally, an AC self-heating solution is introduced for integrated external circuits, motor drive systems, and boost-type power battery modules. Key aspects such as system hardware design, critical parameters, and implementation strategies are analyzed, along with a step-by-step application guide. The practical feasibility of these technologies is demonstrated, reinforcing the role of AC heating in maintaining battery performance, safety, and lifespan in low-temperature conditions. By bridging theoretical modeling with real-world technical applications, this chapter

J. Zhu et al., *Alternating Current (AC) Heating for Lithium-Ion Batteries in Electric Vehicles*, https://doi.org/10.1007/978-981-96-9071-8_6

provides a structured approach to system verification and engineering implementation. The discussion of AC heating solutions not only enhances the understanding of thermal management in LIBs but also lays the groundwork for future advancements in energy storage and electric vehicle technology.

Keywords AC heating system · Practical implementation · Topology · Simulation · Experimental verification

6.1 Current AC Heating Solutions

Due to the inherent poor charge–discharge characteristics of LIBs at low temperatures and the instability at high temperatures, battery management systems in new energy vehicles, such as pure EVs and hybrid EVs, often include battery thermal management functions. A typical thermal management system includes both low-temperature heating and high-temperature cooling functions.

To improve battery performance at low temperatures and enhance the charge–discharge capabilities of LIBs in cold environments, the low-temperature heating system for LIBs mainly uses traditional external heat conduction methods. This typically involves adding one or more high-temperature heat sources to the exterior of the battery pack(BP) or battery module, such as installing heating devices, setting up high-temperature liquid flow channels or introducing high-temperature air. These high-temperature heat sources transfer internal energy to the low-temperature LIBs through heat conduction, thus achieving external-to-internal low-temperature heating for the LIB.

These traditional heating methods generally have certain issues, making it difficult to meet the automotive requirements for convenience, quick heating, simple layout, and low cost in terms of heating effects, device cost, and heating time. In recent years, with the rapid development of the new energy vehicle industry, some new low-temperature heating methods for LIBs have emerged that offer better heating effects and shorter heating times. Among these, AC heating has attracted attention due to its high efficiency.

To advance the application of AC heating in vehicles, enhance the charging and discharging capacity of LIBs at low temperatures and improve their overall performance, researchers have proposed a variety of implementation schemes. These schemes can be classified into two categories based on whether external energy injection is involved: heating solutions with energy injection and heating solutions without energy injection.

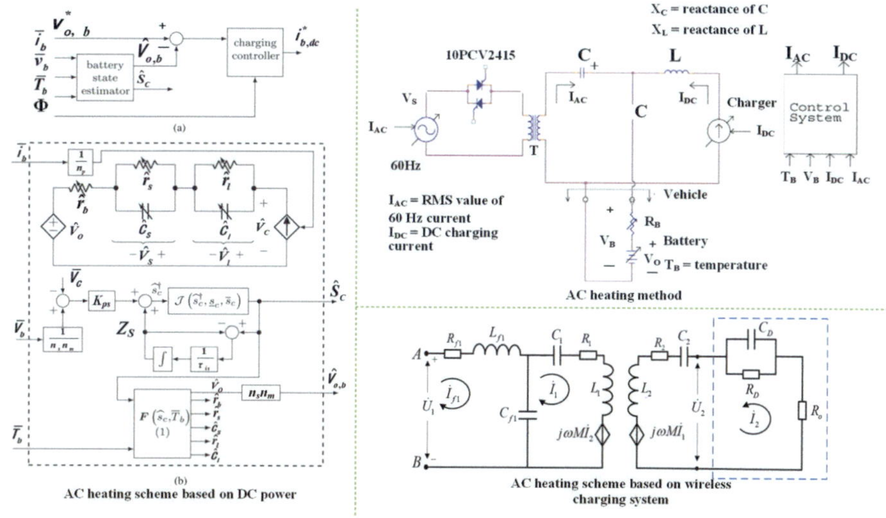

Fig. 6.1 Heating solutions with energy injection [1, 3, 4]

6.1.1 Heating Solutions with Energy Injection

As shown in the Fig. 6.1, in the category of solutions involving external energy injection, Stuart et al. [1, 2] were the first to design an AC heating method for onboard BPs using an AC power source. They successfully heated batteries to suitable temperatures under −40 °C ambient conditions.

In contrast to the use of AC power sources, Guillen Montenegro et al. [3] proposed an AC heating scheme based on DC power. By incorporating an H-bridge between the DC charging port and the BP, they integrated AC heating functionality into the charging topology.

Beyond wired charging-based AC heating solutions, Xiong et al. [4] improved the wireless charging system using a series-compensated topology with inductors and dual capacitors. They developed a novel AC heating scheme that was capable of heating the battery from −20 to 0 °C within 142 s. The discharge efficiency of the BP is relatively low during the use of this solution, which severely impacts the power, lifespan, and safety of the BP.

6.1.2 Heating Solutions Without Energy Injection

In addition to charging scenarios, EVs primarily operate without external power sources. Consequently, solutions that do not rely on external energy injection have garnered significant attention.

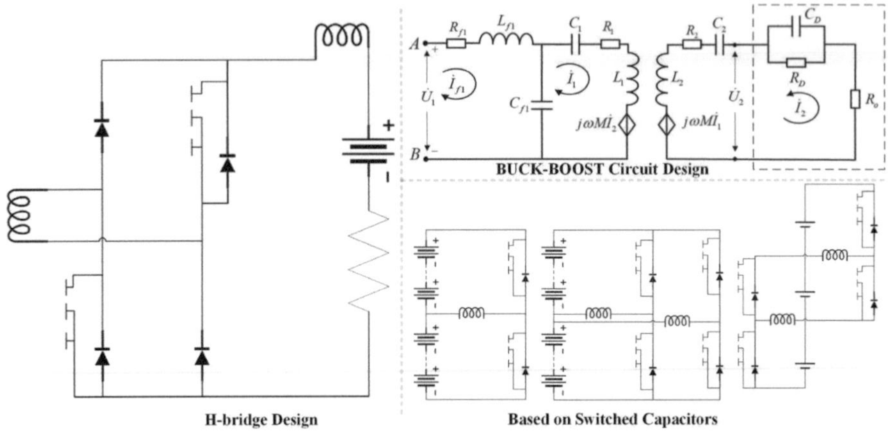

Fig. 6.2 Solutions with topological circuits design and energy storage components [4, 8, 10]

As shown in the Fig. 6.2, these solutions typically involve designing topological circuits and energy storage components around the BP to enable energy exchange between the BP and the storage components, thereby achieving AC heating. The main energy storage components in these solutions include inductors, capacitors, and LC resonance circuits [5]. Shang et al. [6] utilized inductors as energy storage devices and designed a BUCK-BOOST circuit to achieve high-frequency self-heating for onboard BPs. Building on this, they developed an interleaved parallel BUCK-BOOST circuit, which not only provided self-heating functionality but also featured balancing capabilities [6, 7]. Hu et al. [8] designed an H-bridge to control the direction of inductor current flow, enabling flexible control of self-heating frequency and amplitude. This solution achieved heating of the battery from −20 to 5 °C within 7.33 min. Due to the limited energy storage of inductors, Zhang et al. [9] proposed an AC heating circuit based on switched capacitors, which could heat the battery from −20 to 0 °C in 2.2 min.

However, whether using inductors or capacitors, the AC heating frequency is relatively high. Excessively high heating frequencies reduce the internal resistance of the battery, increasing the heating current amplitude and resulting in higher circuit energy consumption. Furthermore, when using inductors and capacitors as energy storage devices, heating topologies are restricted to hard-switching modes, leading to high peak voltages and switching losses in power devices such as MOSFETs. To address these issues, heating designs based on LC resonance circuits have attracted considerable interest. Shang et al. [10, 11] designed several high-frequency sinusoidal AC heating solutions using LC resonance circuits, allowing power switches to operate in soft-switching mode. Similarly, Jiang et al. [12] and Zhu et al. [13] employed H-bridges to control LC resonance circuit currents, enabling low-temperature heating for batteries.

Although the aforementioned methods can effectively heat BPs, the additional heating circuits increase both cost and space requirements. As shown in the Fig. 6.3, in recent years, integrating heating circuits into motor drive systems has emerged as a novel solution. Li et al. [14] modified the circuit arrangement of battery modules and designed a low-frequency square-wave AC self-heating solution based on motor inductance. This method could heat the battery from −7 to 29.1 °C within 300 s. Similarly, Du et al. [15] utilized motor inductance as the energy storage device to design a high- frequency AC heating solution, successfully heating the BP from −30 to −10 °C within 400 s. Zhu et al. [16] restructured the motor inverter to enable energy exchange between the BP and the DC-side capacitor by controlling the inverter switches, achieving self-heating functionality. To enhance heating efficiency, Liu et al. [17] divided the BP into two sections and achieved energy transfer between these sections using the motor drive circuit. This method achieved a temperature rise rate of 3.871 °C/min at −20 °C.

6.2 AC Heating Solution for Integrated Charger Design

In this section, we will focus on discussing an innovative AC heating solution, specifically how this solution can be integrated into charger designs.

To enhance the charge–discharge performance of vehicle-mounted LIBs in low-temperature environments, this design proposes a prototype system based on a full-bridge topology structure and constructs a complete system model to verify the effectiveness of the control strategy. The system is capable of performing AC heating on the lithium-ion(Li-ion) battery module, providing technical support for efficient preheating of batteries in low-temperature environments.

In order to thoroughly investigate the heating effect of AC excitation on Li-ion batteries, a comprehensive simulation model was developed. This model integrates internal battery mechanism analysis and heat transfer characteristics, enabling accurate simulation of the battery's thermal response under different operating conditions. The model is not only suitable for temperature regulation during vehicle operation but can also be applied in battery replacement or preheating scenarios, ensuring system stability and applicability in various environments.

The prototype system uses a current closed-loop control strategy, capable of outputting a sine AC current with a frequency range of 20–200 Hz and a maximum amplitude of 10 A. When connected to an external battery, the ripple of the output current is controlled to below 5%, with an error of approximately 1%, ensuring high-precision current control capability. Experimental results show that the system can raise the temperature of a 18650 battery from −20 to 6 °C within 20 min, effectively addressing the issue of battery performance degradation in low-temperature conditions. Even in extreme environments below freezing, the system can quickly elevate the battery temperature to near zero, significantly improving the battery's usability and charge–discharge efficiency in low-temperature environments.

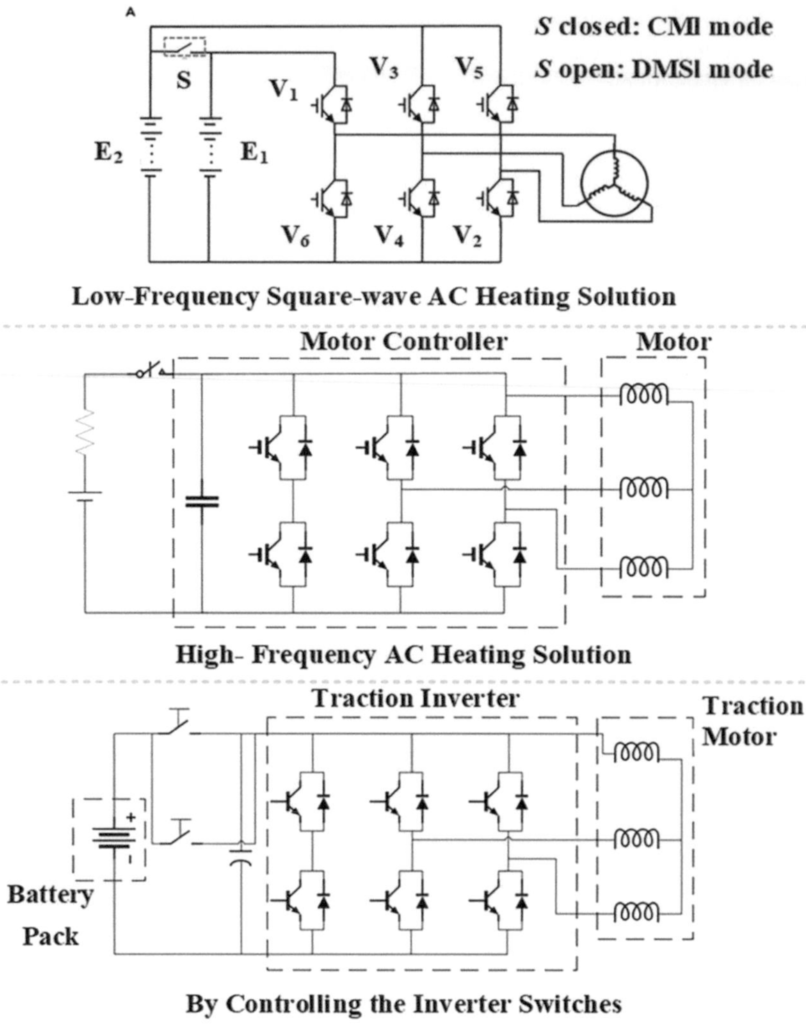

Fig. 6.3 Solutions by integrating heating circuits into motor drive systems [14–16]

Furthermore, the system demonstrates excellent performance in closed-loop current control, with a control response time of only 70 ms, greatly enhancing the precision and dynamic response speed of current regulation. The system supports real-time adjustments to the output current's amplitude and frequency, making it adaptable to different types of Li-ion batteries and various excitation strategies, achieving flexible and efficient heating control. The application of this technology not only optimizes the low-temperature performance of batteries but also provides a reliable solution for temperature control management in new energy vehicles and energy storage systems.

6.2.1 *Topology of the Innovative AC Heating Solution*

To generate a variable excitation current, common circuit topologies include bridge circuits and buck-boost chopper circuits.

1) Full-Bridge Circuit Topology

As shown in Fig. 6.4, this is a traditional full-bridge circuit, where the input is a constant voltage source, which in this case U_{IN} is a 36 V power input. The output is a purely resistive load, and the circuit uses inductors as energy storage components, with the current direction and magnitude controlled by the alternating on/off switching of four switches.

However, in AC excitation heating applications, the traditional full-bridge circuit has a problem: its output is not a purely resistive load, but rather an active load. When the input is a single-output power source (i.e., the power source cannot discharge), the inductor in the circuit does not reduce the current during the discharge cycle. Therefore, an additional energy storage component C_1, is required to store the energy released from the battery side.

Furthermore, the circuit can only charge the battery when the input voltage is higher than the output voltage.

2) Buck-Boost Circuit

Figure 6.5 shows a typical two-potential bidirectional Buck-Boost circuit, which stores energy through an inductor when switch V_1 is on and V_2 is off, and releases energy when switch V_2 is on and V_1 is off. By appropriately controlling the alternating on/off switching of the switches, this circuit can achieve bidirectional control of the output and input voltages, thereby generating a variable excitation current.

However, this solution shares a similar drawback with the full-bridge circuit topology: it can only achieve bidirectional current control when the input allows current to flow. If the input does not accept reverse current, the battery discharge

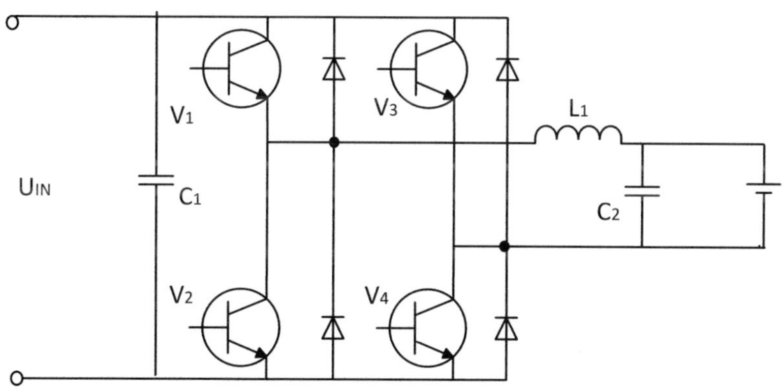

Fig. 6.4 Full-bridge circuit topology [18]

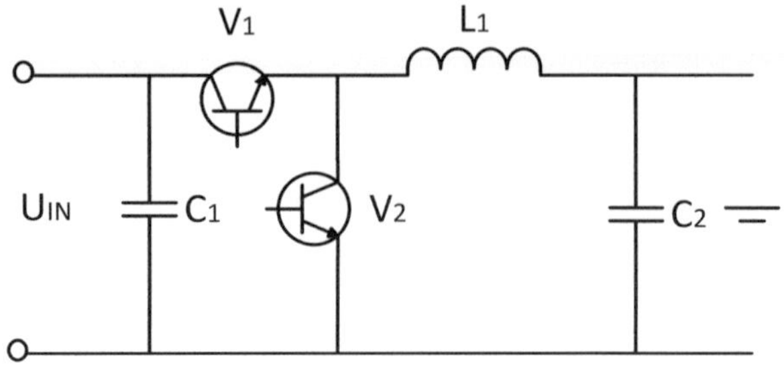

Fig. 6.5 Buck-boost circuit topology [18]

cycle cannot proceed. Therefore, an additional energy storage component C_1, is also needed at the input side.

Similarly, the circuit can only charge the battery when the input voltage is higher than the output voltage.

Both methods theoretically achieve the AC excitation heating function required in this design. However, each has different advantages and disadvantages, as shown in Table 6.1.

Based on the above analysis, this design will adopt the full-bridge circuit as the basic topology for the circuit, for the following reasons:

1) In the application of onboard LIBs in vehicles, due to the large capacity and high rated voltage of the BP, the AC excitation heating system needs to handle high current and high power situations. Under these conditions, since the Buck-Boost circuit works with only one switch transistor active at a time, it places higher performance requirements on the switch transistors. On the other hand, the full-bridge circuit operates with two switch transistors working simultaneously, so the voltage across each switch transistor is only half of that in the Buck-Boost circuit under the same operating conditions. Therefore, when selecting switch transistors, the full-bridge circuit has an advantage. In actual operation, the lifespan of the switch transistors in the full-bridge circuit is also longer than in the Buck-Boost circuit.

Table 6.1 Analysis of advantages and disadvantages of full-bridge circuit and buck-boost circuit

	Full-bridge circuit topology	Buck-boost circuit
Advantage	1. Most onboard chargers have a full-bridge topology, making it easy to integrate with the charger 2. During operation, the switching transistors are subjected to relatively low switching voltage and current	1. The circuit requires fewer components, making it cost-effective and simple in structure
Disadvantage	1. The circuit requires more components, resulting in higher costs and a more complex structure	1. This topology is rarely used in chargers 2. During operation, the switching transistors need to withstand higher switching voltages and currents

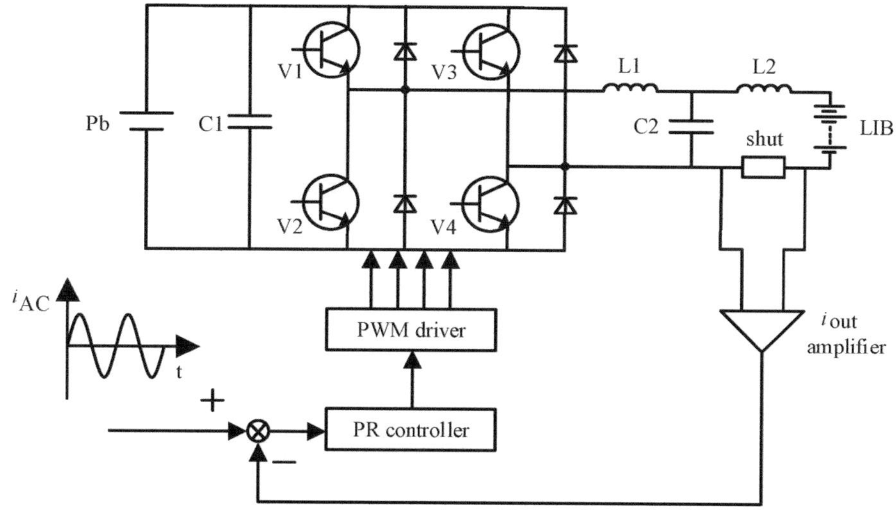

Fig. 6.6 The structure of the circuit system [18]

2) The full-bridge circuit is much more commonly used in onboard chargers than the Buck-Boost circuit. Using the full-bridge circuit as the main circuit topology will allow the AC excitation heating system to integrate better with the onboard charger.

The circuit system structure is illustrated in Fig. 6.6.

6.2.1.1 Topological Structure and Hardware Design

In this design, the main circuit topology utilizes the full-bridge circuit. The control system primarily consists of an output current sampling circuit, an A/D conversion circuit, a full-bridge driver module, a communication module, and an MCU incorporating data storage and closed-loop control. The selected MCU is the DSP digital signal processor TMS320C2812 from Texas Instruments. This study utilizes a four-way isolated power supply and driver IC for separate isolated driving. To convert the 3.3 V amplitude PWM wave into a 12 V amplitude with driving capability, the PWM wave needs to be isolated and output. The isolation chip used is the TLP250 optocoupler model produced by TOSHIBA Company.

The sampling circuit is used to collect the current at the battery end, including both the magnitude of the current and the direction of current flow. The sampling circuit includes an operational amplifier circuit, which is used to convert the large current signals into analog signals within an appropriate range. The purpose of collecting the output current is twofold: first, to monitor whether the current output by the circuit meets the required value; second, to use the detected current value as an input to the closed-loop control system for real-time regulation of the circuit's charge and discharge currents.

The output from the sampling circuit is an analog signal that is difficult for the MCU to process directly. Therefore, an A/D conversion module is required to convert the analog signal into a digital signal, allowing the MCU to process the data. During the design process, based on the required control response speed and precision, parameters such as the resolution and conversion speed of the A/D module need to be carefully selected.

The communication module is used to exchange information between the device and external modules, in order to obtain real-time battery data, such as the battery module's current temperature, voltage, fault information, etc. This helps determine when to start heating, stop heating, or adjust the output excitation current parameters. During the design process, the needs of external communication must be considered in order to determine the type and number of communication modules to be configured.

The full-bridge driver module is a device that converts the PWM control signal output by the MCU into a driving signal capable of controlling the full-bridge circuit. It enables the MCU to control the switching of each transistor in the full-bridge circuit and is a critical component of the full-bridge circuit.

6.2.1.2 Operating Mode Analysis

Firstly, this design uses a fixed frequency and changes the duty cycle to control the output current. The advantage of this control method is that the working frequency is stable, making it easier to select and design electromagnetic interference filters. Additionally, the control algorithm is relatively simple. Based on the charging and discharging situation within one cycle and the variation in current, the operation process of the device can be divided into four modes, as shown in Table 6.2.

In these modes, the direction of the current flowing through the inductor during charging is considered positive, i.e., when the current flows into the positive terminal of the output battery, the current direction in the inductor is positive.

It is important to note that due to the high power and large current requirements of the AC excitation heating device, the inductance required on the output side will be large in practical applications. To reduce the burden on selecting inductors, the design of the output inductor-capacitor structure is similar to an LCL filter circuit. The benefit of using this structure is that by connecting inductors in series, each inductor's inductance can be reduced, preventing the situation where a single inductor becomes

Table 6.2 AC excitation heating hardware circuit input and output target parameters

	$i \geq 0$	$i < 0$
$\frac{di}{dt} \geq 0$	Mode 1	Mode 3
$\frac{di}{dt} < 0$	Mode 2	Mode 4

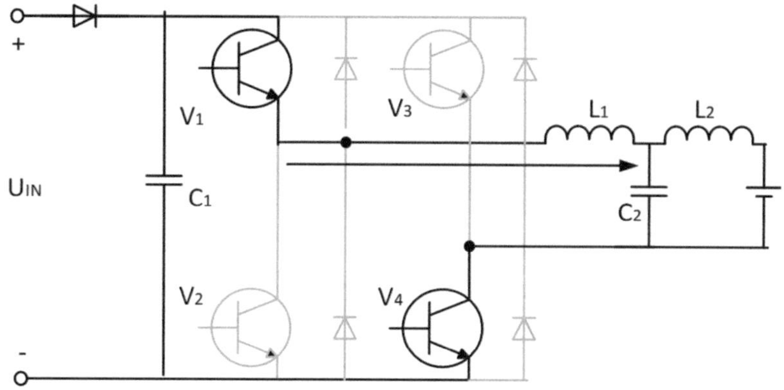

Fig. 6.7 Mode 1

too large in volume. Additionally, this structure helps minimize the current ripple on the output side under a given total inductance value, which not only reduces the working noise of the inductors but also improves the circuit's output efficiency to some extent. The detailed design and parameter selection for this structure will be explained later.

1) Mode 1

In this mode, the battery is in the charging state, and the current is gradually increasing. As shown in Fig. 6.7, switches 1 and 4 are closed, while switches 2 and 3 are open. The current flows out of the positive terminal of the input side, passes through the inductor, and flows into the positive terminal of the output battery. After flowing out of the negative terminal, the current returns to the negative terminal of the input side. In this mode, the inductor stores energy in the forward direction, and the current gradually increases as the inductor stores energy. It is important to note that the input voltage must be designed to be higher than the voltage across the output battery for charging to occur.

2) Mode 2

In this mode, the battery is in the charging state, and the current is gradually decreasing. As shown in Fig. 6.8, switches 1 and 4 are open, while switches 2 and 3 are closed. The input side is disconnected from the circuit, and the inductor, input side capacitor, and output battery form a series loop. At this point, the inductor continues to discharge in the forward direction. Since the voltage difference across the inductor is opposite to the current flowing through it, the inductor starts releasing energy, which is discharged to the battery and input-side capacitor. Consequently, the current gradually decreases.

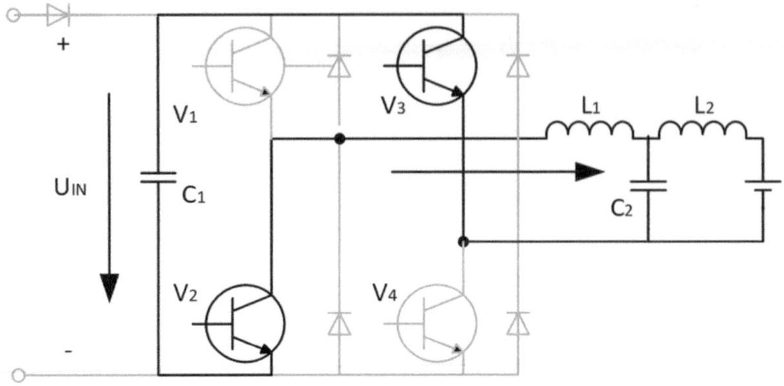

Fig. 6.8 Mode 2

3) Mode 3

In this mode, the battery is in the discharging state, and the current is gradually increasing. As shown in Fig. 6.9, switches 1 and 4 are open, while switches 2 and 3 are closed. The input side is disconnected from the circuit, and the inductor, input-side current, and output battery form a series loop. At this point, the voltage across the inductor is in the same direction as the current, and the inductor stores energy in the negative direction. The current flows out of the positive terminal of the battery, passes through the inductor, and returns to the negative terminal of the battery. The battery releases current, and the current gradually increases in the negative direction.

4) Mode 4

In this mode, the battery is in the discharging state, and the current is gradually decreasing. As shown in Fig. 6.10, switches 1 and 4 are closed, while switches 2

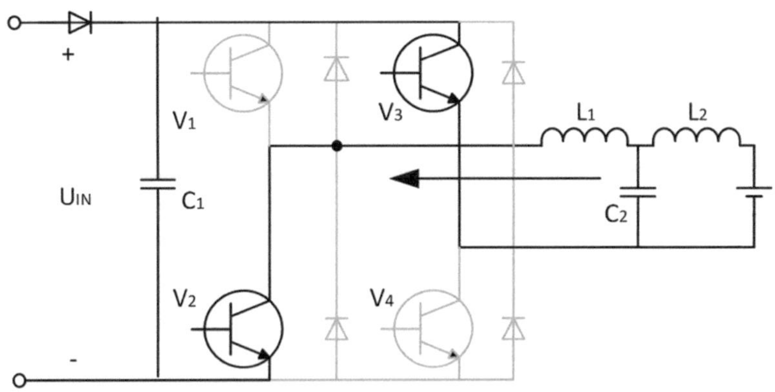

Fig. 6.9 Mode 3

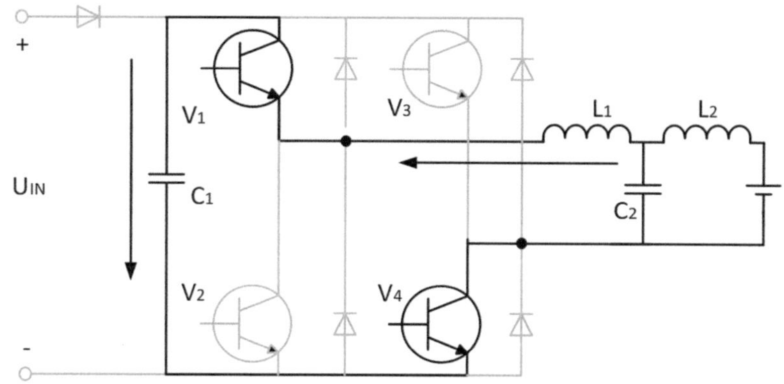

Fig. 6.10 Mode 4

and 3 are open. The current flows out of the positive terminal of the output battery, passes through the inductor, and flows into the positive terminal of the input capacitor. After flowing out of the negative terminal of the capacitor, the current returns to the negative terminal of the battery. In this mode, both the inductor and the battery release energy to the capacitor. Since the voltage across the capacitor is always higher than the battery voltage due to the influence of the input voltage, the voltage across the inductor is opposite to the direction of current flow, causing the current to gradually decrease in the circuit.

From the above operating states, it is clear that the alternating operation of Modes 1 and 2 can control the increase and decrease of current during the charging phase. By reasonably controlling the duty cycle, i.e., the switching times of switches 1, 4 and 2, 3, specific waveforms can be generated during the charging cycle. Similarly, the alternating operation of Modes 3 and 4 can control the current waveform during the discharging phase.

6.2.1.3 Full-Bridge Topology Modeling

To investigate the frequency response characteristics of the system and provide guidance for control design, this study adopts the state-space method to model the full-bridge topology. The modeling process takes into account the presence of parasitic parameters such as inductance, the output-side DC capacitor, and diodes in the switches. The system equivalent circuit used for state-space modeling is depicted in Fig. 6.11.

Because the two H-bridges diagonally opposite the full bridge always switch on and off simultaneously, the current i_{L1}, varies with a period of T, while the current i_1, varies with a period of 0.5 T. Therefore, the analysis focuses on the current variation within the 0.5 T, period to establish the state-space equations for the system. According to Kirchhoff's law, in the system topology depicted in Fig. 6.11 the following Eq. (6.1) holds:

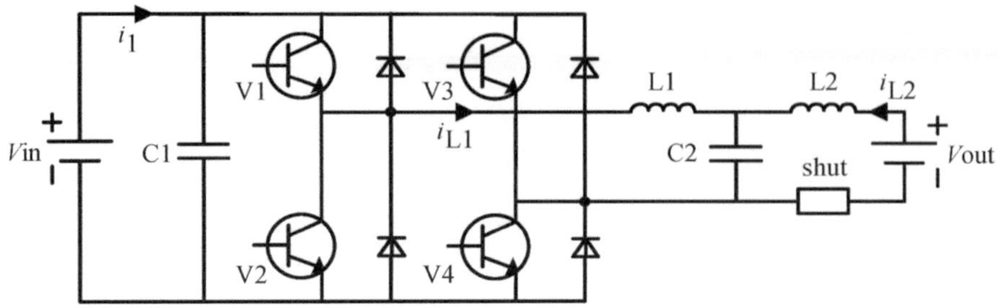

Fig. 6.11 The circuit used for state-space system modeling [18]

$$\begin{cases} u_C = e + z i_{L2} + L_2 \dfrac{di_{L2}}{dt} \\ i_{L1} = C \dfrac{du_C}{dt} + i_{L2} \\ u_S = \mathrm{u_C} + \mathrm{L_1} \dfrac{di_{L1}}{dt} \end{cases} \tag{6.1}$$

where u_c represents the voltage across the filtering capacitor, e denotes the voltage across the output load, z signifies the load impedance, and u signifies the output voltage of the H-bridge.

The variable u_s represents the switching function, which assumes different values depending on the states of the switches, as illustrated in Eq. (6.2).

$$u_S = \begin{cases} 1 \ , & V_1 \ and \ V_4 \ are \ connected \\ -1 \ , & V_2 \ and \ V_3 \ are \ connected \end{cases} \tag{6.2}$$

Define i_{L1}, i_{L2} and u_c as state variables, with the output-side battery voltage treated as an external disturbance, resulting in the following state-space Eqs. (6.3) for the system.

$$\begin{cases} \dot{x} = \begin{bmatrix} \frac{di_{L1}}{dt} \\ \frac{di_{L2}}{dt} \\ \frac{du_C}{dt} \end{bmatrix} = \begin{bmatrix} 0 & 0 & -\frac{1}{L_1} \\ 0 & -\frac{z}{L_2} & \frac{1}{L_2} \\ \frac{1}{C} & -\frac{1}{C} & 0 \end{bmatrix} \begin{bmatrix} i_{L1} \\ i_{L2} \\ u_C \end{bmatrix} + \begin{bmatrix} \frac{1}{L_1} \\ 0 \\ 0 \end{bmatrix} u_S + \begin{bmatrix} 0 \\ -\frac{1}{L_2} \\ 0 \end{bmatrix} e \\ y = \begin{bmatrix} 0 & 1 & 0 \end{bmatrix} \begin{bmatrix} i_{L1} \\ i_{L2} \\ u_C \end{bmatrix} \end{cases} \tag{6.3}$$

Equation (6.3) describes a system with time-invariant coefficients, where the time-varying quantity is the output voltage of the H-bridge, denoted as the control input u. Within one switching cycle, the average value $\overline{\mathrm{u}}$ of u can be given by Eq. (6.4)

$$\overline{u} = ki_s \tag{6.4}$$

where i_s represents the specified reference current control input and k denotes the amplification factor of the reference input after PWM by the H-bridge. For an H-bridge with PWM producing bipolar outputs, k is the ratio of the H-bridge bus voltage to the triangular wave used for PWM, reflecting the H-bridge's ability to "amplify" the given signal.

The transfer function from control input to output current is given by Eq. (6.5)

$$I_{L2}(s) = \frac{\mathrm{kI_s}(s) - (C_{L1}s^2 + 1)E(s)}{CL_1L_2s^3 + CL_1Zs^2 + (\mathrm{L}_1 + L_2)s + Z} \tag{6.5}$$

For a battery, its external characteristics can be approximated using a first-order RC equivalent circuit model. The equivalent capacitance in this model typically has a very large value, and the parallel RC section significantly affects the low-frequency characteristics or the system. At high frequencies, the analysis can approximate the battery as a resistive load with $Z = 0.1\ \Omega$.

6.2.1.4 Determination of Main Parameters

The selection parameters for each component are presented in Table 6.3.

Table 6.3 AC heating hardware circuit input and output target parameters [18]

Component	Model	Parameter
Switching devices	IRF540N MOSFET	f_H-*bridge* = 10 kHz
The filter circuit at the output end	LCL filter	Input voltage $Ua = 36$ V Output current $Is = 10$ A Switching frequency $fs = 10$ kHz Rated output voltage $Us = 20$ V
Input-side inductance	–	$L1 = 300$ μH
Input filter capacitor	–	$C = 470$ μF
Output-side inductance	–	$L2 = 75$ μH
Input capacitance	–	$V = 75$ V $Ci = 1000$ μF

6.2.2 Control and Simulation of the AC Heating Solution

6.2.2.1 System Control Strategy

This design presents a closed-loop control strategy employing the PR control algorithm to regulate the output current. The designed AC heating system demonstrates a notable DC voltage bias in the output load. To tackle this issue, an integration link is integrated at low frequencies to amplify the amplitude gain of the quasi-PR controller and adequately mitigate the influence of the DC voltage. The improved PR controller is formulated as Eq. (6.6):

$$G(s) = K_P + \frac{2K_R\omega_C s}{s^2 + 2\omega_C s + \omega_0^2} + \frac{K_I}{s} \tag{6.6}$$

where K_p represents the proportional term coefficient, K_R denotes the resonance term coefficient, ω_0 stands for the resonance frequency, ω_c is an adjustable parameter that influences both the gain of the PR controller and the bandwidth of the controller cutoff frequency, and K_I indicates the integral term coefficient.

The continuous quasi-PR controller is discretized using the bilinear transformation method. Through the application of this transformation and leveraging the connection between the Z-domain transfer function and the difference equation, the difference equation can be derived as (6.7)

$$y(k) = \frac{1}{3}\left[(K_P + b_0 + c_0)x(k) + c_0 x(k-1) - b_0 x(k-2) + (2y - a_1)y(k-1) - a_2 y(k-2)\right] \tag{6.7}$$

where the parameters ω_0, ω_c, T, K_P and are predetermined, and the values of b_0, a_1 and a_2 can be calculated directly, In the digital controller, discrete quasi-PR control is realized using a difference equation. Considering the H-bridge gain, the capacitance and inductance values are substituted and incorporating the improved quasi-PR controller yields the open-loop and unit negative feedback Bode plots of the system under $K_P = 0.01$, $K_R = 0.5$, $K_I = 0.5$ and $\omega_c = 1$, as shown in Fig. 6.12. It is observed that as the output frequency of the system increases the gain at the specified frequency is greater than 0. At low frequencies, there is a gain of approximately 25 dB. This indicates that the improved quasi-PR controller meets the output characteristic requirements of the AC excitation device.

6.2.2.2 System Simulation

In this design, simulations were conducted using MATLAB R2019b/Simulink, and a complete circuit model was constructed, including the full-bridge circuit and output filtering circuit. The selected components, such as the switching devices in the full-bridge circuit and the filtering circuit, were built according to the specified models

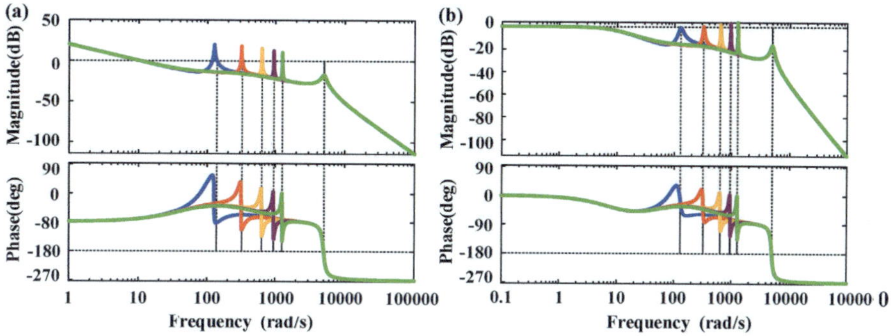

Fig. 6.12 **a** The open-loop Bode plot; **b** The unit negative feedback Bode plot of the system using the improved quasi-PR controller [18]

and parameters in the design. The detailed data for all switches and diodes were sourced from the official data manuals published by IR Corporation (headquartered in Los Angeles, California, USA), and the parasitic components in the model were configured based on the parameters provided in these manuals.

In this simulation module, a PR algorithm was used to calculate the duty cycle of each PWM. The calculation of the duty cycle is based on the error between the current circuit current and the expected current value, with dynamic adjustments made through the PR algorithm to precisely control each PWM modulation signal. The PWM module then generates the corresponding duty cycle PWM signal based on the output from the PR control module, controlling the switching devices in the circuit to turn on or off, ensuring precise current regulation and control.

The main circuit topology of the simulation model includes the full-bridge circuit, LCL filtering circuit, and output battery module. The full-bridge circuit is primarily responsible for energy conversion and current control, the LCL filtering circuit effectively filters out high-frequency noise, ensuring smooth and stable output current, while the output battery module is connected to the battery system for energy storage and output. Through this integrated simulation design, the feasibility of the control strategy and circuit design can be effectively verified, providing a theoretical basis for practical engineering applications.

For the convenience of setting operations, the main input and output parameters of the designed hardware circuit are summarized in Table 6.4.

To verify the circuit's ability to generate excitation currents within the specified range—particularly in terms of output current amplitude and frequency—relevant parameter adjustments were conducted. According to the design specifications, the output current amplitude was ensured to meet the required value while keeping its frequency within the designated range. Through simulation analysis, the circuit's output capability and characteristics under the set limit conditions were validated.

Figure 6.13 presents the simulated system output current waveforms at frequencies of 20, 50, 100, and 200 Hz, with a peak amplitude of 10 A. When the excitation frequency was set to 20 Hz, the simulation system reached a stable output current

Table 6.4 AC heating hardware circuit input and output target parameters. [18]

Target parameters	Value and unit
Input voltage	36 V
Output voltage	25 V
Maximum output current value	10 A
Output current frequency range	20~200 Hz
Output current waveform	Sine
Switching frequency	10 kHz

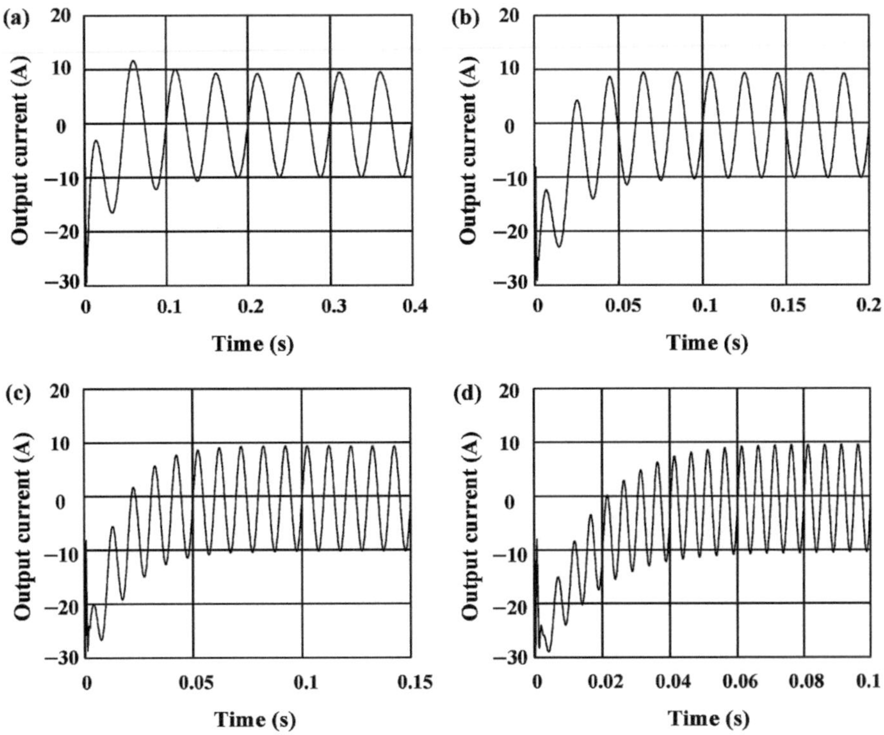

Fig. 6.13 System simulation output current wave forms of **a** 20, **b** 50, **c** 100, and **d** 200 Hz at 10 A [18]

state within 0.1 s. As the frequency increased to 50 Hz, the system's stabilization time significantly reduced to approximately 0.06 s. Further increasing the frequency to 100 and 200 Hz maintained the stabilization time at around 0.06 s, demonstrating the system's ability to respond rapidly and maintain stable operation even at higher frequencies.

Notably, at 20 Hz, the system achieved stable current output within the first three cycles, indicating a fast response capability that allows it to quickly adapt to load variations and provide stable output. Under different frequency settings, the simulation results confirmed that the system effectively achieved optimal waveform control. Moreover, as the frequency increased, the response speed of the control system improved, further enhancing control accuracy and real-time performance.

Based on these simulation results, along with the improved quasi-PR controller closed-loop control method and the full-bridge circuit topology, the system effectively meets the performance requirements of the Li-ion battery module outlined in this design. This validation not only confirms the circuit's stability and accuracy under different frequency conditions but also provides a critical foundation for optimizing control systems in practical applications.

6.2.3 *Experimental Validations*

In this section, we mainly focus on the Experimental Section, Verification of Current Control Effect, and Heating Effect Verification. The Verification of Current Control Effect mainly validates the closed-loop control performance of the system, checking whether it can meet the output requirements at different frequencies and amplitudes, and analyzing its overall performance. On the other hand, the experiment verifies whether it can operate for extended periods and whether it can heat the Li-ion battery module quickly and without damage. The Heating Effect Verification primarily analyzes the energy loss during the AC excitation heating process.

6.2.3.1 Experimental Section

The output load of the AC heating system comprised six 18650 LIBs arranged in series. For the input, a 36 V DC-regulated power supply was employed, which consists of three 12 V lead-acid batteries and a diode connected in series. Due to the absence of a battery management system that can interface with the system, the amplitude and frequency of the output current were fixed during the verification of the efficacy of the AC heating system. Changes in battery temperature were monitored and recorded using an independent temperature-monitoring device. The influence of the AC heating system on the performance of the target LIB during operation was assessed by analyzing the variation in the battery's internal resistance before and after multiple AC heating experiments. Figure 6.14 illustrates the prototype of the AC heating system and the test platform, with thermocouples affixed to the battery surface using specialized stickers.

High-voltage input source: Fengfan 46B24L-MF lead-acid battery. Low-voltage input source: DF17315C3A, a dual-channel constant-current constant-voltage controllable power supply manufactured by Ningbo Zhongce Company. Oscilloscope: Tektronix DP03054model. Voltage probe: Tektronix P6139B high-voltage probe. Current probe: TektronixTCP0020 current probe. Thermostat: Shanghai Dongda Company's T-TH-225-C high- and low-temperature alternating wet and heat

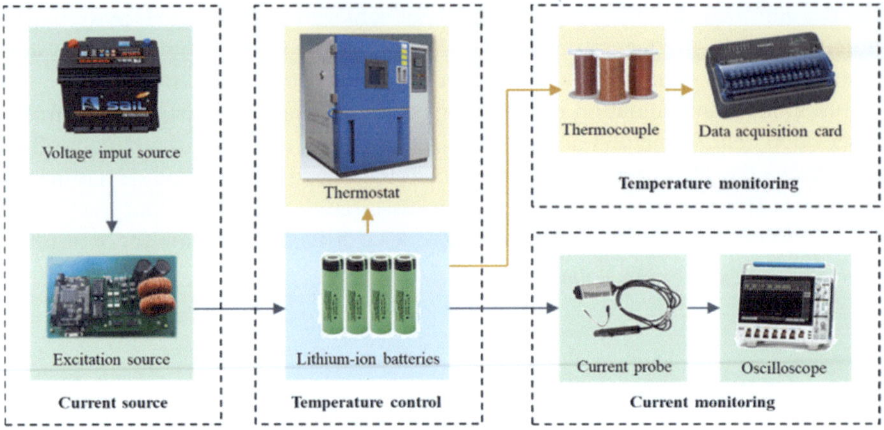

Fig. 6.14 AC heating system and test platform [18]

test chamber. Temperature sensor: American Omega company's TT-T-36-SLE thermocouple wire. Data acquisition card: japan Nippon Company's LR8510 wireless voltage and temperature unit data acquisition instrument. Current pulse excitation source: Arbin BT2000 electronic load. Host computer: Texas Instruments' CCStudio, a dedicated DSP debugging software version v3.3.

To simulate a low-temperature environment of −20 °C, the Li-ion battery module was placed within the thermostat, gradually reducing its temperature to match the ambient −20 °C. A thermocouple was attached to the surface of a single battery within the module with the temperature changes monitored and recorded by the sampling monitoring module. The thermocouples were placed on the surface of the third cell in the BP. Since the six cells were connected in series, the current flowing through each cell was the same. Additionally, the entire system was placed in a temperature chamber, ensuring consistent environmental conditions. Therefore, the temperature of a single cell was measured to represent the temperature of the entire battery module. By applying 100 Hz, 10 A excitation to the module with the designed prototype, it was observed whether the battery module could raise the temperature of the LIB to above 0 °C within 20 min. The experiment was repeated 20 times, and the internal resistance of the battery was measured at room temperature at the beginning and end of the test to verify the effect on the life of the LIB.

6.2.3.2 Verification of Current Control Effect

Closed-loop regulation analysis is mainly used to verify the control effect of closed. loop control, including current closed-loop response analysis. In the closed-loop response analysis of the current test, the initial current amplitude is set to 2 A at a frequency of100 Hz, and subsequently adjusted to 10 A. As depicted in Fig. 6.15, the actual change in current during the test reveals that the closed-loop response time for

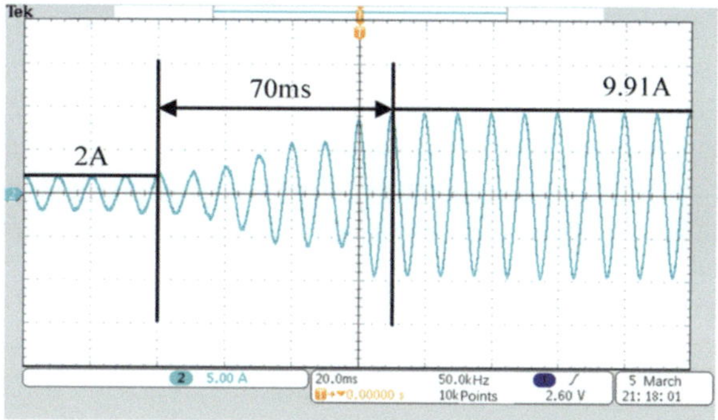

Fig. 6.15 Current closed-loop response analysis [18]

current is 70 ms with a calculated steady-state error of 1%. Because the AC heating of the battery is a long-timescale process relative to the response time, and its time can generally reach several minutes or even tens of minutes, the requirement for its adjustment time is not high-the amplitude accuracy of its output current is mainly required and the steady-state error in the test is 1%, which can fully meet the needs of practical applications.

6.2.3.3 Heating Effect Verification

Figure 6.16a illustrates the variation in surface temperature of the LIB during a test. It is evident that the prototype successfully applies AC excitation continuously to the LIB even at low temperatures. The LIB exhibits an increase in temperature from −20 to 6 °C within the 20-min duration. At this temperature, the LIB satisfies the requirements for charge and discharge operating temperatures.

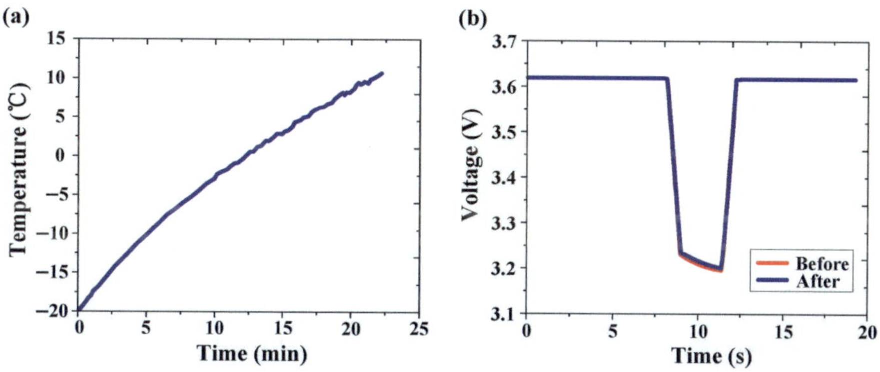

Fig. 6.16 **a** Diagram of rise in battery module AC heating test temperature; **b** Battery pulse excitation response before and after AC heating test [18]

Figure 6.16b presents the changes in the internal resistance of the LIBs detected by applying a short-time pulse current to the Li-ion battery module before and after 20 tests, thereby characterizing the degree of battery aging. These internal resistance tests were conducted at room temperature and an SOC of 50%. The red curve represents the initial state of the battery, while the blue curve depicts the test results after subjecting the battery to the same excitation following 20 AC heating tests. The experimental findings indicate that after 20 tests, the internal resistance of the LIBs exhibits almost no change, suggesting no loss of life. Consequently, the prototype devised in this study effectively heats the Li-ion battery module with AC excitation fulfilling the design requirements.

6.3 AC Self-Heating Solutions for Battery Packs

In this section, we will explore the self-heating solution for BPs. Self-heating technology, as an innovative thermal management method, can generate heat spontaneously through built-in circuits or systems, effectively addressing the impact of low-temperature environments on battery performance. Compared to traditional external heating solutions, self-heating not only offers higher energy efficiency but also allows for more precise control of the battery's temperature distribution, thereby improving the overall reliability and safety of the system.

6.3.1 AC Self-Heating Solution for Integrated External Circuitry

6.3.1.1 Structure Design

As shown in Fig. 6.17, this design proposes a LIB self-heating system and method that enables BPs to achieve self-heating. The system primarily consists of a temperature sensor (2), a single-cell status monitoring circuit (3), a high-frequency charge–discharge control circuit (4), and a control unit (5).

Structurally, the temperature sensor is directly connected to the Li-ion BP (11) to monitor battery temperature in real time. The single-cell status monitoring circuit is responsible for detecting the operating state of each individual cell and transmitting the collected data to the control unit. One end of the high-frequency charge–discharge control circuit is connected to the positive and negative terminals of the Li-ion BP, while the other end is linked to the control unit. Based on the battery status information transmitted by the single-cell status monitoring circuit, the control unit intelligently regulates the operating mode of the high-frequency charge–discharge control circuit, allowing the battery to undergo high-frequency charge–discharge

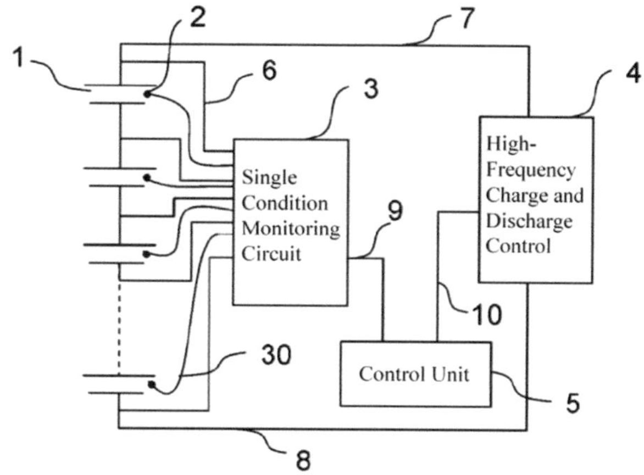

Fig. 6.17 Schematic diagram of the structure [19]

cycles. This process enables the self-heating function of the BP, thereby improving battery performance and extending service life in low-temperature environments.

As shown in Fig. 6.18, The high-frequency charge–discharge control circuit comprises multiple core components, including a first control switch (14), a second control switch (12), a third control switch (15), a fourth control switch (13), a first diode (19), a second diode (20), a first inductor (16), a second inductor (17), and a capacitor (18). Specifically, one end of the first control switch is connected to the positive terminal of the Li-ion BP via the battery positive branch (21), while the other end is connected to the negative terminal of the first diode and the first inductor. One end of the second control switch is connected to the negative terminal of the Li-ion BP via the battery negative branch (22), while the other end is connected to the first inductor, the capacitor, and the fourth control switch. The positive terminal of the first diode is connected to the capacitor and the third control switch. One end of the second inductor is connected to the third control switch and the positive terminal of the second diode, while the other end is connected to the fourth control switch and the negative terminal of the Li-ion BP. The fourth control switch is connected to the negative terminal of the Li-ion BP via the battery negative branch (24). Meanwhile, the negative terminal of the second diode is ultimately connected to the positive terminal of the Li-ion BP via the battery positive branch (23). The operating frequency of this circuit is higher than the lowest frequency point of the electrochemical reaction process in the electrochemical impedance spectroscopy (EIS) test of the battery, ensuring that the battery can achieve self-heating under efficient and safe conditions.

Additionally, the Li-ion BP consists of multiple Li-ion cells connected in series. Each cell is connected to the battery status monitoring circuit through voltage sampling signal lines (6) to precisely acquire the working state of individual cells. The number of temperature sensors matches the number of battery cells, and they are evenly distributed on the surface of the Li-ion battery cells. These sensors are connected to the battery status monitoring circuit through separate lines to ensure accurate battery temperature monitoring.

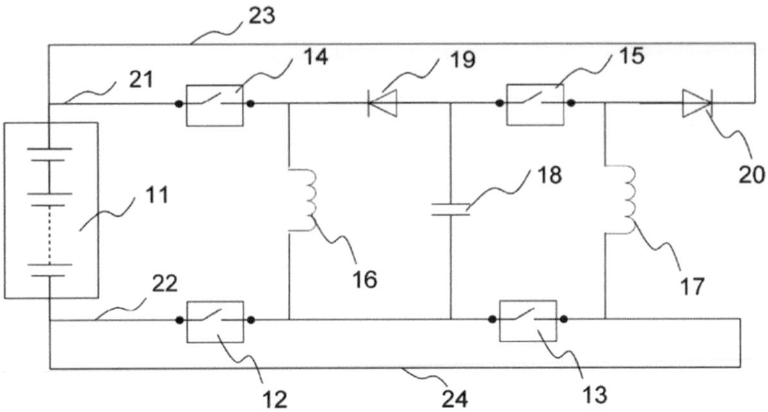

Fig. 6.18 The schematic diagram of the high-frequency charge–discharge control circuit [19]

Compared with existing technologies, this design utilizes the heat generated by the Li-ion battery during its cyclic charging and discharging process for self-heating, offering the following advantages:

This design uses high-frequency charge and discharge control to heat the battery itself. Due to the internal resistance of the Li-ion battery and the mechanisms of electrochemical reactions, heat is generated during the battery's cyclic charge and discharge process, which heats the battery from within, leading to more uniform temperature distribution. In contrast, traditional methods rely on external devices to heat the battery, transferring heat through the battery casing, resulting in higher energy loss and lower efficiency compared to this invention.

Under low temperature conditions, charging and discharging, especially at high rates and for extended periods, may cause dendrites to form on the negative electrode, which affects the battery's safety. The frequency of charge and discharge control in this invention is higher than the lowest frequency point of the electrochemical reaction process in the EIS test, which effectively prevents the formation of dendrites during low-temperature charge and discharge.

During the self-heating process of this design, the battery's temperature and voltage information are monitored in real-time. High-frequency charge and discharge control is then adjusted based on the voltage and temperature information, further ensuring the safety of the battery.

6.3.1.2 System Operation Control Process

Step 1: The single-cell state monitoring circuit collects voltage and temperature data of each Li-ion cell in real time via voltage sampling lines and temperature sensors, transmitting this information to the control unit.

Step 2: The control unit detects over voltage or under voltage in the cells. If detected, the system halts (Step 9). Otherwise, proceed.

Step 3: The control unit checks whether the temperature of any cell falls below a preset threshold. If so, proceed to Step 4; otherwise, execute Step 9.

Step 4: The control unit retrieves stored frequency and preset current values.

Step 5: The control unit initiates high-frequency charge–discharge cycles in the BP based on the set parameters.

Step 6: The control unit rechecks for over voltage or under voltage in the cells. If detected, it reduces the charge–discharge current amplitude as per set rules (Step 7). If not, continue with Step 8.

Step 7: If the current drops below the set limit, execute Step 9. Otherwise, return to Step 5.

Step 8: The control unit monitors whether the battery temperature has reached the target level. If achieved, proceed to Step 9; otherwise, return to Step 5.

Step 9: The control unit stops the high-frequency charge–discharge cycles (Fig. 6.19).

6.3.1.3 Determination of the Operating Frequency Setting for the High-Frequency Charge–discharge Control Circuit

A detailed analysis of the EIS test results for the Li-ion single cell shown in Fig. 6.20 is conducted to determine the boundary line (27) between the high-frequency region (28), the mid-frequency region, and the low-frequency region (29). Based on this boundary line (27), the frequency differentiation points between the high-frequency region (28), the mid-frequency region, and the low-frequency region (29) are further identified, providing a basis for subsequent battery management and optimization control.

To accurately interpret the test results, the equivalent circuit model shown in Fig. 6.21 is used to fit the EIS test data from Fig. 6.20. In this equivalent circuit model, the voltage source (31) represents the open-circuit voltage associated with the battery's state of charge. The resistor (32) represents the internal resistance of the battery and approximates the high-frequency impedance characteristics related to the SEI layer. Meanwhile, the loop consisting of resistor (34) and capacitor (33) is used to simulate the impedance semicircle characteristics of the high-frequency and mid-frequency regions shown in Fig. 6.20. The low-frequency region (29) is described by a model composed of resistor (36), capacitor (35), and the Warburg impedance element (37).

By fitting the EIS test data shown in Fig. 6.20, the key parameters of the equivalent circuit model in Fig. 6.21 can be accurately obtained. The loops formed by resistors and capacitors correspond to specific time constants, which can be calculated. The

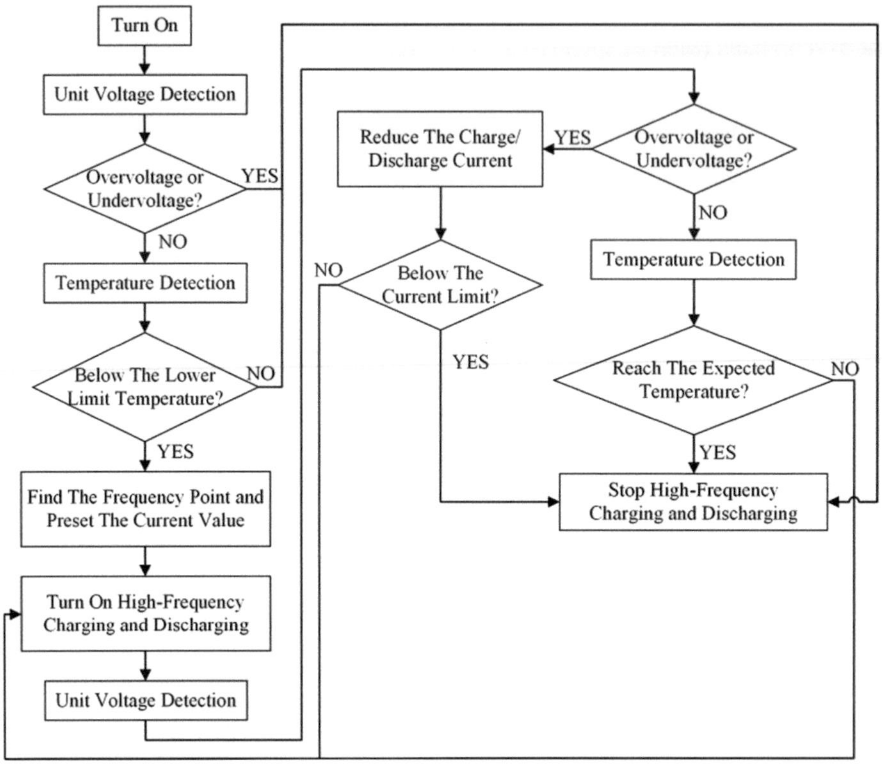

Fig. 6.19 System operation control flowchart [19]

Fig. 6.20 Typical EIS test graph of Li-ion battery [19]

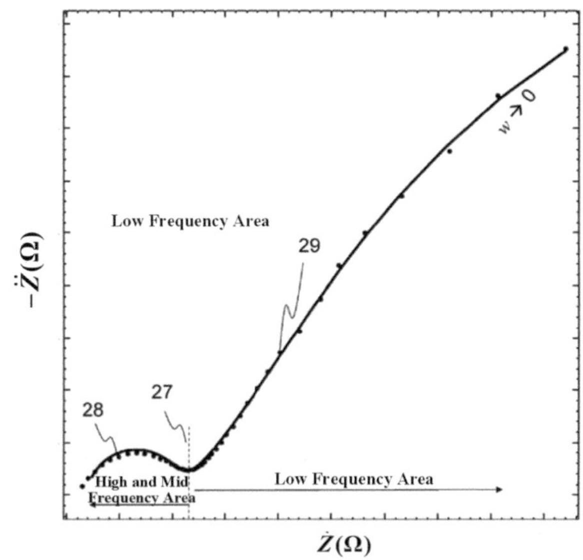

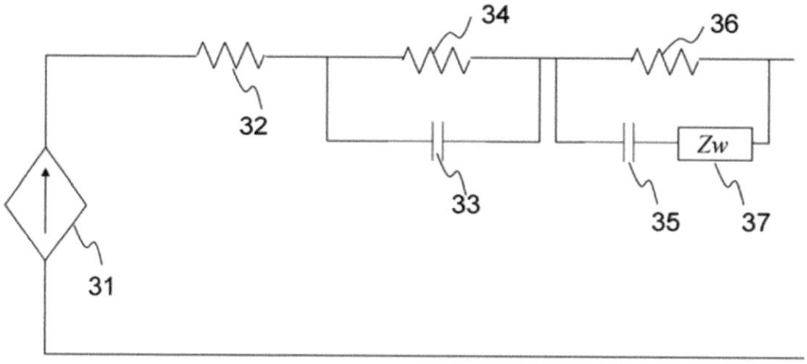

Fig. 6.21 The structure of the battery equivalent circuit model [19]

reciprocal of these time constants is then used as the operating frequency of the high-frequency charge–discharge control circuit. This operating frequency is stored in the non-volatile memory of the control unit, such as ROM, to ensure long-term data validity. Based on this stored value, the control unit precisely regulates the operating frequency of the high-frequency charge–discharge control circuit, optimizing the Li-ion battery's charge–discharge process and enhancing its performance and stability under various operating conditions.

6.3.1.4 Operating Cycle of the High-Frequency Charge–discharge Control Circuit

1) Working Principle

Step 1: Connection of BP to High-Frequency Charge–Discharge Control Circuit
The BP is connected to the high-frequency charge–discharge control circuit to achieve periodic high-frequency charge–discharge cycles, thereby optimizing battery performance.

Step 2: High-Frequency Control Switch Turned On (Phase 25)
In this phase, the high-frequency control switch is turned on for a conduction time of t12. At this moment, the BP enters a discharge state, with its positive terminal connected to the positive branch of the output circuit and its negative terminal connected to the negative branch. Simultaneously, the high-frequency control switch charges the inductor, forming a current loop that enables efficient energy transfer.

Step 3: High-Frequency Control Switch Turned Off
After the high-frequency control switch is turned off, the inductor continues to release the stored energy. The current flows through the diode, charging the capacitor and allowing it to accumulate charge in preparation for subsequent energy transfer.

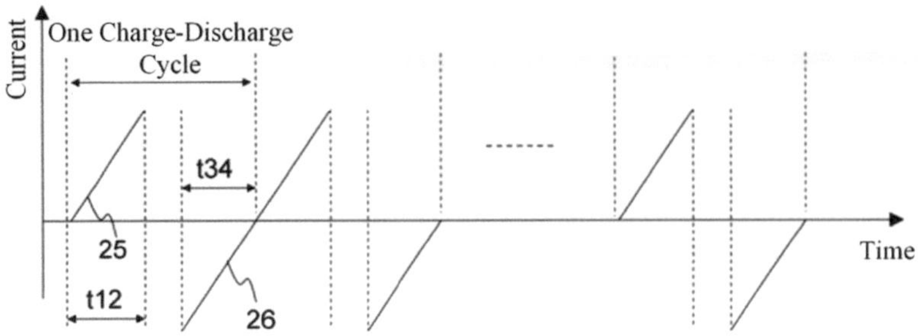

Fig. 6.22 Schematic diagram of high-frequency charge–discharge current of the battery [19]

Step 4: High-Frequency Control Switch Turned On Again (Phase 26)
The high-frequency control switch is turned on again for a conduction time of t34. At this point, the capacitor releases the stored energy, and the current discharges through the high-frequency control switch, the inductor, and the circuit loop. Simultaneously, the inductor is recharged, maintaining continuous energy transfer.

Step 5: High-Frequency Control Switch Turned Off, Completing Battery Charging
After the high-frequency control switch is turned off, the energy stored in the inductor continues to be released. The current flows through the diode, the positive terminal of the battery, and the output circuit branch, allowing current to enter from the negative terminal and thereby completing the charging process of the BP.

Throughout the entire operating cycle, the BP's current waveform is shown in Fig. 6.22. The duration of this cycle corresponds to the reciprocal of the lowest operating frequency of the charge–discharge circuit, which is determined based on the EIS analysis results.

2) Waveform Analysis

Phase 25 (Corresponding to Step 2):
The battery gradually discharges into the external circuit, with the current progressively increasing, reflecting the energy transfer process from the battery to the load.

Phase 26 (Corresponding to Step 5):
The battery absorbs current from the external circuit, with the current gradually decreasing, indicating the charging process of the battery.

Between Phases 25 and 26 (Corresponding to Steps 3 and 4):

The inductor and capacitor work in coordination to complete current conversion and energy storage, ensuring that the high-frequency charge–discharge process remains stable and efficient.

By employing this high-frequency charge–discharge control strategy, the BP can achieve efficient energy management under different operating conditions, enhancing charge–discharge stability and optimizing battery lifespan.

6.3.2 *AC Self-Heating Solution for Integrated Motor Drive System*

6.3.2.1 Structure Design

As shown in Fig. 6.23, this design involves a low-temperature AC heating device for Li-ion batteries used in EVs, intended for AC heating of power batteries. The heating system consists of a control unit, energy conversion circuit, energy storage inductor, capacitor, pre-charge circuit, and start switch. The power battery, energy storage inductor, energy conversion circuit, and capacitor are sequentially connected to form a loop. The pre-charge circuit and start switch are connected in parallel between the power battery and the capacitor. The control unit is connected to the energy conversion circuit, pre-charge circuit, and start switch.

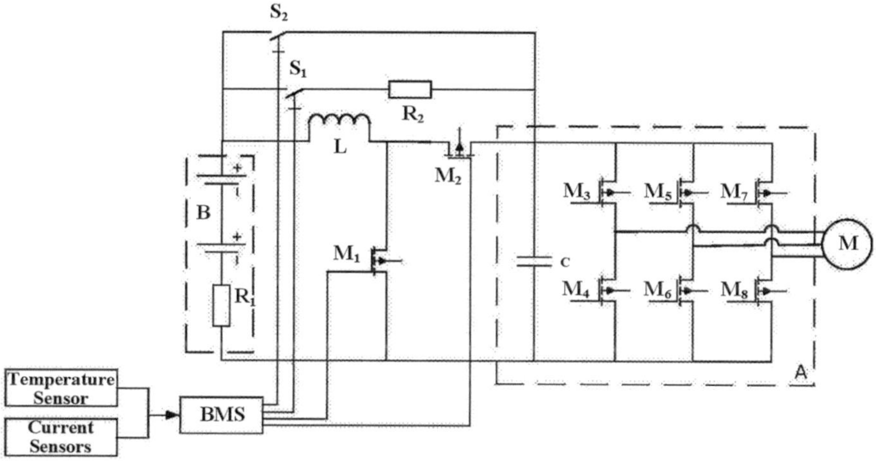

Fig. 6.23 The overall circuit topology diagram [20]

The control unit operates by real-time monitoring of the battery temperature and controlling the activation or deactivation of the start switch, thereby switching between the AC heating and normal power supply modes. Upon entering the AC heating mode, the control unit adjusts the working state of the energy conversion circuit based on real-time battery current information, enabling internal heating during the AC charge–discharge process. The energy conversion circuit includes switching transistors M_1 and M_2, which are connected to the control unit. The power battery, energy storage inductor, and switching transistor M_1 are sequentially connected to form a loop, with one end of M_1 connected to the inductor and the other end to the capacitor. The control unit adjusts the duty cycle of M_1 and M_2 based on battery current information, enabling AC charge–discharge with arbitrary waveforms. The switching transistors M_1 and M_2 are semiconductor devices such as MOSFETs, IGBTs, silicon carbide devices, or other power semiconductor components. The pre-charge circuit utilizes the pre-charge resistor and relay in the BMS, with the pre-charge resistor connected to the pre-charge relay, which is in turn connected to the control unit. The control unit includes a controller linked to temperature and current sensors, which may be a dedicated heating controller or a repurposed BMS controller. The energy storage inductor is either a high-frequency magnetic core inductor or a coreless inductor, while the capacitor may be the bus capacitor of the motor inverter circuit or a combination of external and bus capacitors. The AC charge–discharge process involves two main modes: battery discharging and capacitor charging, or battery charging and capacitor discharging. The control unit dynamically adjusts the duty cycles of M_1 and M_2 to modify the AC current waveform and amplitude. The operating frequency of the control unit exceeds the lowest frequency of the battery's electrochemical impedance spectrum response.

This design achieves internal heating of the battery through AC charging and discharging of the power battery. As the electrochemical reaction mechanism of Li-ion batteries indicates, the internal resistance of the battery is a known factor. During the battery's cyclic charging and discharging process, the internal resistance generates heat, enabling uniform internal heating of the battery. Traditional external heating methods rely on the battery casing to transfer heat, which results in low heating efficiency and uneven temperature distribution. In contrast, internal heating of the battery ensures more uniform heating. This design, therefore, has lower energy loss, higher efficiency, and more consistent temperature rise.

In low-temperature environments, high-rate and prolonged charging of Li-ion batteries can cause lithium deposition and dendrite formation on the negative electrode, which negatively impacts the battery's lifespan and safety. The frequency of AC charging and discharging in this design can be controlled by the control unit, and the operating frequency of the control unit is higher than the lowest frequency point of the electrochemical reaction process in the EIS test, effectively preventing lithium deposition and dendrite formation.

This design reuses resources from the battery management system and the bus capacitors in the vehicle's inverter circuit, requiring few additional components, which effectively saves costs. Therefore, the space required for additional components is minimal, and it is possible to implement the AC heating function while only slightly increasing the design burden of the BP structure. The reused BMS controller only needs to control the duty cycle of two switches, which does not consume excessive controller resources.

By collecting the battery's input and output currents and adjusting the duty cycle of the switches in real-time, this design allows for any desired AC charging and discharging current waveform under average energy conditions, making it suitable for different types of batteries and battery heating under various operating conditions. By monitoring the current and temperature information, the invention can terminate the AC heating process when the battery current is too high or the voltage is too low, enhancing safety.

This design enables fast low-temperature heating of the power battery without the need for an additional power source and prevents lithium metal deposition during the charging process.

6.3.2.2 AC Self-Heating Method Steps

As shown in Fig. 6.23, a Li-ion battery for EVs for AC heating of power battery B, including control unit, energy conversion circuit, energy storage inductance L, capacitor C, pre-charging road and start switch S_2, power battery, energy storage inductance, energy conversion circuit and capacitor in turn form circuit, pre-charging road and start switch in parallel between the power battery and capacitor, the control unit connected energy conversion circuit, pre-charging circuit and start switch respectively. Energy storage inductor L includes high frequency magnetic core inductance or no core inductor. The power BP to be heated includes a BP consisting of n-cell power batteries in series and parallel, where the equivalent AC impedance of the BP is indicated.

The energy conversion circuit is a buck-boost circuit, including the switch pipe M_1 and M_1 connected to the control unit, the switch pipe M_1, the power battery B and the energy storage inductor L are connected successively to form a loop. One end of the switch pipe M_1 is connected to the energy storage inductor L and the capacitor C; the control unit adjusts the duty cycle of the switch pipe M_1 and M_2 according to the current information of the power battery B to realize the AC charge and discharging of the power battery under any AC waveform (such as, triangular wave, trapezoidal wave, sine wave, etc.). Switswitch tube M_1 and M_2 are semiconductor switch tube devices, including MOSFET, IGBT, silicon carbide devices or other power semiconductor devices.

The pre-charging road adopts the pre-charging resistance R_2 and the precharged relay S_1 in the battery management system, the pre-charging resistance R_2 is connected to the precharged relay S_1, and the precharged relay S_1 is connected to the control unit. The control unit includes a controller and a temperature sensor and a current sensor connected to the controller, respectively. The controller can reuse the battery management system controller BMS or set up a dedicated heating controller. The temperature sensor is a thermocouple or other heat-electric conversion sensor.

The current sensor is a Hall sensor, sampling resistance, or other current sampling sensor.

Capacitor C is the combination of drive motor inverter circuit bus capacitor (as part of motor inverter) or drive motor inverter circuit and drive motor inverter circuit. In this embodiment, the capacitor C multiplexes the bus capacitors of the drive motor inverter circuit. In the low-temperature AC heating device of Li-ion battery for EVs, the control unit controls the opening and turning off of the starting switch according to the temperature information of the power battery, and realizes the conversion of the power battery. In the control unit according to the current information of the power battery, reasonably controls the switch tube duty ratio and the switching state of the starting switch, and realizes the internal heating in the AC charging and discharging process of the power battery.

6.3.2.3 Four Modes of AC Self-Heating

1) Battery Discharge and Capacitor Charging Modes: Figs. 6.24 and 6.25 illustrate the operational diagrams of the battery discharge and capacitor charging modes during AC heating as described in this invention. In this mode, the inductor current flows from left to right, charging the inverter capacitor using the battery system. The BMS controller regulates the duty cycles of the switching transistors M_1 and M_2, thereby controlling the rise and fall of the battery current within a single PWM cycle. Figure 6.24 shows the current path when M_1 is on and M_2 is off, while Fig. 6.25 illustrates the current path when M_2 is on and M_1 is off. By integrating BP temperature and current information, the BMS dynamically adjusts the duty cycle of M_1 and M_2 in real time to modify the waveform and amplitude of the battery discharge current.

2) Battery Charging and Capacitor Discharging Modes: Figs. 6.26 and 6.27 illustrate the operational diagrams of the battery charging and capacitor discharging modes during AC heating as described in this design. In this mode, the inductor current

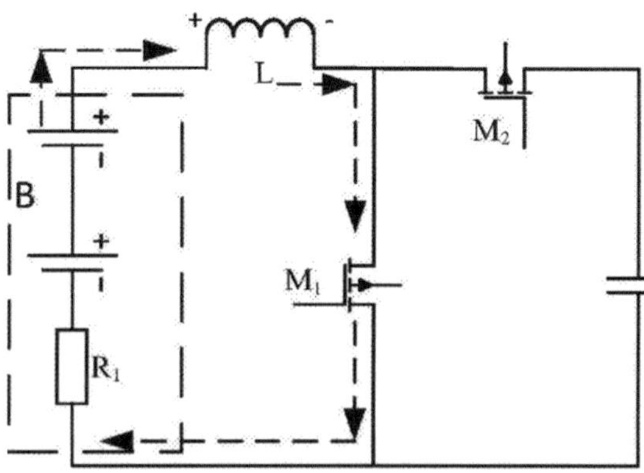

Fig. 6.24 Schematic diagram of battery discharge during AC heating [20]

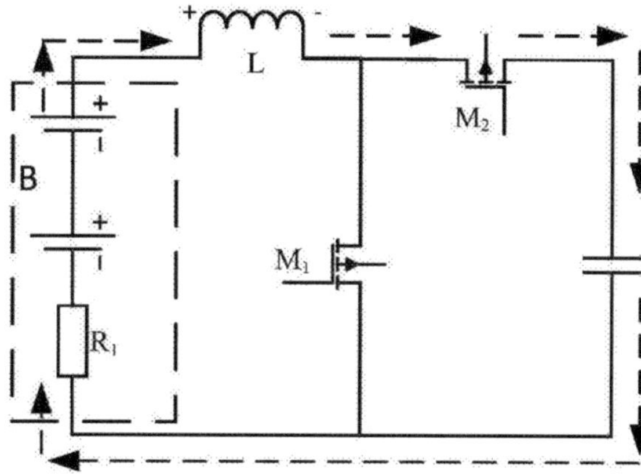

Fig. 6.25 Schematic diagram of capacitor charging during AC heating [20]

flows from right to left, charging the battery system from the inverter capacitor. The BMS controller regulates the duty cycles of the switching transistor M_1 and M_2, thereby controlling the rise and fall of the battery current within a single PWM cycle. Figure 6.26 shows the current path when M_2 is on and M_1 is off, while Fig. 6.27 illustrates the current path when M_1 is on and M_2 is off. By integrating BP temperature and current information, the BMS dynamically adjusts the duty cycle of M_1 and M_2 in real time to modify the waveform and amplitude of the battery charging current.

3) Capacitor Pre-Charging Mode: Before entering the AC heating mode, the traction battery needs to charge the bus capacitor C of the drive motor inverter circuit, ensuring the voltage across capacitor C is close to or higher than the battery voltage. As shown in Fig. 6.28, the diagram illustrates the current flow during the pre-charging process of capacitor C prior to AC heating. Using the pre-charging circuit, the relay switch S_1 is closed, allowing the battery to charge capacitor

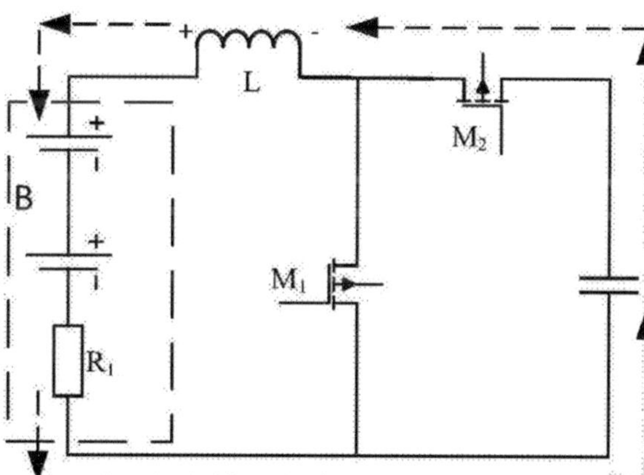

Fig. 6.26 Schematic diagram of capacitor discharging during AC heating [20]

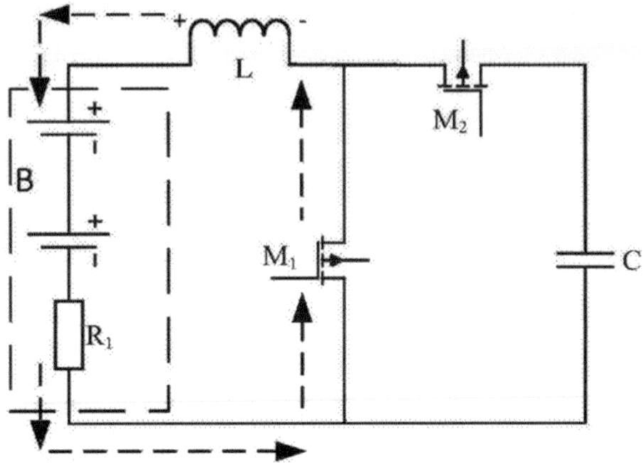

Fig. 6.27 Schematic diagram of battery charging during AC heating [20]

C through S_1 and the pre-charging resistor R. The pre-charging resistor R is employed to prevent excessive pre-charging current. Once the capacitor voltage approaches the battery voltage, the relay switch S_1 is opened.

4) Normal Battery Power Supply Mode: This design enables real-time activation or deactivation of the AC heating process based on commands from the upper-level controller and the current battery temperature status. When the BMS turns off switches S_1 and S_2, and simultaneously disables the drive motor inverter, the AC heating process is activated by controlling the states of switching transistors M_1 and M_2. The BMS initiates the capacitor C pre-charging process by turning on switch S_1, while turning off S_2 and switching transistors M_1 and M_2. To deactivate the AC heating function, the BMS turns on S_2, turns off S_1, and

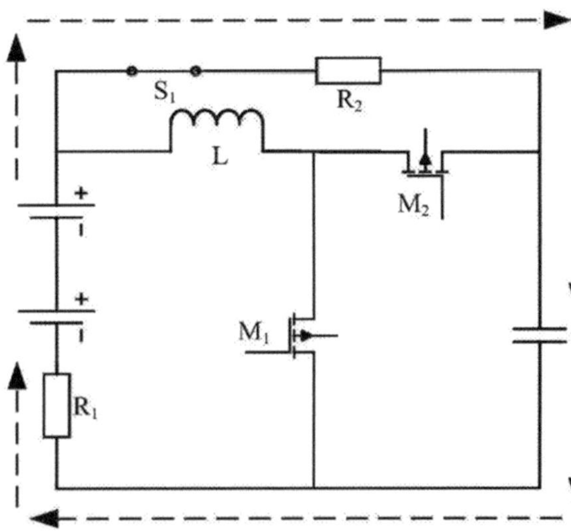

Fig. 6.28 Schematic diagram of current flow during the pre-charging of the bus capacitor of the motor drive inverter before AC heating operation [20]

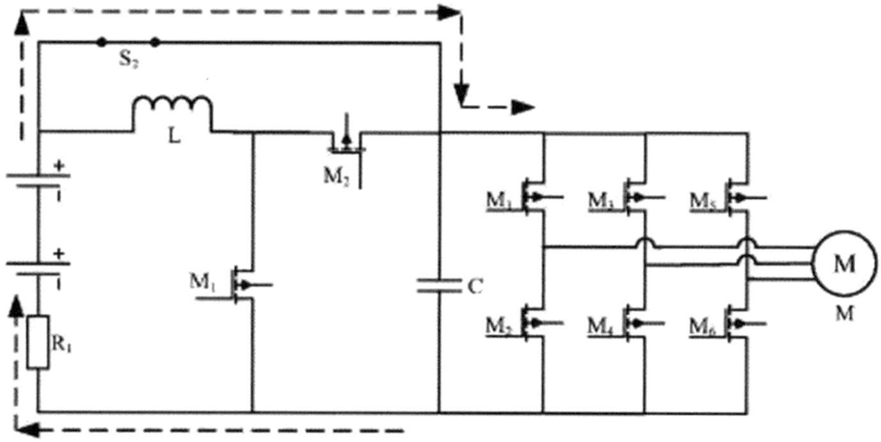

Fig. 6.29 Schematic diagram of current flow when AC heating is not required [20]

disables switching transistors M_1 and M_2. At this point, the drive motor inverter circuit can be enabled, allowing the battery to supply energy to the drive motor to operate it. The current flow during this process is shown in Fig. 6.29.

6.3.3 *AC Self-Heating Solution for Boost Type Power Battery Module*

6.3.3.1 Structure Design

As shown in Figs. 6.30 and 6.31, a boost-type AC low-temperature heating circuit for a power battery module is provided. The power battery module includes battery modules B1 and B2, each formed by connecting n individual Li-ion battery cells in series. The low-temperature heating circuit includes switching transistors M1, M2, M3, M4, a transformer T, diodes D1, D2, D3, D4, D5, D6, D7, D8, a switching transistor driver circuit, a temperature sensor TS, and a microcontroller MCU. The switching transistors M1, M2, M3, and M4 are MOSFET or other types of semiconductor switching devices. The transformer T and diodes D1, D2, D3, D4, D5, D6, D7, D8 form a forward and reverse boost structure.

The temperature sensor is connected to the battery modules B1 and B2 and the microcontroller MCU. It can be a thermocouple, infrared sensor, or other types of sensors, which collect the temperature of the battery modules B1 and B2, convert the temperature into electrical signals, and send them to the microcontroller MCU. The microcontroller MCU can receive the signals from the temperature sensor in real-time and select the working mode based on the temperature. When the temperature is high, the system enters a low-power mode, and the microcontroller MCU enters sleep mode, but it can be woken up at low temperatures, with all peripherals powered down.

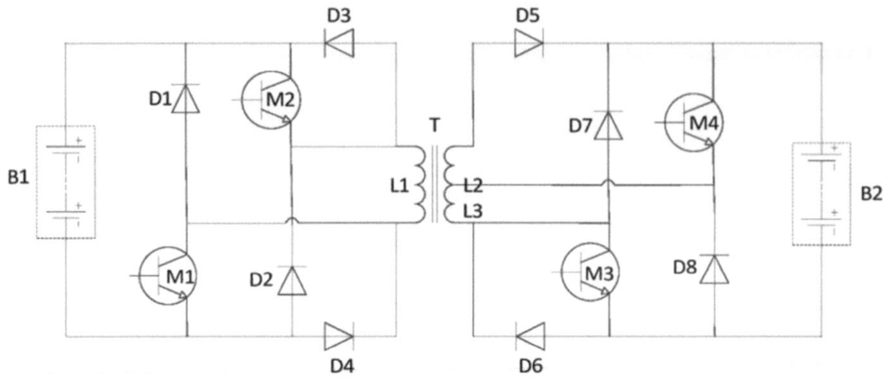

Fig. 6.30 Overall structural block diagram of the design (a) [21]

Fig. 6.31 Overall structural block diagram of the design (b) [21]

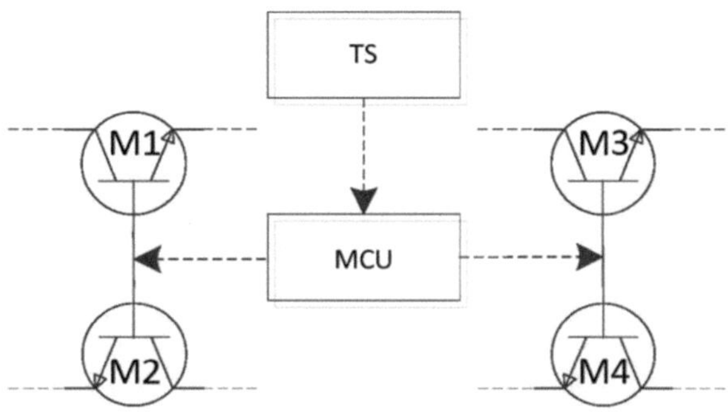

When the temperature drops below a specific value Tmin, the system is awakened and enters normal working mode, with peripherals powered on. The microcontroller MCU can control the generation, stopping, PFM (Frequency Modulation), and PWM (Pulse Width Modulation) of output rectangular pulses. When the Li-ion battery module requires low-temperature heating, the microcontroller generates rectangular pulses with specific duty cycles and frequencies based on the temperature, which are sent to the switching transistor driver circuit to control the turning on or off of the switching transistors M1, M2, M3, and M4.

The transformer T includes the primary coil L1 and secondary coils L2 and L3. One end of the primary coil L1 is connected to M2, the anode of D3, and the cathode of D2, while the other end is connected to M1, the anode of D1, and the cathode of D4. One end of secondary coil L2 is connected to the anode of D5, while the other end is connected to M3, the cathode of D6, and the anode of D7. One end of secondary coil L3 is connected to M4 and the cathode of D8, while the other end is connected to M3, the cathode of D6, and the anode of D7.

The number of turns of primary coil L1 (n_1), secondary coil L2 (n_2), and secondary coil L3 (n_3) satisfies the relationship: $n_2 > n_1 > n_3$. Switch transistors M1, M2, M3, and M4 are all semiconductor switch devices, and the semiconductor switch devices include MOSFET. The temperature sensor is either a thermocouple or an infrared sensor.

This design uses high-frequency charge and discharge control to heat the battery itself. Due to the internal resistance of the Li-ion battery and the mechanism of electrochemical reactions, heat is generated during the battery's charge and discharge cycle, which heats the battery from within, resulting in a more uniform temperature distribution. In contrast, traditional methods rely on external devices to heat the battery, where heat is transferred through the battery shell, leading to higher energy loss, lower efficiency, and uneven temperature distribution. This design, however, has lower energy loss, higher efficiency, and more uniform temperature rise.

In low-temperature conditions, charging the battery, especially at high rates or over extended periods, may cause the formation of dendrites on the negative electrode, which can compromise the safety of the battery. The charge and discharge control frequency in this design is higher than the lowest frequency point in the electrochemical impedance spectrum of the battery's electrochemical reaction process, effectively preventing the formation of dendrites during low-temperature charge and discharge.

During the self-heating process, this design continuously monitors the battery's temperature and voltage information and adjusts the high-frequency charge and discharge control based on this data, ensuring the safety of the battery.

6.3.3.2 AC Self-Heating Method Steps

The switching transistors in the low-temperature heating circuit are in an off state when not in operation. After the system enters normal working mode, when the Li-ion battery module requires low-temperature heating, switching transistors M1 and M2 are first turned on. Battery module B1 discharges, and the current flows from the positive terminal of battery module B1, through switching transistor M2, the primary coil L1 of transformer T, and switching transistor M1, to the negative terminal. This completes the discharge cycle for battery module B1. Meanwhile, the secondary coil L2 of the transformer generates an induced current, which flows through diode D5 to the positive terminal of battery module B2, then through the negative terminal of battery module B2, diode D6, and returns to the secondary coil L2 of transformer T, completing the charging cycle for battery module B2. The current flow is shown in Fig. 6.32.

Next, switching transistors M1 and M2 are turned off. Since the energy stored in the primary coil L1 and secondary coil L2 of transformer T cannot dissipate instantly, the energy stored in the primary coil L1 is used to charge battery module B1, while the energy stored in the secondary coil L2 is used to charge battery module B2. At this point, the current flows out of the primary coil L1, through diode D1 to the positive terminal of battery module B1, and then through the negative terminal of battery

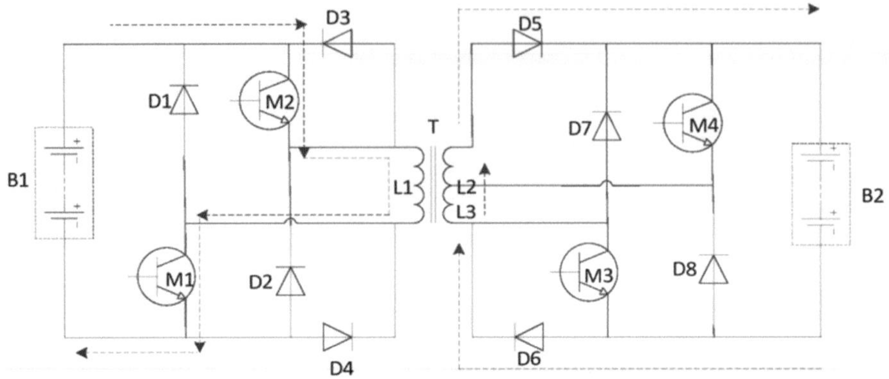

Fig. 6.32 Current flow schematic diagram when switch transistors M1 and M2 are on, and switch transistors M3 and M4 are off during the discharging phase of battery module B1 and the charging phase of battery module B2 [21]

module B1, through diode D2, and back to the primary coil L1, thus completing the charging process for battery module B1 using the stored energy in the coils. Meanwhile, the current flows out of the secondary coil L2, through diode D5 to the positive terminal of battery module B2, then through the negative terminal of battery module B2, through diode D6, and returns to the secondary coil L2, thus completing the charging process for battery module B2. The current flow is shown in Fig. 6.33.

Then, switching transistors M3 and M4 are turned on, utilizing transformer T for reverse boosting. At this point, the L3 coil of transformer T becomes the primary coil, and the L1 coil of transformer T becomes the secondary coil, completing the discharge of battery module B2 and the charge of battery module B1. The current flows from the positive terminal of battery module B2, through switching transistor

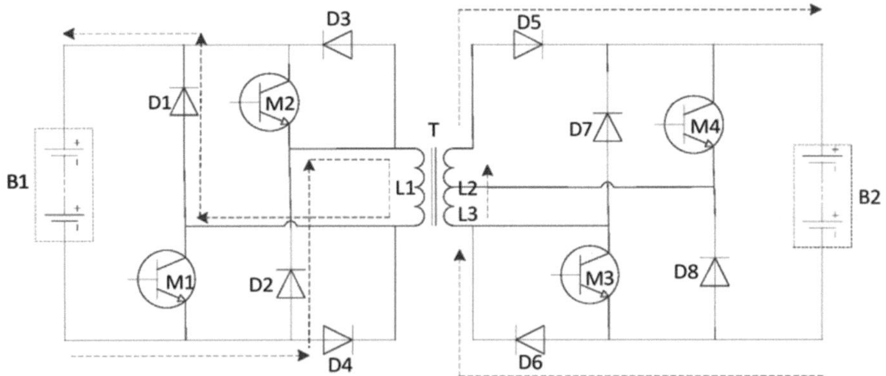

Fig. 6.33 Current flow schematic diagram when switch transistors M1, M2, M3, and M4 are off during the stage of charging battery modules B1 and B2 using the energy stored in the primary coil L1 and secondary coil L2 of transformer T [21]

M4, the primary coil L3 of the transformer, and switching transistor M3, returning to the negative terminal of battery module B2, completing the discharge cycle for battery module B2. Simultaneously, the secondary coil L1 of transformer T has current flowing out, through diode D3, the positive terminal of battery module B1, the negative terminal of battery module B1, and diode D4, ultimately flowing back to the secondary coil L1 of transformer T, thus completing the charging process for battery module B1. The current flow is shown in Fig. 6.34.

Then, switching transistors M3 and M4 are turned off. Since the energy stored in the primary coil L3 and secondary coil L1 of transformer T cannot dissipate instantly, the energy stored in the primary coil L3 is used to charge battery module B2, while the energy stored in the secondary coil L1 is used to charge battery module B1. At this point, the current flows out of the primary coil L3, through diode D7, to the positive terminal of battery module B2, then through the negative terminal of battery module B2, through diode D8, and returns to the primary coil L3, thus completing the charging process for battery module B2. Simultaneously, the current flows out of the secondary coil L1, through diode D3, to the positive terminal of battery module B1, then through the negative terminal of battery module B1, through diode D4, and back to the secondary coil L1, thus completing the charging process for battery module B1. The current flow is shown in Fig. 6.35. This completes one charge–discharge cycle. The current variations in battery module B1 during the AC charge–discharge process are shown in Fig. 6.36, and the current variations in battery module B2 during the AC charge–discharge process are shown in Fig. 6.37.

The self-heating circuit operates in continuous high-frequency charge–discharge cycles, and the frequency is higher than the highest frequency point where electrochemical reactions occur during electrochemical impedance testing. The battery modules experience minimal external energy loss, with heat generated only by the

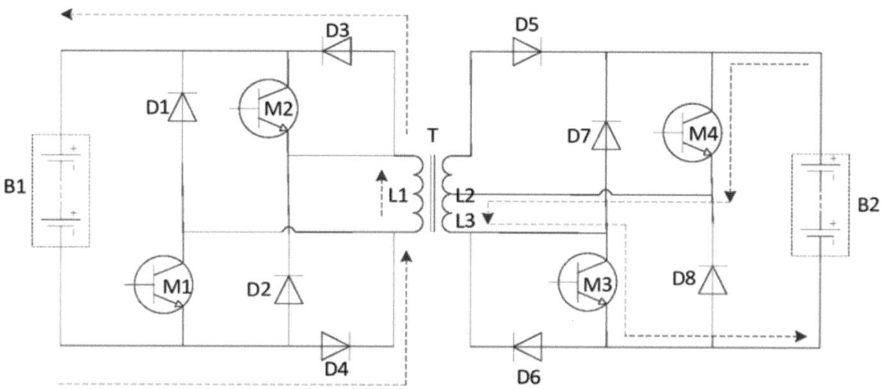

Fig. 6.34 Current flow schematic diagram when switch transistors M3 and M4 are on, and switch transistors M1 and M2 are off, during the stage of discharging battery module B2 and charging battery module B1 [21]

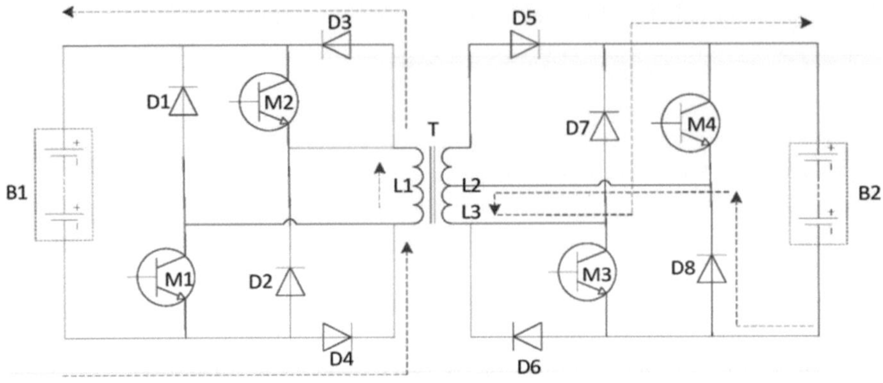

Fig. 6.35 Current flow schematic diagram when switch transistors M1, M2, M3, and M4 are off, during the stage of charging battery modules B1 and B2 using the energy stored in the primary coil L3 and secondary coil L1 of transformer T [21]

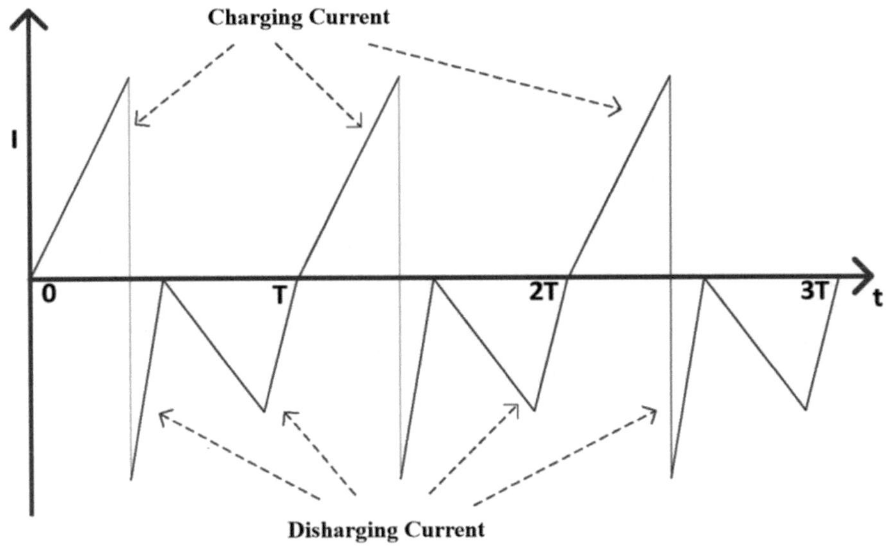

Fig. 6.36 Current variation diagram of battery module B1 during the AC charge and discharge process in this design [21]

battery's internal resistance and the Li-ion liquid-phase migration resistance, thus achieving efficient self-heating of the battery modules.

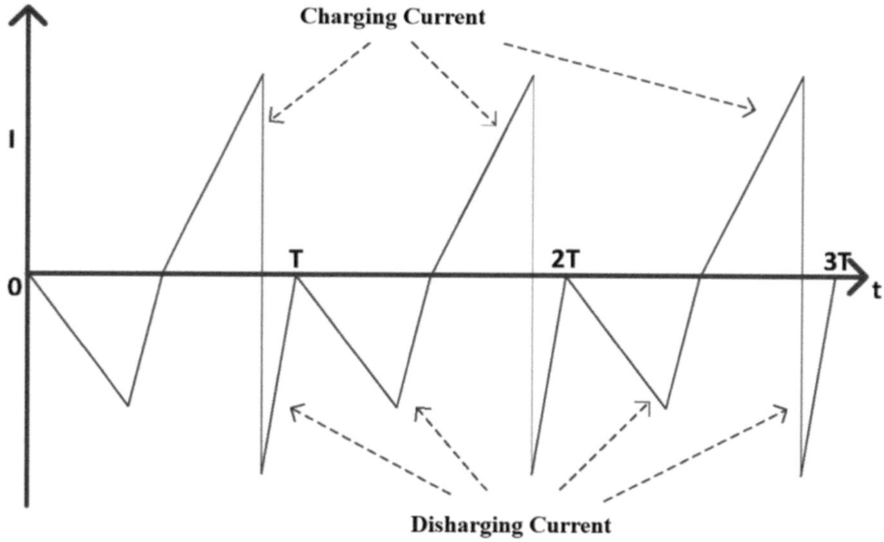

Fig. 6.37 Current variation diagram of battery module B2 during the AC charge and discharge process in this design [21]

6.4 Conclusions

This chapter begins with the implementation and application of AC heating technology, providing a detailed introduction to current research on AC heating. Using an innovative integrated charger design as an example, we highlight a practical, feasible battery-integrated AC heating technology and methodology through a detailed description of its topology, control strategies, and simulation processes. By presenting three AC self-heating solutions for integrated external circuits, motor drive systems, and boost-type power battery modules, we strengthen the reader's understanding of the thermal management performance of Li-ion batteries in EVs and other energy storage systems.

Through an in-depth study and analysis of existing AC heating technologies, we categorize them into energy injection-type and non-energy injection-type heating methods, understanding their respective advantages and disadvantages. The former relies on external energy injection to generate heat, while the latter utilizes internal electrochemical processes within the battery to generate heat. Although traditional methods are still in use, the limitations of these solutions, particularly regarding energy efficiency and thermal distribution, have spurred the development of more integrated and effective solutions. The comprehensive study and analysis of existing AC heating technologies provide a reference background for subsequent AC heating system designs.

Based on this, we propose an integrated AC heating solution with energy injection. The design introduces a prototype system based on a full-bridge topology, aimed at enhancing the charge and discharge performance of Li-ion batteries under low-temperature conditions. To verify the effectiveness of the control strategy, a full-bridge topology system model was also developed and implemented. This system is capable of AC heating for Li-ion battery modules. To analyze the heating effect of AC excitation on Li-ion batteries, a comprehensive simulation model was developed, combining heat transfer and battery mechanism analysis, applicable to both normal usage and battery swapping scenarios. By employing a current closed-loop control strategy, the prototype system can raise the temperature of a 18650 battery from −20 to 6 °C in just 20 min, fully meeting the heating system's requirements. The system rapidly increases the temperature of the Li-ion battery from sub-zero to near zero degrees, with a control response time of only 70 ms. Additionally, it allows real-time adaptation to different battery types and excitation strategies by adjusting the output current amplitude and frequency through closed-loop control.

Furthermore, we discuss three AC self-heating solutions without energy injection specifically developed for BPs, including integrated external circuits, motor drive systems, and boost-type power battery modules. These solutions are particularly suited to address the challenges faced by EVs, such as the reduced charging efficiency, increased internal resistance, and potential dendrite formation risks associated with Li-ion batteries at low temperatures. Each solution offers unique advantages. For instance, the integrated external circuit self-heating solution uses high-frequency charge–discharge control for self-heating of the battery, enabling real-time monitoring of temperature and voltage, and adjusting the high-frequency charge–discharge control based on this data to ensure battery safety. The motor drive system self-heating solution allows rapid low-temperature heating of the power battery without the need for an external power source, avoiding lithium metal deposition during charging. The boost-type power battery module self-heating solution also utilizes high-frequency charge–discharge control for self-heating, offering advantages such as low energy loss, higher efficiency, and more uniform temperature rise. These self-heating methods not only help extend the lifespan of Li-ion batteries but also enhance the overall safety and performance of energy storage systems. By effectively managing temperature and preventing thermal runaway, these solutions ensure the efficient operation of batteries in various environmental conditions.

Through the content of this chapter, we extend from theoretical modeling and heating methods to specific technical implementation and system verification. We provide concrete and feasible solutions for the practical application of AC heating systems. The proposed solutions not only offer practical methods for improving battery performance in low-temperature environments but also lay the foundation for future innovations in battery technology. As EVs and renewable energy storage systems continue to develop, integrating these innovative heating solutions will be key to ensuring Li-ion batteries remain safe, efficient, and reliable, ultimately contributing to the sustainable development of the global energy landscape.

References

1. Stuart T A, Hande A. HEV battery heating using AC currents [J]. J Power Sources, 2004, 129(2): 368–378.
2. Hande A, Stuart T A. AC Heating for EV/HEV Batteries [Z]. Power Electronics in Transportation. Auburn Hills, MI, USA; IEEE. 2002: 119–124.https://doi.org/10.1109/PET.2002.1185559
3. Guillen Montenegro L E, VILLEGAS PICO H N. Ac-Heating and Fast-Charging Power Requirements of EV Battery Packs in Subzero Temperature [J]. IEEE Trans Transp Electrification, 2023, 9(3): 3936–3946.
4. Xiong R, Zhang K, QU S, et al. A fast pre-heating method for lithium-ion batteries by wireless energy transfer at low temperatures [J]. eTransportation, 2023, 16: 100227.
5. Huang X, Meng J, Jiang W, et al. Alternating current heating techniques for lithium-ion batteries in electric vehicles: Recent advances and perspectives [J]. J Energy Chem, 2024, 96: 679–697.
6. Shang Y, Xia B, Cui N, et al. An Automotive Onboard AC Heater Without External Power Supplies for Lithium-Ion Batteries at Low Temperatures [J]. IEEE Trans Power Electron, 2018, 33(9): 7759–7769.
7. Zhu C, Shang Y, Lu F, et al. Core Temperature Estimation for Self-Heating Automotive Lithium-Ion Batteries in Cold Climates [J]. IEEE Trans Industr Inform, 2020, 16(5): 3366–3375.
8. Hu Z, Li Y, Liu F, et al. Thermal Characteristics Investigation of Lithium-Ion Battery Under High-Frequency AC Excitation in Low-Temperature Environment [J]. IEEE Trans Transp Electrification, 2022, 8(1): 407–419.
9. Zhang Y, Yang Y, Shang Y, et al. A high frequency AC heater based on switched capacitors for lithium-ion batteries at low temperature [J]. J Energy Storage, 2021, 42: 102977.
10. Shang Y, Liu K, Cui N, et al. A Sine-Wave Heating Circuit for Automotive Battery Self-Heating at Subzero Temperatures [J]. IEEE Trans Industr Inform, 2020, 16(5): 3355–3365.
11. Shang Y, Liu K, Cui N, et al. A Compact Resonant Switched-Capacitor Heater for Lithium-Ion Battery Self-Heating at Low Temperatures [J]. IEEE Trans Power Electron, 2020, 35(7): 7134–7144.
12. Jiang J, Ruan H, Sun B, et al. A low-temperature internal heating strategy without lifetime reduction for large-size automotive lithium-ion battery pack [J]. Appl Energy, 2018, 230: 257-66.
13. Zhu C, Cao Y, Zhang H, et al. Comprehensive Design and Optimization of an Onboard Resonant Self-Heater for EV Battery [J]. IEEE Trans Transp Electrification, 2021, 7(2): 452–463.
14. Li Y, Gao X, Qin Y, et al. Drive circuitry of an electric vehicle enabling rapid heating of the battery pack at low temperatures [J]. iScience, 2021, 24(1): 101921.
15. Du C, Peng Q, Chen F, et al. Investigation on the method of battery self-heating using motor pulse current [J]. Proceedings of the Institution of Mechanical Engineers, Part D: Journal of Automobile Engineering, 2021, 236(10-11): 2399–2409.
16. Zhu C, Han J, Zhang H, et al. Modeling and Control of An Integrated Self-Heater for Automotive Batteries Based on Traction Motor Drive Reconfiguration [J]. IEEE Journal of Emerging and Selected Topics in Power Electronics, 2023, 11: 384–395.
17. liu Z, Liu F, Lu S, et al. Amplitude-Frequency Decoupled Heater and Integrated Strategies for Automotive Batteries Based on Inverter and Motor [J]. IEEE Trans Power Electron, 2023, 39: 1565–1576.
18. Qian Xu, Xueyuan Wang, Wenjun Fan, Xuezhe Wei, Haifeng Dai. Design and Implementation of a Non-Destructive AC Heating System for Lithium-Ion Battery Modules[J].Batteries,2024,Vol.10(9): 300
19. Dai Haifeng, Wei Xuezhe, Zhu Jiangong, Sun Zechang. A kind of lithium ion battery self-heating apparatus and method [P]. CN104282965B,2015.01.14
20. Sun Zechang, Wei Xuezhe, Dai Haifeng. A kind of lithium ion battery for electric vehicle low temperature exchanges heater [P]. CN107666028A,2018.02.06

21. Wei Xuezhe, Liu Yaofeng, Zhu Jiangong, Sun Zechang, Dai Haifeng. A kind of boost type power battery module alternating current low-temperature heat circuit [P]. CN203722291U,2014.07.16